Agriculture and Animal Husbandry

Agriculture and Animal Husbandry

Gaurav Tyagi

RANDOM

RANDOM PUBLICATIONS

NEW DELHI (INDIA)

Agriculture and Animal Husbandry

ISBN 978-93-5111-783-4

Published in 2016 in India by

RANDOM PUBLICATIONS

4376-A/4B, Gali Murari Lal, Ansari Road
New Delhi-110 002
Phone : +9111-43580356, 011-23289044, 011-43142548
e-mail: sales@randompublications.com,
info@randompublications.com, randomexports@gmail.com

Reprinted 2018

Type Setting by : Friends Media, Delhi-110089
Digitally Printed at : Replika Press Pvt. Ltd.

Preface

A large number of farmers in India depend on animal husbandry for their livelihood. In addition to supplying milk, meat, eggs,wool and hides, animals, mainly bullocks, are the major source of power for both farmers and drayers. Thus, animal husbandry plays an important role in the rural economy.

Agriculture in India has a significant history. Today, India ranks second worldwide in farm output. Agriculture and allied sectors like forestry and fisheries accounted for 16.6% of the GDP in 2009, about 50% of the total workforce. The economic contribution of agriculture to India's GDP is steadily declining with the country's broad-based economic growth. Still, agriculture is demographically the broadest economic sector and plays a significant role in the overall socio-economic fabric of India.

The situation of the Thar Desert has become critical for many rural farmers as they struggle with poverty and malnourishment. Overall, India is home to approximately half of those living with hunger worldwide. The arid climate and unpredictable droughts make food security a serious challenge for subsistence farmers.

While agriculture is the primary occupation in the Thar Desert, animal husbandry is the secondary occupation for over half the families. Livestock provide stability in the face of crop failure and drought and villagers are able to provide wool, milk, and other dairy products for personal consumption or to sell in the market .

Keeping in mind this nature of the subject this book has been written to advance the fascination and alleviate the challenges. Animal Husbandry and Farming Management will prove to be and interesting guide into the world of animals, farming and agriculture.

– Author

Contents

1

Understanding Farmers and Land Owners

FARMERS AS ACTORS IN A CAPITALIST POLITICAL ECONOMY

Whereas the perspectives of Mann-Dickinson and Friedmann on one hand, and of de Janvry, Friedland et al., and others working in the Lenin (and, to a lesser extent, the Kautsky) traditions on the other, are, in a sense, in diametrical opposition, much of the most provocative work in the Marxist tradition in the "new sociology of agriculture" has revolved around formulations that explicitly or implicitly are attempts at synthesis. One of the most noteworthy attempts has been by Mooney.

Mooney's principal contribution has been to cast doubt as to whether the existence of conventional capital-labor relations on farms is an adequate benchmark for gauging the existence of capitalist penetration of agriculture. Following Wright, Mooney has developed a model of agrarian class structure involving "contradictory class locations" such that class positions other than family labor farmer (unity of capital and labor in the farm household), capitalist farmer, and agricultural wage labor are seen to exist. In particular, Mooney sees that there are several "detours" that can be taken by farmers in order to avoid proletarianization (e.g., either being forced from agriculture or becoming hired agricultural laborers).

These detours involve tenancy, contract farming, part-time farming, and debt. In each, whereas there is no capital-labor relation at the point of agricultural production, farmers are exploited by some fraction of nonagricultural capital (in tenancy, by landlords; in contract farming, by agribusiness; in part-time farming, by off farm capitalists; and in debt, by finance capital). Thus, Mooney argues that the exploitation of farm wage workers by agrarian capitalists is only one form that capitalist penetration of agriculture can take.

Moreover, Mooney sees that these detours — that is, the alternative ways in which capital acts to "strip simple commodity producers of ... surplus value other than through the extension of wage labor" — may be more significant than full-blown capital-labor relations at the point of agricultural production. It is significant, however, that Mooney's explanation of why these contradictory class locations in agriculture tend to emerge has a subjectivist component, based on Weber's distinction between formal and substantive rationality. Mooney sees that many farmers are motivated more by forms of substantive rationality (e.g., the desire for autonomy over their work) than they are by formal capitalist rationality. Accordingly, these farmers tend to be tenacious in holding onto their farms and farm lifestyles and will often tend to take one of the four "detours" to capitalist development in order to remain in agriculture.

Mann and Dickinson, however, replied vigorously to Mooney's neo-Weberianism with two major arguments: first, that so-called contradictory class locations are really based on the same social relations of production grounds that Mann and Dickson use and, second, that Mooney has misconstrued their notion of "obstacles to capitalist development" in agriculture to be immutable barriers, so that Mooney was unable to recognize that his notion of "detours" to capitalist penetration of agriculture is similar to their own views.

Mann and Dickinson thus argue that subjectivism, and the explanatory ambiguities it involves, does not yield insights beyond those afforded by a structural, neo-Marxist theory. They were also critical of Mooney's project of synthesizing Marxian and Weberian approaches on the ground that these approaches are incompatible, except on an ad hoc, eclectic basis. Mooney added further to the debate by arguing that Mann and Dickinson have exaggerated his departure from a neo-Marxist framework-that his perspective retains Marxist insights by incorporating key Marxian categories as ideal-types within a Weberian framework.

He argued that this approach overcomes the incompatibilities of the Marxian and Weberian frameworks — unless, of course, one adopts a mechanistic Marxist or Leninist perspective. His key position was that the presumption of a predictably patterned and necessary differentiation of farmers into the capitalist class and proletariat is erroneous and that subjectively meaningful human action affects the process of differentiation substantially.

With the 1980s farm crisis coming under increasing scrutiny by rural sociologists, Mooney has extended his analytical framework to an historical examination of the state-sponsored farm credit system (Farmers Home Administration [FMHA], Commodity Credit Corporation [CCC], and the Farm Credit System [FCS]) and tax policy incentives for increasing the capitalization of agriculture. Mooney conceptualizes credit and tax policies as means of dealing with the crisis of legitimacy posed by the increasing prevalence of tenant farming produced by the Great Depression. Mooney argues that this credit and policy system created more indebted farmers who ultimately became vulnerable to financial crises in the form of rising real interest rates and declining agricultural commodity prices, a situation such as that which occurred in the 1980s. Financial crises expose to perceptive actors the role of the state in creating the crisis, thus creating another crisis of legitimacy and providing a basis for affected farmers to mobilize against the state.

Pfeffer has taken a considerably different tack than Mooney, though Pfeffer's work has some similarities to that of Mooney in that he stakes out middle ground on the question of whether the predominant feature of agriculture in advanced capitalism is the persistence of the family labor farm on one hand, or its demise into capitalist relations of production on the other.

Pfeffer conducted an historical analysis of three systems of agricultural production in the United States (industrial agriculture in California, family farming in the Great Plains, and the sharecropping system in the South prior to World War II) and demonstrated that recruitment and maintenance of access to a suitable labor force was a crucial factor shaping the disparate organizational structures of these three systems. Only in California, largely because of preexisting land concentration and of state actions to recruit and restrict the residential, educational, and occupational mobility of ethnic minority farm workers, was capitalist agriculture able to set root prior to World War II.

Noting that the U.S. South had a high degree of land concentration during the entirety of the nineteenth century, Pfeffer goes on to explain why southern planters were unable to recruit a wage labor force after the Civil War and why the tying of sharecroppers to plots on the landlord's plantation through the crop-lien system was the only possible route to recruiting the labor necessary to operate a large plantation. Finally, Pfeffer demonstrates that a combination of labor scarcity, the unwillingness of immigrants to work for wages on farms, and severe declines in the price of wheat in the late nineteenth century led to the establishment and persistence of family farming in the Great Plains. Pfeffer thus suggests that while capitalist agriculture can emerge under particular circumstances, there have been several alternative forms of agricultural organization — sharecropping and family farming, in particular — that have become established due to regional and commodity variations in the ability of farmers to recruit a dependable labor force.

Still a different approach was taken by Whatmore et al., who have developed a theoretically based typology of farms in Britain. Their goal was to understand the effects of external pressures by nonfamily and nonfarm capitals to penetrate agriculture and how these pressures articulated with internal processes on farms. By cross-classifying levels or degrees of subsumption of internal and external relations, they developed a typology with four main categories of farm types, ranging from relatively unsubsumed ("marginal closed unit") to the "subsumed unit." The internal relations variable pertained to ownership and control of farm capital and land, control over management, and the balance of family and hired labor.

The external relations variable pertained to level of dependence on industrial capitals for production inputs and purchase of farm outputs and to the degree of involvement in credit relations, especially with finance capitals. Applying this theoretically based typology to the data from three areas in southern England — an urban fringe area, an agricultural area undergoing commercial and industrial development, and a primarily agricultural area — they found that the process of subsumption of farm production relations, and thus their integration into the "wider circuits of capital," varied over time and space. Whatmore et al. concluded that farm families need to be understood as actors in the process of subsumption rather than its passive victims.

The neo-Marxian theoretical enterprise (and that of the modernization school as well) on the conversion of traditional peasant economic relations to commodity relations under capitalism has also been scrutinized by Vandergeest. Among his major criticisms were that neo-Marxist approaches tend to be too unilinear, too macro-structural, too denying of human agency, too inattentive to the role of the state in the process, and too unconnected with practice. He has argued, from a neo-Weberian position rooted in the work of Pierre Bourdieu, that all categories and theories are in the last analysis historically contingent, ideological, and interpretive social products.

The world is a complex whole which can only be investigated empirically and understood through theory. It cannot be reduced to the simple working out of a model derived through deductive theory — whether that of "simple commodity production" or of "capitalism". There is not a single deductive logic (such as the logic of exchange-value or accumulation) underlying or determining all relations in a capitalist formation, but there are different, historically contingent principles which we can only investigate through empirical research.

There has recently been a tendency in the neo-Marxist and neo-Weberian literatures toward a deemphasis on "deductivist"-functionalist theoretical postures. Subjectivist perspectives such as those of Vandergeest and Mooney have been influential in leading to this reorientation, though this tendency is by no means limited to those persuaded by neo-Weberianism.

The three postures in the Marxist political economy of agricultural tradition — those emphasizing the persistence of petty commodity production, those emphasizing the inevitable differentiation of petty commodity producers into antagonistic social classes, and those seeking synthetic positions — have yielded a continuing, lively debate in the rural sociology literature. Particularly instructive is the debate over the Mann-Dickinson hypothesis.

Also of importance is Goodman and Redclift's commentary on the work of Friedmann in which Goodman and Redclift critique Friedmann for taking an overly deductive approach to petty commodity production that ignores the role of historical conjuncture and ideology. Bernstein has made a comparable argument — that generic theories of simple commodity production under advanced capitalism tend to overemphasize its "functional" aspects and downplay contradictions and the diversity of household forms of production.

It should be emphasized that theoretical and empirical treatments of the political economy of North American agriculture have not been made only by rural sociologists. Many of the pioneers in this tradition of scholarship have been nonsociologists or sociologists who have had little or no connection with the Rural Sociological Society or the institutionalized form of rural sociology in U.S. land-grant universities. Nonetheless, rural sociologists have made influential contributions to this literature.

The emerging political economy tradition in the sociology of agriculture has led to a provocative and stimulating literature. It should be mentioned, however, that those working from this perspective are just beginning to establish a distinctive research program; much of the existing literature, in fact, often tends to involve superimposing a new vocabulary on already-established data, such as Goss et al. set out to do. Establishment of a distinctive research tradition will probably occur slowly since conventional data sources such as censuses and sample surveys are often inapplicable to theoretical issues in the political economy of agriculture. There have, nonetheless, been several research papers in the political economy tradition that have utilized innovative new methodologies to generate research findings that bear directly on specific hypotheses.

Subculture and Agricultural Structure

It is useful to note that the growing interest in subjectivist approaches in the new sociology of agriculture has not been confined to those who work within the neo-Marxist and neoWeberian traditions. Anthropologically-oriented researchers have also contributed to this literature through the examination of the effects of ethnic background on state and local level farm structure. Salamon and associates have done the pathbreaking studies in the anthropological tradition in Illinois communities, though recently this work has been taken up by others.

While their work was not intended to contribute to the "structure versus agency" debate that has raged between Mann and Dickinson on one hand, and Mooney on the other, and actually predated much of this debate, it is clearly relevant to the debate. In particular, the work of Salamon and associates pertains to the role of subjective motivations in farm decision making, especially motivations conditioned by persisting subcultural variations based in differences in ethnic origins of farm families.

The key finding of this line of research has been that farm families with German ethnic backgrounds tend to view farming as a way of life and hold a strong value for keeping the family farm intact. These Salamon has labeled "yeoman" farmers. In contrast, farm families with British ethnic backgrounds tend to be more entrepreneurially oriented, viewing farming as a way to make profits and having little attachment to farming or to particular farms. These

Yankee farmers are more likely to seek growth in scale and less likely to support their local communities. This research has led to understanding differentials in farm size as a partial function of differences between these subcultural variants. Flora and Stitz found that yeoman farmers in a Kansas county generally expanded less than did Yankee farmers. Foster et al. also found support for the Salamon hypothesis in a study of a sample of Illinois farms that had been in the same families for 100 years or more.

Several other anthropological studies in the tradition of Salamon and colleagues have been reported in Chibnik. Of particular importance is Barlett's article in which she criticizes notions of the demise of the family farm and of the "disappearing middle" from an anthropological perspective.

AGRICULTURE AND THE RURAL COMMUNITY

Perhaps in no other area of the "new sociology of agriculture" is there so much apparent continuity with the pre-Korean War tradition of rural sociological research as in research on farm structure and rural communities. The early rural sociologists primarily conceived of agriculture as one of the major institutional ensembles affecting rural community life and, especially during the 1930s, devoted much of their research effort relating to agriculture to understanding the implications of agricultural structural changes for rural community life and viability.

The conceptualization of agriculture and rural community interactions and the methodologies for exploring these interactions in pre-Korean War and post-1970 rural sociological research, however, exhibit dramatic differences. Whereas most of the early rural sociologists tended to be less interested in agriculture per se and were principally concerned with understanding rural community institutions, contemporary researchers have tended to be more interested in structural change in agriculture and have deemphasized understanding the totality of the fabric of rural social life.

The research tools and data sources of these two generations of rural sociological researchers are also dramatically different. The early rural sociologists relied on rural social surveys, while the bulk of contemporary research is based on county-level census data. A further difference between the early and contemporary rural sociologists of agriculture in their work relating to rural communities concerns the pivot of the latter studies around Goldschmidt's "As You Sow". This is somewhat ironic in that Goldschmidt's study was anthropological and qualitative, whereas the research conducted since 1970 that has drawn on Goldschmidt's work has largely been quantitative and has involved far less "anthropological" detail on rural communities than did Goldschmidt's.

Rodefeld is generally credited with the "rediscovery" and popularization of Goldschmidt's pioneering work among rural sociologists, though Heffernan

(1972) made important contributions to elevating "Goldschmidt-style" research to a prominent position in rural sociology. Nonetheless, after Rodefeld's refocusing of rural sociological attention on the Goldschmidt thesis, Goldschmidt's book has been virtually an obligatory citation in research articles on this topic. Despite some trenchant criticisms of Goldschmidt's perspective and research methodology, Goldschmidt's general hypotheses remain accepted theoretically and thus continue to shape, if not dominate, research on farm structure and rural communities within the "new sociology of agriculture."

Briefly, Goldschmidt's thesis was that there are differences in the quality of rural community life depending upon the social organization of agriculture and its attendant occupational structure. Goldschmidt observed that the quality of rural life in two carefully matched California communities in the San Joaquin Valley differed dramatically and that these differences were largely accounted for by the organization of agriculture.

He argued that Dinuba, a community surrounded by "family" farms, had a greater number of businesses, a larger retail sales volume, a higher per capita income, and a greater diversity of social, educational, recreational, and cultural institutions than did Arvin, a community surrounded by large, industrial, corporate farms. These differences were principally accounted for by an intervening variable — community occupational composition — that was seen to be strongly influenced by the social organization of agriculture and which directly contributed to the dramatic variations in indicators of the quality of life. Goldschmidt reported that the high ratio of laborers to other occupational groups in Arvin accounted for its low level of community viability relative to Dinuba. From these observations Goldschmidt inferred that the rise of large-scale corporate agriculture then underway in California and elsewhere would have adverse implications for rural communities.

Goldschmidt's study received essentially no further attention until it was referred to in a 1968 hearing on corporation farming before the Senate Select Committee on Small Business. LaRose and Sonka and Heady soon reported further data consistent with the Goldschmidt study. In addition to Rodefeld's and Heffernan's efforts to resurrect Goldschmidt's work in rural sociological circles, there was a major research effort by the Small Farm Viability Project in California, drawing largely on Isao Fujimoto's unpublished research, which presented a larger restudy of the Goldschmidt thesis. The results largely supported Goldschmidt's findings of over three decades earlier. Flora and Conboy also reported results consistent with the thesis.

Since that time there have been numerous studies in the new sociology of agriculture tradition that have explored the Goldschmidt thesis in a number of contexts. Several observations can be made on the conduct of recent research in the Goldschmidt tradition, what we might call "neoGoldschmidt" research. First, it must be stressed that contemporary rural sociological researchers have taken considerable liberties with Goldschmidt's formulations. As noted earlier, Goldschmidt's method was a comparative anthropological case study of two communities and eschewed quantitative analysis of the sort that has predominated in the subsequent rural sociological literature.

Also, Goldschmidt's analysis focused on several aspects of the property relations and control over labor, such as absentee ownership and use of ethnic minority laborers, rather than farm size, which has been the most important farm structural indicator in the neo-Goldschmidt tradition. In that regard it is important to stress that the "family" farms of Dinuba were by no means "small" farms; in fact, the family farms of Dinuba were generally larger in gross sales terms than most large farms in the Midwest today.

Second, neo- Goldschmidt research has largely focused on a single direction of causality — the impact of agricultural structure on rural communities — and has largely ignored other types of interrelations between agriculture and rural communities. Third, most studies have been based on Census of Agriculture data and have employed countylevel data as a proxy for community characteristics have utilized survey data to infer community quality of life consequences of changing agricultural structures. Fourth, despite obvious major regional variations in farm structures and in rural communities and their economies, there has, until very recently, been a virtual lack of a comparative perspective on farm and community structures.

Goldschmidt and Harris and Gilbert are partial exceptions to this generalization, since they have done research using states as the unit of analysis, though this procedure compounds the aggregation bias problems. Very recently, nonetheless, there has appeared a series of regionally based studies of the effects of farm structure on rural community viability and quality of life commissioned by the Office of Technology Assessment. These studies are Buttel et al., Skees and Swanson, Flora and Flora, MacCannell, and van Es et al. as published in abridged form in Swanson. These five studies were very briefly summarized in the main report. Together they indicate strongly that there are pronounced regional variations in how changing farm structures affect rural communities.

The impacts of farm structural changes on rural communities are generally most pronounced in the Great Plains and industrial agriculture states, less so in the South and Midwest, and very little in the Northeast. There is evidence that the degree to which farm structure affects rural communities is directly associated with the dependence of the rural community on agriculture.

The general model pursued in these neo- Goldschmidt studies has been to estimate coefficients for a causal chain in which farm structural change indicators are seen to affect the composition of the farm population (e.g., as measured by indicators of the size of the farm population, the proportion of full-time hired workers) and the size and composition of the rural population.

These clusters of variables are then seen to affect indicators of the quality of life in rural-agricultural communities (e.g., proportion of families in poverty, median family income, per capital retail sales, unemployment rate).

This general model is sometimes supplemented by technological factors as antecedents of farm structure, since technological change is considered to be one of the major factors influencing farm structural change. More recently, studies in this genre have gone beyond the cross-sectional census data designs that predominated from the mid-1970s to the early 1980s and have included measures of change, where appropriate, in variables in the model. One limit to studies that use census and other data based on single-year assessments of farm income is that net farm incomes are highly variable over the short-run; for example, as many as 20 percent of farms with annual sales of over $250,000 had poverty-level incomes in 1986.

Heffernan has effectively summarized the general configuration of results in the neo-Goldschmidt literature by noting that "all relevant research to date suggests that a corporate type of agriculture results in a reduction in the quality of community life for at least some people, especially the hired workers in rural communities." As such, the results continue to lend support to the Goldschmidt thesis. But there have recently been several expressions of dissatisfaction with Goldschmidt's perspective and with the generalizations that have been made from research in this tradition.

A major theoretical concern is that Goldschmidt and many researchers working in this tradition have taken large-scale and corporate agriculture to be basically synonymous when, in fact, not all large-scale farms are corporate or industrial-type farms. Much more important than industrial-type farms in farm numbers and shares of sales, assets, and profits are larger-than-family farms. Empirically, as Goldschmidt-style research has been done in the Midwest and other traditional family farming areas, there have been admonitions that large-scale agriculture, especially of a nonindustrial character, does not necessarily lead to adverse impacts on rural communities, though some studies note that increased scale of production could have detrimental impacts on community services.

A further area of dissatisfaction with research on farm structure and rural communities is that the treatment of farm structure in this literature has generally failed to incorporate new concepts relating to agricultural dualism. Typically farm size or scale is operationalized as a single variable representing a central tendency measure of farm size, which was a suitable procedure when there was a singular process of differentiation among farms. But with the emerging dualism in U.S. farm structure since the late 1970s, there has been little change in mean acres per farm since the 1969 census. This measure of central tendency thus ignores the processes of dualism that have contributed to increased concentration with apparent stability in average acres per farm since the late 1960s. The neo- Goldschmidt conceptualization of an agricultural structure in which particular communities can be characterized along a continuum of farm sizes ranging from small to large has also been questioned by Wimberley who, using multiple indicators from county-level Census of Agriculture data and a longitudinal design, found that there are instead three continua relating to degree of "corporate-commercial" agriculture, "largefarm-area" agriculture, and "small-farm" agriculture. Using Wimberley's measures and including sources of off-farm employment as well, Reif found that large-farm-area agriculture and core-industry structures were related both to family poverty and income inequality in counties.

Thus, while studies of the impacts of changing farm structures on rural communities have been a productive area of research in the new sociology agriculture, a number of research frontiers for the future are suggested by recent criticism. There is a need to diversify the methodological approach, consider alternative directions of causality, introduce more sophistication in terms of the measurement of change in farm structure, and expand recent efforts at comparative regional and interdisciplinary research.

The work of Vail represents an encouraging thrust in this direction. Vail has conducted extensive research on the relationships between agriculture and rural communities in Maine in which a variety of factors of demonstrated importance — the reciprocal relations between agriculture and rural communities, the role of an increasingly dualistic agrarian structure, nonfarm aspects of agriculture, and the historically specific nature of farm-community interactions — have been incorporated into a single study.

SMALL FARMS AND PART-TIME FARMING

The nature, social bases, and socioeconomic consequences of "small" farms and part-time farming are, in a sense, among the most longstanding foci in rural sociological research on agriculture. Nevertheless, sociological analysis of small-scale agriculture and part-time farming received substantially increased attention and exhibited considerable theoretical reformulation in the 1970s and 1980s for several reasons. First, spearheaded by the writings of Hightower and others on the demise of "small" farms, there was a startling renaissance of concern with the status and prospects of small family farms, which had greatly subsided a few decades earlier in the aftermath of the apparent resolution of the depression-induced increase in tenancy.

Second, quite ironically, several observers began to document the fact that the 1970s had witnessed an unprecedented trend toward the stability and persistence of small farms; in most regions the absolute number of farms with fewer than 50 acres has increased over the past one to two decades. Third, it became apparent that much of the reason for the enhanced viability of small farms was the growing access of farm households to off-farm jobs and income;

indeed, the U.S. Department of Agriculture began to note that, in most years, the majority of farm households' aggregate income came from nonfarm sources and that there had been a slow but steady increase in the prevalence of off-farm labour market participation among farm families since World War II.

Small and part-time farms are treated here as a particular focal point in the new sociology of agriculture, a tack that has been taken by many others. Whereas there is a key connection between small-scale agriculture and part-time farming — the fact that part-time farming has in recent decades been much more common among "small" than among "medium-size" or "large" farms — the two phenomena, however, are by no means equivalent. In particular, some "small farms"-which provisionally can be referred to here as low-sales-volume farms, with about $50,000 or less in gross annual sales volume-have little or no off-farm income and, accordingly, have long been overrepresented among rural poverty groups in the United States and elsewhere.

Moreover, in all advanced capitalist societies, especially the UK, part-time farming occurs across the whole spectrum of sizes of family-proprietor farms; in the UK it has been quite common for the operators of very large farms to have off-farm occupations, especially the operation of small businesses related to agriculture. Thus small-scale and part-time farming, although tending to be significantly related, are by no means coterminous. Further, the analysis of neither small nor part-time farming has been dominated or influenced by neo-Marxist postures nearly as much as other areas in the new sociology of agriculture. In the huge literature on these topics there have been theoretical and empirical neo-Marxist works, but these approaches have been a distinct minority component of the literature.

It is useful to begin with a very generalized model of the recent evolution of U.S. agriculture to situate the theoretical and empirical issues relating to small and part-time farms. Over the past decade or so there has been growing acceptance of the notion that U.S. agriculture has followed a dualistic or bimodal pattern of development since the early 1970s. There is evidence that prior to the late 1960s (and thus preceding the phase of dualistic development) there was a singular pattern of increased concentration and differentiation in U.S. agriculture in which there was a direct association between size of farm and its persistence or viability; thus large farms increased in numbers, acreage, assets, and so on at the expense of small farms, particularly small tenant farms without the means to acquire off-farm income. Farm operators with access to off-farm income were thought to be largely in a transitional status, either into or out of agriculture.

It is now suggested that a more complicated or dualistic pattern of concentration and differentiation has emerged. The stylized facts of this association are, first, that large farms — including large family-type, larger-than-family, and industrial farms in Rodefeld's terms — are, much as in previous decades, exhibiting a demonstrable increase in their shares of aggregate production, sales, assets, profits, use of hired labour, and so on.

Second, however, it is argued that small farms are increasing in numbers and holding their own in terms of acreage and aggregate production. This bolstered status of small farms is considered to be made possible largely by access of their operators and other family members to off-farm wage income. U.S. small farmers have been noted to have, on average, total family incomes in excess of those of medium-sized farmers and the U.S. median family income because of the higher than average educational attainments of members of small farm households and their consequent ability to acquire highly remunerative off-farm work. While outside of the South small farmers have tended to have higher educational levels than larger, full-time operators for several decades, the educational disparities between small, part-time and larger, full-time operators are suggested to have increased. This has often been explained in terms of the 1970s "population turnaround" in which well-educated persons, many of urban origin, were attracted to farm residential situations for lifestyle reasons.

Finally, it is asserted that there has been a "disappearing middle" — that medium-size (traditional family-type) farms have decreased in numbers and in shares of aggregate output, assets, profits, and the like. It has been suggested that these medium-size farms enjoy neither the advantages of bigness or of smallness. In particular, "middle farmers," because they have operations too large to be farmed on a part-time basis, tend to lack the labour resources (and often the educational qualifications) to participate effectively in off-farm labour markets. Thus "the middle" is tending to "disappear," albeit slowly. Approximately 5, 70, and 25 percent of U.S. farms are seen to lie, respectively, in the "large," "small," and "medium" categories.

These stylized facts appear to be largely accurate, although Ehrensaft *et al.*, and Barlett have raised empirical and theoretical objections to the disappearing middle thesis. Salamon and Davis-Brown's findings suggest that the "disappearing middle" may have come more from certain segments of the farm population (the more risk-accepting, entrepreneurial farmers) and less from other segments (the "yeoman" farmers whose key goals are preserving their family farms). Similarly, this conclusion may not hold for black farmers whose limited access to credit for land purchases and mechanization led to rapid declines in the total number of black farmers. Nonetheless, there has been general confirmation of this empirical model for the U.S. farm sector, which provides a preliminary framework for placing the phenomena of small and part-time farming in historical context.

As Coughenour and Wimberley have noted, "most American farm families live on small and part-time farms." Thus in the sense of understanding the composition of the U.S. farm sector, it is essential to account for the status of and trends in small-scale and part-time agriculture. A dramatically different context of research on small farms emerged in the mid- to late 1970s. This context can now be seen as contradictory; at the same time that there had been a rekindling of concern with the status and prospects of the small-scale farm that had been unprecedented since the Great Depression, there was a correspondingly unprecedented increase in the numbers of small farms in most regions of the country.

The changing context of the decline and persistence of smallscale agriculture strongly influenced the conduct of research on small farms. There emerged a number of studies examining the factors leading to small farm decline, persistence, survival, adaptation, and so on. The dominant finding in this research was that small farmers pursue multiple strategies for survival — access to off-farm income; minimizing debt; direct marketing of high-value, specialty crops; and so on — and that the choice of strategies was related to the socioeconomic characteristics of the farm, farm operator, and farm household members.

Recognition of the diversity of small farmers and their survival strategies set forth a related line of inquiry: documenting the internal differentiation among small farmers and generating typol ogies of small farms and small farming households. Major studies in this genre include Buttel and Gertler, Schulman *et al.*, Mufioz, Schroeder *et al.*, and Heffernan *et al.*. These studies have typically found that gross sales (or other indicators of scale), age, education, motivations for farming, household composition, off-farm occupation, and off-farm income are important dimensions of internal stratification among small farmers.

A third cluster of research activities relating to small farms has been to explore small farmers' sociopolitical attitudes. One major area of inquiry has been in documenting the motivations of small scale farm operators and their family members.

Another focus of attitudinal research has been on the political orientations of small farmers. Research to date has generally not reported that small farmers have dramatically different political orientations than large farmers on either general dimensions of political ideology or orientations toward agricultural politics. It has been demonstrated, however, that there are major differences in political orientations among small farmers that are primarily accounted for by off-farm occupation, total family income, and race.

Finally, there has been a nagging concern among rural sociological researchers as to how small farms should be defined. Gross farm sales has for a very long time been the principal scale variable used by the Census of Agriculture in collecting and reporting data, and our macro-sociological knowledge base on trends in the structure of small-scale farming is largely based on gross sales data. Nonetheless, it is widely recognized that gross farm sales have significant limitations as a measure of scale or "smallness."

For example, many economists would prefer value-added over gross sales as a measure of scale. But in the absence of a convenient measure of value-added (or some other concept preferable to gross annual sales) for census and survey research purposes, some arbitrary criterion based on gross sales, sometimes combined with a total family income criterion, has continued to be the measurement convention in small-farm research. Nonetheless, as Carlin and Crecink have shown, alternative definitions of small farms have significant implications for the numbers and characteristics of farmers and farm households included in that category.

Part-time farming research in the United States was initiated during the Great Depression of the 1930s, mainly by agricultural economists and to a lesser extent by rural sociologists, and was part of the concern about "agricultural adjustment" that prevailed in USDA and land-grant university circles. It was observed that many farmers were attempting to survive depression conditions and avoid (or extricate themselves from) tenancy by acquiring off-farm income. As was also to become the case in the 1980s, part-time farming came to be widely advocated as a partial solution to the difficult problems of the agricultural depression of the 1930s.

In the decades up until the early and mid-1970s, however, it was recognized that "agricultural recovery," which typically was conceptualized in terms of increased owner-operation of farms, did not correspond to a decrease in part-time farming. Indeed, coincident with agricultural recovery from the late 1930s to the 1960s was an unprecedented exodus of farmers from the land — a new "agricultural adjustment" problem born of expansion, dynamism, and what latter-day neo-Marxists would refer to as differentiation, rather than depression and decline — that again called for part-time farming as an adjustment mechanism.

Whereas in the 1930s farm size and part-time farming did not appear to be strongly correlated, by the 1950s it was widely recognized that there had emerged a strong inverse association of gross farm sales and various indicators of the involvement of the operator in off-farm work. Among many researchers, especially agricultural economists, the principal research concern, on which there was a decade of conflicting results, was with whether part-time farmers were as efficient as full-time farmers.

Whereas rural sociologists were a minority among researchers on part-time farming during this period, Fuguitt in particular made several important contributions to the literature that continue to be drawn upon today. Fuguitt first demonstrated that part-time farming represented a combination of "push-pull" factors — that the prevalence of part-time farming was positively related to off-farm employment opportunities and inversely related to opportunities in agriculture. Fuguitt then established that there were categories of part-time farmers (based on farm and nonfarm origin and on farmers' aspirations) with distinctly different social characteristics. Fuller has noted that after 1975 research attention to part-time farming increased dramatically and that several new issues occupied the attention of rural sociologists and kindred social scientists. Definitional matters received major attention after 1975; in particular, there came to be dissatisfaction with conceptualizing part-time farming in terms of the presence or absence, or the number of days, of off-farm work by the operator — the way that off-farm employment data have traditionally been collected and reported in census sources. A consensus began to emerge on conceptualizing part-time farming in terms of the activities of the farming household rather than the farm operator. There also continued to be typological research, following Fuguitt's efforts.

Further, there was a significant amount of research on mobility into and out of part-time farming, on the significance of differences in the off-farm occupations that are held, on the implications of part-time farming for rural communities, and especially on the motivations of part-time farmers. Finally, there has emerged a significant political economy tradition in small and part-time farming research that has explored issues such as the rooting of part-time farming in the concentration and differentiation process, the roles that part-time farming plays in accumulation and legitimation in the agricultural and larger economy, the role of off-farm work in the sexual division of farm household labour, and the "embourgeoisiement" thesis — that part-time farming, by enabling smaller farmers to avoid proletarianization, may blunt the emergence of radicalism by subordinate groups in the agrarian class structure.

In light of excellent review articles on the status of part-time farming research, there is a less pressing need to summarize what has been a voluminous literature. Instead, we will comment on some "metathemes" in this literature that have not yet been the subject of thorough reviews.

First, there is a general tendency for the literature on part-time farming to be polarized into highly structural (if not structuralist in the Althusserian-Marxist sense) and highly voluntarist accounts. Structuralist accounts see part-time farming in almost teleological fashion — that the dynamics of capitalism lead inexorably to agrarian proletarianization in (among others) the form of part-time farming, which provides a mechanism for sustaining profitable capital accumulation in farm, and especially nonfarm, industry by providing a low-wage labour reserve. The subjective motivations and experiences of part-time farming households are thus ignored and viewed to be epiphenomena of larger laws of motion of capitalist development. This approach has historically been prominent in Italian rural sociology circles.

Voluntarist accounts, on the other hand, have largely focused on the motivations of members of part-time farming households — especially their desire to remain in farming and to fulfill lifestyle aspirations. To the degree that nonsubjective factors are considered, they are treated as an "opportunity structure" and not as a social structure that is rooted in the historical and material dynamics of agriculture. Fuller has made a related diagnosis of the state of research on part-time farming — namely that despite widespread social science research on part-time farming, as yet there exists no satisfactory explanatory theory.

Fuller attributes this to the fact that there has been disproportionate attention to the "farming" aspects of part-time farming and too little attention to the family and other nonfarming aspects of rural life. We would suggest that the frontier of research in part-time farming should lie in integrating objective and subjective aspects of these phenomena, which should help to stimulate the theoretical thinking that Fuller has found lacking.

A second metatheme that can be distilled from the part-time farming literature concerns the need for a truly comparative approach to part-time farming research. This observation is, in a sense, ironic, since there has probably been more cross-national comparative observation of part-time farming than in any other subject matter in the sociology of agriculture. But while cross-national variations in part-time farming are widely recognized, these divergent systems of parttime farming have yet to be explained theoretically.

This suggests the need for a comparative, integrative perspective on agricultural and rural development, which, like de Janvry's rendering of Lenin's classical schema of "roads of agricultural development," would examine the interrelations between agriculture and the larger, nonfarm rural economy and society. A major step in this theoretical direction has been taken by Bonanno, though his focus has been more on the persistence and role of small farms than on part-time farming per se. Nonetheless, much work remains to be done to explore part-time farming and other aspects of agrarian structure from the vantage point of the dynamics of agrarian and regional-development dualism.

GENDER AND AGRICULTURE

The role of women in U.S. agriculture was by no means a new area of inquiry when interest in this topic soared in the late 1970s and early 1980s. Wilkening, in particular, had sustained a major research programme on the work roles of farm women and men since the early 1950s. And for several decades there had been a substantial amount of attention paid to rural and agricultural family structures. Nonetheless, rural sociological research on gender and agriculture was to embrace distinctly different themes in the era of the "new sociology of agriculture". Whereas there have been a number of foci in this new tradition of research, the dominant cast of roughly 20 recent studies has been to explore the "invisibility" of women in U.S. agriculture, to paraphrase the title of Sachs' recent book-length work in this area.

Research on gender and agriculture has fallen into four major categories, which basically have represented stages in the development of knowledge on the role of women on American farms. The first category is that of cross-sectional descriptive studies that have sought to document the nature and extent of farm women's involvement in on-farm, household, and off-farm tasks. The second is a set of studies that have attempted to assess changes in women's involvement in on-farm, household, and off-farm tasks by utilizing longitudinal research designs. The third category is that of studies exploring the interrelations between the on-farm, household, and off-farm roles of farm men and women. The final category is that of comprehensive theoretical and empirical approaches that attempt to understand how the institutions of capitalism and patriarchy interact to shape the involvement of men and women in various work roles.

The first descriptive phase of research on gender relations in agriculture was initiated by Joyce and Leadley, who summarized the existing literature on rural women, both farm and nonfarm. Joyce and Leadley lamented the lack of research on rural farm and nonfarm women and set forth an agenda for research. Soon thereafter a number of studies would be conducted in which many of the basic parameters of women's participation in farm, off-farm, and household activities were clarified.

Because census data have included little information on farm women and thus have seriously underestimated the economic contribution of women to farming operations, the knowledge base on the productive activities of farm women has been greatest and most longstanding in terms of off-farm employment. It had been recognized for quite some time that farm women were increasing their participation in off-farm labour markets, and that women's off-farm earnings were often crucial in ensuring the viability of farms. It was also recognized, however, that farm women's off-farm jobs typically were concentrated in low-skill, low-wage sectors of the economy and tended to involve less-than-full-time or otherwise intermittent employment.

Subsequent descriptive studies on farm women's work roles focused primarily on their on-farm and household work and its relationships with off-farm employment. One of the most extensive studies was by Jones and Rosenfeld, who used 1980 national survey data to profile farm women's involvement in farm work, off-farm employment, community activities, and several USDA programmes. They found a high incidence of women's involvement in most agricultural production tasks, including those typically considered to be "men's work". Over three-quarters of farm women were found to play a role in bookkeeping and other financial management tasks on the farm. Slightly over 30 percent of the farm women (60 percent of whom worked less than 40 hours per week) reported off-farm employment.

A further major descriptive study was Salant's survey of farm households in 29 rural counties in Mississippi and Tennessee. She concluded that farm women are extensively involved in onand off-farm work; over half of the women reported employment either on-, off-, or both on- and off-farm, about 75 percent of whom worked off-farm. Women's off-farm labour force participation was found to be related to age and family life cycle. The off-farm income earned by farm women made a major contribution to the income of households in the survey — in many cases enabling the family to rise above poverty-level status — despite the fact that farm women's employment was largely in low-wage jobs. Women who worked on-farm primarily did so as unpaid family laborers. Only 3 percent of the women in the sample were principal farm operators. Salant did not, however, find any support for the hypothesis that women who work on-farm tend to do so to free the labour of men in order to acquire more lucrative off-farm jobs.

Another important descriptive study is that of Fassinger and Schwarzweller, which was based on a 1979 sample survey of 124 farm households in Michigan. Fassinger and Schwarzweller found that farm size was positively related to the amount and proportion of time spent by farm men and women in on-farm work. Farm size was inversely related to men's off-farm employment but had no association with women's off-farm employment.

They found that men's time devoted to on-farm work was positively associated with women's on-farm work. Fassinger and Schwarzweller found a very high degree of division of labour by gender for both farm and household tasks. On the farm, women were more likely to assist with financial and recordkeeping aspects of the business and to plant and maintain a vegetable garden. Men, on the other hand, were primarily involved in field operations, the purchase of farm inputs and machinery, and the repair of machinery and farm buildings. The lion's share of household tasks was performed by women regardless of the scale of the farm.

Kalbacher's study was the first comprehensive report profiling the role of women farmers in American agriculture. Kalbacher used 1978 census results and data from a 1979 Farm Finance Survey to present a statistical profile of women farm operators. About 5.2 percent of all U.S. farm operators, or 128,000 farmers, were women in 1978 — representing a near-doubling of the number of female farmers since 1970 and a tripling since 1950. Femaleoperated farms were located disproportionately in the South, and tended more often to be owner-operated, part-time, and smaller than male-operated farms.

Female farm operators were also older and more likely to be in a racial-minority group than their male counterparts. But whereas women farmers were less likely than men to report fanning as their principal occupation, their average off-farm income was significantly less, resulting in a much lower average household income for women farmers, though femaleoperated farms typically had lower debt and more favorable debtasset ratios than did male-operated farms. A further examination of the role of farm women in the United States was conducted by Tigges and Rosenfeld in their study of differences between men and women among "independent" farmers using a national sample. The independent farmer was defined as "a woman or man who has managerial responsibility for the farm operation and lacks a spouse's regular farm labour". They found that, unlike independent farming men who were similar to men whose wives participated in farming, independent farming women tended to have lower net farm incomes and total family incomes than did other women. Part of these differences were explained by differences in type of farm, class, and life-cycle situation variables. Tigges and Rosenfeld concluded that independent women farmers tended to enter farming by default (as upon the death of a spouse) rather than seeking to enter farming on their own.

After several years of descriptive studies documenting the variety of contributions of women to U.S. farm households, the attention of several researchers shifted to changes in these roles over time. The most important study in this regard has been by Wilkening in which he presented data from a 1978 mail survey of Wisconsin farm families, personal interviews with farm couples completed in 1979, and data from a 1962 study carried out in the same region. The data reveal that families remain the most important source of labour on Wisconsin farms, particularly on dairy farms that dominate in Wisconsin.

Nonetheless, Wilkening found that both farm men and women now contribute less time to the farm operation than they did in 1962. Off-farm employment by one of the spouses was significant in the late 1970s on about one-quarter of the farms, and off-farm work was associated with low levels of gross farm and high levels of total family income, but was unrelated to stage in the family life cycle. While Wilkening found that farm women have not played an increased role in farm production activities since 1962, the data did show that farm women have begun to assume major responsibilities for recordkeeping and other financial aspects of Wisconsin farm operations. Women were reported to be equal partners in making important farm decisions on about half of the farms, a slight increase since 1962. Farm women remained largely responsible for the vast majority of household work during the study period.

Testing part of a theory about the sexual division of labour and authority in farm households, Bokemeier and Garkovich found that farm women's self-concepts relating to their identification with their farms' operations were complex in their sources, including scale of farm, farm background, women's human capital, and family characteristics. They identified four roles for farm women, ranging from little or no involvement and authority to full partnership: homemaker, "ag" helper, business manager, and full "ag" partner.

A quite different study oriented toward assessing change in women's and men's work roles was conducted by Gladwin in North Florida. Gladwin was concerned with exploring an extension of the Boserup hypothesis that the capitalization and intensification of agriculture are inversely associated with the level of participation of women in farm activities. Specifically, Gladwin examined the reverse hypothesis — that a decrease in intensification, or marginalization, of agriculture will be associated with an increase in female labour with survey and census data. Gladwin reported that women are now involved to a greater extent in farm production than in the 1920s and 1930s and that the majority of farm women think of themselves as "farmers" rather than farm wives.

But she also reported that off-farm work has become more important for both men and women on North Florida farms. About 88 percent of the farms were part-time farms with significant off-farm income supporting the farm family. These part-time farms tended to produce crops or livestock requiring relatively little labour. Gladwin concluded that there is an association between labour intensification and the growing participation of women in farm work, since the few highly capitalized farms in the region received comparably less labour from women in the farm family than did the more numerous part-time, less-capitalized farms.

The third category of research on gender and agriculture has been that of assessing the mutual interrelations of men's and women's on-farm and off-farm work and how these associations are affected by the scale of the farm. One such study, by Fassinger and Schwarzweller, found that men's and women's onfarm work were positively associated, and each was positively related to farm size. Also, men's off-farm work was found to be positively associated with farm scale, while women's off-farm work was not related to the size of the farm.

Coughenour and Swanson examined 1979 Kentucky farm survey data to explore the associations between farm size and the work status of farm men and women. They did so, however, in a manner different from the previous literature; whereas previous studies have seen men's and women's work statuses as a function of farm size, Coughenour and Swanson suggest that the opposite direction of causality may be important. They explored this hypothesis by developing a four-category farm-type typology based on men's and women's work statuses. They found that there was a modest association between farm types and size of farm (measured by acreage and gross sales). They explored this association further in terms of "exchanges" between men's and women's onand off-farm labour.

They found, for example, that farms in which the woman has off-farm employment forfeited some of the woman's on-farm labour (and hence its contribution to farm size and sales) but gained in income relative to other farms. They also found that the allocation of men's labour — either on-farm, which was positively associated with farm size and sales, or off-farm, which was inversely associated with farm size and sales — was a stronger determinant of the size structure of farms than was the allocation pattern of women's labour. Nonetheless, they did find that the women's role in farming had significant effects — in particular, that women's off-farm work, which reduced or eliminated her on-farm labour input, tended to be inversely associated with the size and sales position of the farm despite the fact that women's off-farm work provided additional capital for the farm operation.

Buttel and Gillespie examined the interdependencies of average weekly hours of on-farm and off-farm work by farm men and women, the use of hired labour, and the size of farm with 1979 New York farm survey data. They found that farm men and women tend to specialize mutually in either on-farm or off-farm work, with these interdependencies being greatest among small farm households. It was also found that hired labour tends to be substituted for women's on-farm labour input, with this inverse association between hired labour and women's on-farm work being greatest among larger, commercial-scale farms. The results of the ButtelGillespie study were largely corroborated by a replication and extension using data from tobacco-, peanut-, and soybean-producing farms in North Carolina by Simpson *et al.*

The extension pertained to distinguishing between production of commodities that is "organized around" individual units ("small batch") and those whose production is coordinated as a "continuous process". Dairy and tobacco were presented as examples of the former type and soybeans and peanuts as examples of the latter type. Simpson *et al.* found that crop production organized as continuous processes tended to make farm women's work more independent of farm men's than was the case for crop production organized around small batches, and thus technological change in the production of "continuous process" crops had different effects on men and women.

Rigorous analysis of gender relations on U.S. farms is a sufficiently new focus of research that, save for the pioneering book by Sachs, there has been little effort to incorporate gender relations phenomena into a comprehensive theory of farm structural change. Sachs' book thus is the single representative in the fourth category of research on farm gender relations and the gender division of farm household labour.

Sachs' book is a wide-ranging treatment of research on farm women from a variety of disciplines and in both First World and Third World contexts. For present purposes, Sachs' treatment of U.S. farm women receives principal emphasis. As indicated in the title of her book, one of Sachs' key arguments is that while women have historically made significant contributions to agricultural (and, increasingly, off-farm) labour, these contributions have been overlooked and undervalued.

Whereas Sachs roots her analysis in the new political economy tradition referred to at the beginning of this chapter, she contends that the invisibility of the women's role in agriculture, and the subordination of women in the agricultural labour process, cannot be explained with reference to the dynamics of capitalist development alone. She argues that the simultaneous contribution of women to agriculture along with their invisibility and subordination must be viewed through the dynamics of capitalism and patriarchy, which is defined as a "set of social relations which has a material base and in which there are hierarchical relations between men, and solidarity among them, which enable them to control women". She emphasizes, moreover, that patriarchical relations preceded (although they have been transformed by) capitalism and thus cannot be seen as a direct consequence of capitalist development.

Sachs begins by demonstrating that at each major stage of agricultural development from the seventeenth century to the present women have played important roles in facilitating the viability of U.S. farms. Until the twentieth century the importance of women was typically seen to be in producing and processing food for home consumption. There was a traditional division of labour between men and women in which men did the majority of the field work, but this division of labour would often break down during busy seasons when women would do "male" tasks. With the industrialization of U.S. society and further commercialization of agriculture, women's subsistence production activities diminished in importance. Women accordingly tended to increase their on-farm and/or off-farm work. These patterns, however, varied substantially by region, with women's direct involvement in farming being greatest in the South and West. Sachs notes that one constant historical feature of women's roles in farm households has been in their performing the bulk of household tasks. Women have become somewhat more involved in farm decision making, but Sachs suggests that their decision-making roles have been secondary to their labour contributions to the farm operation and household.

Sachs' analysis then turns to the historical development of a "domestic ideology" — that women are naturally "domestic" beings and more suited to household work than men — in U.S. society in general and in agriculture in particular. She argues that keeping women in domestic roles has facilitated the development of capitalism by maintaining a reserve labour force and creating a consumptive domestic class. Sachs suggests that the family farm, particularly because it is based on the nuclear family, has been one of the most pervasive mechanisms through which patriarchal relations have been maintained on U.S. farms.

She then explores how this patriarchal order has been reflected in the sexual division of labour on farms and in the social relations of women-operated farms. Sachs' book has been exemplary in identifying how gender relations on U.S. farms have been shaped by the twin forces of patriarchy and capitalist development. Her work remains the most comprehensive theoretical statement and overview of empirical evidence on farm gender relations that has yet appeared.

Sachs' book is wide-ranging but brief, suggesting that further elaborations and critiques will increase our theoretical and empirical understanding of farm gender relations and the gender division of farm and rural labour. In particular, several ambiguities remain in the literature. At an empirical level, while the past decade of research has established a number of frequently supported generalizations about the structuring of farm women's work roles, there are a number of conflicting results.

Some observers, for example, argue that women make major contributions to farm work, whereas others see a clear division of labour such that women participate very little in farm tasks. Other empirical controversies may well arise from the fact that most of the major studies in the field have been based on data from a single state.

Thus further research is needed to clarify these empirical patterns and to incorporate them in a larger theoretical perspective — a major component of which will likely revolve around establishing and explaining differential patterns of women's work roles across farming regions. Some of this theorizing and comparative research is, however, taking place in Europe, for example, has analyzed in detail the gender relations on farms in France and has made some initial notes toward a theory of variations in gender differences in the household, child-rearing, and farm work.

LAND OWNERSHIP AND FARM STRUCTURE

OWNERSHIP OF U.S. LAND

The land surface of the United States covers 2.3 billion acres. Private owners held 61 percent in 2002, the Federal Government 28 percent, State and local governments 9 percent, and Indian reservations 3 percent. Virtually all cropland is privately owned, as is three-fifths of grassland pasture and range and over half of forestland. Federal, State, and local government holdings consist primarily of forestland, rangeland, and other land. Most land in Federal ownership—largely in the West—is managed by the Department of the Interior (68 percent) and the Department of Agriculture (28 percent).

Farm operators do not own all the land used in agriculture. According to the 1999 Agricultural Economics and Land Ownership Survey (AELOS), farmers held 58 percent of the land in farms in 1999. These landowning farmers also made up 58 percent of the 3.4 million farmland owners. Nonoperator landlords accounted for the remaining 42 percent of land in farms. Ninety-five percent of nonfarm landlords were individuals/families or partnerships. Of these unincorporated landlords, 55 percent were at least 65 years old. Many nonfarm landlords have a historic connection to farming. Among the people who have exited farming or inherited farmland since the number of farms peaked during the Great Depression, a number have retained ownership of some or all their land.

Farm Numbers, Farm Types, and Conservation Programmes

The number of farms has declined dramatically since its peak of 6.8 million in 1935, with most of the decline occurring during the 1940s, 1950s, and 1960s.

The decline in farm numbers has levelled off since the 1970s. By 2002, 2.1 million farms remained. The remaining farms have a much larger average acreage, but averages mask differences among farms. Today's farms range from very small retirement and residential farms to industrialized operations with sales in the millions. Part of this diversity stems from the very low sales threshold ($1,000) necessary for an operation to qualify as a farm for statistical purposes.

One way to address the diversity of farms is to categorize them into more homogeneous groups. The farm typology developed by ERS identifies five groups of small family farms (sales less than $250,000): limited-resource, retirement, residential/lifestyle, farming-occupation/low-sales, and farming-occupation/high-sales. The typology also includes large family farms, very large family farms, and nonfamily farms. In addition, very small farms (sales less than $10,000) make up more than half of all farms. Very small farms account for a particularly large share of farms in the limited-resource (72 percent), retirement (76 percent), and residential/lifestyle (76 percent) groups. Production, however, is concentrated among larger farms; small farms account for only 27 percent of the total value of production.

The smallness of most farms has implications for conservation and the environment. An ERS study found that smaller corn farms are less likely to use conservation tillage than are larger farms. The practice is more practical for larger farms because they have more acres over which to spread the cost of new or retrofitted equipment necessary to adopt conservation tillage. Small farms whose operators are retired or farm part-time are also less likely to adopt conservation tillage, possibly because of hesitancy to change familiar production practices. Small farms, however, participate widely in the Conservation Reserve Programme (CRP) and the Wetlands Reserve Programme (WRP).

DISTRIBUTION OF CONSERVATION PROGRAMME PAYMENTS BY TYPE OF FARM

These farms harvest most of the land planted to programme commodities and therefore receive three-quarters of commodity programme payments. However, CRP and WRP—the two major conservation programmes—are targeted at particular types of land, not commodities. Since small farms own 70 percent of the land held by farms, they play a large role in natural resource and environmental policy. Retirement, residential/lifestyle, and low-sales farms account for nearly two-thirds of conservation payments and a similar share of the land farmers enrolled in the CRP and WRP. Participating farmers in each of the three groups tend to enroll large shares of their land in these programmes: 46 percent of the land operated on retirement farms, 28 percent on residential/lifestyle farms, and 23 percent on low-sales farms. In contrast, enrollment ranges from 5 to 9 percent for participating high-sales, large, and very large farms.

Because their main job is off-farm, residential/lifestyle operators are limited in the amount of time they can spend farming. As a result, residential/lifestyle farmers find CRP and WRP attractive, since these programmes require little time. Given their life-cycle position, many retired farmers have land available to put into conservation uses. The same forces may also be acting on low-sales operators, who average 57 years of age and may be scaling down their operations. If an off-farm job and advanced age are major determinants of land going into conservation uses, it may be relatively easy to get smaller farms to enrol land in the programmes. Getting larger farms to enrol more of their land might require higher payments, if the opportunity cost of idling their land is higher.

Land Tenure

Farm operators leased 38 percent of their total farmland in 2002, down from 40 percent in 1997 and 43 percent in 1992, according to the census of agriculture. This decline may reflect increasing rental costs as parcels of land become smaller. Parcels of farmland available to rent tend to become subdivided with time due to division among heirs. Smaller parcels increase transaction costs to operators assembling land to expand their operations. Still, rented land as a share of total farmland is higher than the 35-percent rate that prevailed in the 1950s and 1960s.

About 38 percent of all farms rented land in 2003, 32 percent as part owners and 6 percent as tenants. Land leasing has changed from a way for beginning farmers to enter agriculture to a way for established farmers to access additional land. Renting allows farms to expand without the debt and commitment of capital associated with ownership.

In fact, about 17 percent of very large family farms are tenants, a larger percentage than in any other group. Conventional wisdom holds that farmland owners have a long-term interest in their land and thus are more likely than renters to adopt conservation practices. Soule and others (1999 and 2000) found this to be true among corn farmers, at least in the adoption of conservation practices that provide only long-term benefits, such as grassed waterways and strip cropping.

The situation was different for conservation tillage, which can increase profits in the short run by maintaining or increasing yields while reducing machinery, fuel, and labour costs. Cash-renters are less likely than owner-operators to use conservation tillage, but share-renters appear to act like owner-operators in adopting conservation tillage. Share-renters may have an incentive to adopt conservation tillage, if the landlord bears some of the costs that may increase under conservation tillage, such as herbicide expenditures. Share-landlords are also more likely to be involved in management decisions than cash-landlords, which may make share-renters act more like owners.

LAND USED FOR RECREATION AND WILDLIFE AREAS

Special uses include rural transportation; rural parks and wildlife; defence and industrial uses; and farmstead, farm roads/lanes, and other onfarm uses. These special uses increased from 85 million acres (4 percent of the land area of the contiguous 48 States) in 1945 to 153 million acres (8 percent) in 2002. Land in transportation uses (highways and roads, railroads, and airports in rural areas) increased by 4 million acres (17 percent) between 1945 and 1982. Transportation uses declined by about 0.5 million acres from 1982 to 1992 due to the abandonment of railroad facilities and rural roads, and the classification of some transportation uses as urban areas.

Land used for recreation and wildlife areas (Federal and State parks, wilderness areas, and wildlife refuges) expanded 344 percent from 1945 to 2002 (an increase of 78 million acres). The increase came mostly from conversion of Federal lands, previously in forest and grassland pasture and range. Land in defence and industrial uses declined by 10 million acres (40 percent) from 1945 to 2002. Farmsteads, farm roads, and other farm uses declined by 4 million acres (29 percent) between 1945 and 1997. This decline reflects trends toward fewer farms and larger, more consolidated farms, as well as an increasing tendency for farm households to live off the farm. In response to expanding U.S. population, land in urban uses—including homes, schools, office buildings, shopping sites, and other commercial/industrial uses—increased from 15 million acres in 1945 to 25 million acres in 1960, 47 million acres in 1980, and 59 million acres in 2002. While the U.S. population nearly doubled, the amount of land urbanized quadrupled. However, urban uses still comprise only 3 percent of the total land area of the contiguous States.

Miscellaneous other land uses decreased from 1945 to 1964, and have since trended upward, showing a 54-percent increase from 1964 to 2002 reflecting improved data and reclassification of grazing and forest lands. These uses include marshes and open swamps not included in other major land uses, bare rock areas, deserts, some rural residential areas, and other uses not inventoried. Wetlands are defined by soil and hydrological characteristics and may occur on land in many different uses.

REGIONAL CHANGES IN LAND USE

While land in every use occurs in all 10 regions of the contiguous States, some uses are more concentrated in some regions than in others. Regions with the largest cropland acreage are the Northern Plains, Corn Belt, and Southern Plains. Grassland pasture and range is concentrated in the Mountain and Southern Plains regions. Acreage in forest use, special and miscellaneous other uses is highest in the Mountain region.

The Northeast, Appalachian, Southeast, Delta States, and Lake States regions lost cropland between 1945 and 2002. The largest increases occurred in the Northern Plains and Mountain regions, with smaller increases in the Corn Belt, Southern Plains, and Pacific regions. Western increases may have resulted in part from federally subsidized irrigation water.

Nine of the 10 regions lost grassland pasture and range between 1945 and 2002. While grassland pasture and range increased 11 million acres (10 percent) in the Southern Plains, the Northeast region lost about 70 percent of its grassland pasture and range, and the Appalachian and Lake States regions lost more than 50 percent. The Northeast and Appalachian regions saw the reforestation of grassland, loss of some grassland to urbanization, and concentration of the dairy industry. Decreases in the Corn Belt, Northern Plains, and Mountain regions were likely associated with the conversion of some grassland pasture and range to cropland.

Cropland Use and Federal Programmes

While total cropland acreage has varied up and down and generally declined since 1969, greater shifts have occurred between cropland used for crops and cropland idled, mostly because of Federal programmes. Cropland used for pasture has exhibited less variation in acreage than cropland idled. Most cropland used for crops is harvested, but typically 2-3 percent experiences crop failure and 5-10 percent is cultivated summer fallow. In 2002, farmers harvested one or more crops on an estimated 307 million acres of cropland, down 4 percent from 1997. About 8 million acres of the total harvested were double-cropped. When double-cropped land is counted twice, total acres harvested rise to 315 million acres.

Table. Major uses of cropland, 48 contiguous States, 1992-2004

Cropland	1992	1997	2002	2004[1]
			Million acres	
Cropland used for crops[2]	337	349	340	336
Cropland harvested[3]	305	321	307	312
Crop failure	8	7	17	9
Cultivated summer fallow	24	21	16	15
Cropland idled by all Federal programs[2]	55	33	34	35
Annual programmes	19	0	0	0
Conservation Reserve Programme	35	33	34	35
Total, specified uses[2,4]	392	382	374	372

Note:

[1]Preliminary, subject to revision.

[2]Breakdown may not add to totals due to rounding.

[3]A double-cropped acre is counted as 1 acre.

[4]Does not include cropland pasture or idle land not in Federal programmes that is normally included in the total cropland base.

Cropland used for crops was at a record high of 387 million acres in 1949, when no acres were idled by Federal programmes. In 1972, cropland used for crops was near a record low of 334 million acres as Federal programmes idled 61 million acres. Cropland used for crops climbed to 387 million acres in 1981, when Federal programmes idled no cropland, and dropped to 333 million in 1983, when Federal programme set-asides reached a historic peak of 78 million acres under the Payment-in-Kind (PIK) programme. The Federal Agricultural Improvement Act (FAIR) of 1996 eliminated all Federal acreage reduction programmes other than the Conservation Reserve Programme (CRP)

Between 1983 and 2002, cropland used for crops increased overall, while total acreage idled by Federal programmes decreased to 34 million acres. From 1997 to 2002, cropland used for crops declined from 349 to 340 million acres, while acreage idled under CRP increased by about 1 million acres. Between 2002 and 2004, acreage in CRP increased by an additional 1 million acres, while cropland used for crops declined by about 4 million acres.

The 14-million-acre drop in harvested cropland between 1997 and 2002 was coincident with a decrease in cultivated summer fallow and an increase in failed acres due to widespread drought. Crop failure occurred on 17 million acres, over 5 percent of the acreage planted, in 2002. This failed acreage was the largest since 1956. The use of summer fallow has been decreasing since the late 1960s, and stood at 15 million acres in 2002, down from 42 million acres in 1969. Four crops—corn for grain, soybeans, wheat, and hay—accounted for 80 percent of all crop acres harvested in 2002. The additional 17 "principal" crops accounted for another 15 percent of harvested area. Vegetables, fruits, nuts, melons, and all other crops accounted for 4.5 percent of crop area harvested in 2002.

URBANIZATION OF AGRICULTURAL AND OTHER RURAL LAND

Cropland conversion to urban uses is largely irreversible, so it is important to know the rate of conversion and how much of the loss is replaced from other land uses. Excessive loss of cropland to urban uses could lessen the production of food and fibre and the supply of rural amenities, such as open space, watershed protection, and rural lifestyles. A variety of Federal, State, local, and private programmes address such concerns.

Although urban land constituted less than 3 percent of the U.S. land area in 2000, 79 percent of the population lived there. Even large percentage increases in urban area would amount to small decreases in rural area since it is so vast. The rate of expansion (by decade) of urban area has declined from 39 percent during the 1950s, to about 36 percent during the 1960s and the 1970s, and to 18 percent in the 1980s. According to the Census Bureau, urban area was 59 million acres in 2000, just 7 percent above the previous estimate for 1990 (DOC/BOC, 2002). However, the Census Bureau adopted a new definition of urban area for the 2000 Census, improving the precision of urban area measurement but making it more difficult to compare urban area before and after this year. If urban area for 1990 is recalculated using the 2000 definition, the Census Bureau's estimate falls to 51 million acres, implying a 15-percent increase in urban area from 1990 to 2000.

The National Resources Inventory (NRI), conducted by USDA's Natural Resources Conservation Service (NRCS) in cooperation with Iowa State University, is an alternative source for estimates of urban and rural areas. The NRI uses a consistent definition for built-up areas, though it differs from the definition used by the Census Bureau. According to the NRI, "developed land," which includes large and small urban and built-up areas as well as rural transportation land, totalled 107 million acres in 2002 in the contiguous United States. The NRI indicates that developed land increased by 14 million acres (19 percent) over 1982-92 and 21 million acres (24 percent) over 1992-2002.

Land converted to urban uses comes from several different major land uses. The NRI indicates that 20 percent of new developed land came from cropland between 1997 and 2002. Prime cropland— land that has the best combination of physical/chemical characteristics for agricultural production—is converted to developed uses at about the same rate (5 percent per year) as nonprime cropland. About 21 percent of rural non-Federal land is prime and 26 percent of crop land converted to urban uses over 1997-2001 was prime. Rural land, defined as all land that is not urban, contains rural residential land, consisting of houses and associated lots. Nonfarm rural residential area was estimated to be about 94 million acres in 2002, up from 56 million acres in 1980. The average rate of increase in rural residential land was 1.7 million acres per year from 1980 to 2002. Combining both rural and urban residential land, the total increase in residential area was about 2 million acres per year during this period.

2

Animal Husbandry and Agricultural Animal Health Care

The principal scientist or animal management supervisor should make all animal care personnel aware of their responsibilities during both normal work hours and emergencies. A programme of special husbandry procedures in case of an emergency should be developed. It is the reserach facility management's responsibility to ensure that personnel caring for agricultural animals used for research or teaching are appropriately qualified or trained. This responsibility may be delegated to an IACUC. Qualification by experience and training must be documented.

The animal facility manager must ensure that all animal care personnel are aware of their responsibilities during and outside normal work hours. Protocols for emergency care must be developed and made available to all personnel.

OBSERVATION

Animals in intensive accommodations should be observed and cared for daily by trained and experienced caretakers. Illumination must be adequate to facilitate inspection. In some circumstances, more frequent observation or care may be needed. Under extensive conditions, such as range or pasture, observations should be frequent enough to detect illness or injury in a timely fashion, recognize the need for emergency action, and ensure adequate availability of feed and water. A disaster plan must be developed for observing animals and providing care during emergency weather or health situations. Regardless of accommodations, animal observations should be documented and husbandry or health concerns reported to the animal facility manager or attending veterinarian as appropriate.

EMERGENCY, WEEKEND, AND HOLIDAY CARE

There must be a means for rapid communication in case of an emergency. In emergencies, facility security and fire personnel must be able to contact staff members responsible for the care of agricultural animals. Names and contact information for those individuals should be posted prominently in the animal facility and provided to the security department or telephone center. If posting names and contact information poses privacy or security issues, a contact number for a security or command center should be used instead.

The institution must ensure that emergency services can be contacted at any time by staff members. The institution must assure continuity of daily animal care, to encompass weekends, holidays, unexpected absences of assigned personnel, and emergency situations. Staff assigned to weekends and holidays must be qualified to perform assigned duties. Cross-training of staff and establishment of standard operating procedures is encouraged to assure consistent, high-quality care.

Emergency veterinary care must be readily available after daily work hours, on weekends, and on holidays. In the event that weather conditions or natural disasters make feeding temporarily impossible, every attempt should be made to provide animals with a continuous supply of water. Absence of feed for up to 48 h should not seriously endanger the health of normal, well-nourished juvenile or adult cattle, sheep, goats, horses, poultry, or swine. Feed should be provided within 24 h to very young animals that are not nursing their dams.

EMERGENCY PLANS

A site-specific emergency plan must be developed to care for agricultural animals that are used for research and teaching. The goal for a plan should be to provide proper management and care for the animals regardless of the conditions. However, some conditions may be so unusual and extreme that it will not be possible to provide immediate care for the animals and to simultaneously ensure employee safety.

Thus, emergency plans should define proper animal management and care and parameters to ensure employee safety. Emergency plans should name employees or positions that are considered essential for providing proper animal management and care. Those employees should a priori understand that responding to emergencies is a condition of employment and that they will be held accountable should they fail to care properly for the animals.

Plans should focus on emergencies that are most likely to occur in the specific geographic area or the research or teaching facility. Emergency plans should include animal evacuation plans specific to the research or teaching facility and actions to be taken if transportation is interrupted.

ANIMAL IDENTIFICATION AND RECORDS

Animals should be permanently identified by a method that can be easily read. Identification of individual animals is desirable, but, in some circumstances, it is acceptable to identify animals by group, cage, or pen. Individual birds may be wing-banded or leg-banded. Ear-notching, ear tattooing, electronic transponders, and branding may be used for individual identification of other species, and each has its advantages and disadvantages. Ear notches and tattoos are permanent and effective, but notching constitutes elective surgery and tattoos generally cannot be read without restraining animals.

Electronic transponders require special sensor units or stations, but should be considered when possible. Cattle and horses are most consistently identified using freeze-branding on the hip, shoulder, rear leg, or side. In addition, when freeze branding is used on more than one breed of horse, branding is performed under the mane. Some states require that cattle be permanently identified by branding with a hot iron; however, this procedure is more stressful than freeze-branding.

Ear and neck chain tags, although readable at some distance, can become lost and are therefore not necessarily permanent. In addition, neck chains and straps should be avoided in situations in which the animal could become entangled in a fence, rock outcropping, or other feature of the environment. Any associated pain and distress should be considered when determining the method of identification. In some cases, it may be necessary to identify animals in multiple ways. Individual records are needed for most animals.

These records should include information about the animal, its source and location, its productivity, its reproductive performance, protocols the animal is assigned to, and its ultimate disposition. Records for individual animals or groups should also include dates of vaccination, parasite control measures used, blood testing dates and results, and notations as to whether castration, spaying, or other elective procedures have been performed.

Applicable veterinary data to be recorded include dates of examination/ treatment, clinical information/diagnosis, names of medications and amounts and routes of administration, descriptions of surgical procedures, and resolution of surgical procedures or illnesses. Principal scientists or animal facility managers may wish to record nutritional information. Research protocols often dictate that additional information be recorded.

VERMIN CONTROL

Programmes should be instituted to control infestation of animal facilities by vermin. The most effective control in facilities prevents entry of vermin into the facility by screening openings and ceilings; sealing cracks; eliminating vermin breeding, roosting, and refuge sites; and limiting access of vermin to feed supplies and water sources. Building openings should be screened with 1.3-cm mesh, and ceilings with ridge vents should be screened with 1.9-cm mesh to minimize rodent and bird entry. Smaller mesh sizes are recommended where they will not interfere with airflow. Mesh may need to be installed along foundations below ground level, especially with wood foundations. Pesticides should be used only as approved. Particular caution should be exercised with respect to residues in feedstuffs, which could injure animals and eventually pass into the meat, milk, or eggs. Pesticides should be used in or around animal facilities only when necessary, only with the approval of the scientist whose animals will be exposed to them, and with special care.

A pesticide applicator or a commercial service may be used. In some regions, wildlife and stray cats and dogs may spread zoonotic diseases, including rabies, to agricultural animals. In highrisk locations, institutions should implement an educational programme that includes training scientific and animal care personnel to recognize the signs of rabies in both wildlife and agricultural species and to handle and report potentially rabid animals.

Inoculation may be advisable for humans who may come into contact with animals in regions where rabies is endemic. Many agricultural institutions keep cats for pest-control purposes. Although the use of free-roaming cats is a traditional form of pest control for agricultural facilities, cats may limit the ability for baiting and may present hygiene or accident risks or serve as disease vectors. However, when cats are present, proper veterinary care and oversight should be provided to these animals. Veterinary care should include vaccinations, parasite control, and neutering.

AGRICULTURAL ANIMAL HEALTH CARE

Agricultural animal health care involves proper management and husbandry as well as veterinary care. Proper management is essential for the well-being of animals, the validity and effectiveness of research and teaching activities, and the health and safety of animal care personnel.

Sound animal husbandry programmes provide systems of care that permit the animals to grow, mature, reproduce, express some species-specific behaviour, and be healthy. Specific operating procedures depend on factors that are unique to individual institutions. Well-trained and motivated personnel can often achieve high-quality animal care with less than ideal physical plants and equipment.

NEW HEALTH PRODUCTS FOR FARM ANIMALS

Animal diseases cost U.S. agriculture $17 billion each year, according to the Congress Office of Technology Assessment. To help prevent these losses, animal scientists are using biotechnology to develop an array of products to diagnose, treat, and prevent disease in farm animals. An area where animal biotechnology has already had a profound impact is diagnostic testing. Many animal diseases are difficult to diagnose. A veterinarian often has to wait hours or days for laboratory results to confirm a diagnosis. In the meantime, the veterinarian must either withhold treatment or risk using an inappropriate therapy.

Using biotechnology, scientists are developing fast, accurate diagnostic tests for many of the most common farm animal diseases. In some cases, a veterinarian can conduct the diagnostic tests right on the farm and immediately start the best therapy. Many of these new tests use monoclonal antibodies. Antibodies are proteins an animal's immune system produces in response to invasion by bacteria, viruses, or parasites. Each kind of antibody binds specifically with a characteristic part of a particular kind of invader. Scientists make monoclonal antibodies by fusing two kinds of cells. One is an immune system cell that produces antibodies, which bind to part of a particular disease-causing microbe; the other is a cancer cell. The cell that results from the fusion inherits from the immune system cell the ability to produce antibodies, and from the cancer cell the ability to reproduce indefinitely.

This new hybrid cell, or hybridoma, acts as an antibody factory. The antibodies are called monoclonal because they are produced by identical "clones" of the original hybridoma-all will bind with the identifying structure on the surface of the microbe in question. One application of monoclonal antibodies is as a test for brucellosis in cattle. This bacterial disease often causes cows to abort pregnancies, and it can be spread to farmers and people who drink milk from infected cows. Vaccines protect pregnant animals from abortion, but vaccinated cows may become carriers of the disease. Conventional diagnostic tests cannot distinguish between the disease-causing microbe and the vaccine, which is made from the microbe. Thus, these tests cannot help the farmer determine which animals are carriers of the disease and which are not. But diagnostic tests using monoclonal antibodies are so specific, they distinguish between cattle that carry the disease-causing bacterium and those that carry only the vaccine. With the aid of these new tests, farmers can isolate carriers to prevent the spread of infection. Diagnostic tests are not just for diseases.

A monoclonal antibody test to detect pregnancy in animals is so simple that a farmer can read the test right on the farm. Monoclonal antibodies are also used in a test to detect a disease called scours in piglets and calves. Scours is a bacterial disease of newborn animals that causes diarrhea. Many of the afflicted animals die of dehydration. By providing quick, accurate diagnosis, the monoclonal antibody test for scours permits immediate administration of therapy. The therapy for scours in calves is also based on monoclonal antibodies. Infected calves are fed monoclonal antibodies that coat the offending bacteria, preventing the microbes from causing diarrhea.

Some other therapies for animal diseases are based on recombinant DNA technology. Genetic engineering allows scientists to insert one or more new genes into an animal, plant, or microbial cell. In developing treatments for an animal disease, for instance, an animal gene is inserted into a microbe or an animal cell that can be grown in a laboratory culture. The inserted gene is one that produces a natural disease-fighting protein. The microbes or animal cells become minute factories, producing large quantities of the therapeutic protein. For example, genetic engineering is used to produce interferons and interleukin-2, natural proteins that fight viruses.

Veterinarians long have used antibiotics against bacterial diseases, but they have had few drugs to treat viral infections. Interferons and interleukin-2 not only kill viruses directly, they also stimulate the animal's immune system, improving the immune system's ability to fight disease naturally. One target of research regarding interferons and interleukin-2 is shipping fever, a disease of cattle caused by several viruses infecting an animal simultaneously.

Although the animal's immune system normally keeps the viruses in check, outbreaks of the disease may result from stress during transport to feedlots or market. Shipping fever costs the beef industry more than $250 million a year. Injections of interferons or interleukin-2 before shipping may help to augment the cattle's natural immunity and prevent shipping fever.

BETTER VACCINES TO PREVENT DISEASE

In addition to producing new diagnostic tests and therapeutic proteins, animal scientists are using biotechnology to develop vaccines to prevent disease. A safe, effective vaccine for swine pseudorabies, a fatal herpes-virus, is already in use. An advantage of the swine pseudorabies vaccine, and other vaccines made through recombinant DNA technology, is that they use only a small portion of the original microbe. Conventional vaccines use killed or weakened forms of the disease-causing microbe.

Sometimes, when the microbes are not completely killed or sufficiently weakened, a conventional vaccine may cause the disease it was supposed to prevent. Disease-causing genes are not included in the genetically engineered vaccines. Therefore, the recombinant vaccines build up the body's immunity without the risk of causing disease. Another advantage of recombinant vaccines is the speed with which they can be developed.

"Conventional vaccine development can take 20 or 30 years, maybe even 100 years," says Karen Jacobsen, D.V.M., University of Georgia College of Veterinary Medicine. "This is one reason why we don't have vaccines for a lot of animal diseases today. " Recombinant vaccines are being developed for foot-and-mouth disease, a highly contagious viral disease that infects cattle, sheep, and other animals. Although this disease has been eradicated in North America, it still causes substantial livestock productivity losses elsewhere, particularly in developing countries. Infected herds must be slaughtered, and contaminated ranches must be left idle for months to prevent new outbreaks of the disease. Conventional foot-and-mouth vaccines are made by weakening the virus that causes the disease. These vaccines sometimes revert to the virulent state, and they have caused outbreaks of the disease in Europe. The foot-and-mouth vaccine made with biotechnology cannot cause the disease because, as in other biotech vaccines, the disease-causing genes have been removed. The biotech vaccine does not need to be kept cold and, therefore, can be used in developing countries. In addition to being used in treating swine pseudorabies and foot-and-mouth disease, recombinant DNA technology is being used to develop vaccines against dozens of livestock and poultry diseases, such as scours and a variety of respiratory disorders. Some vaccines may be injected into eggs so that chickens are immunized before they hatch. Most vaccines protect against viral or bacterial infections, but using genetic engineering, researchers in New Zealand and Australia have developed a vaccine against a parasitic disease. Farmers must destroy the meat from sheep infested with a species of tapeworm common in those two countries. The new vaccine, which protects sheep from the tapeworm, may save farmers millions of dollars annually.

PROTEINS FOR ANIMAL NUTRITION

Natural proteins called somatotropins, or growth hormones, help animals convert feed into muscle or milk. Using recombinant DNA technology, scientists have been able to develop bacteria that produce commercial quantities of somatotropins.

They have found that administering small additional amounts of these proteins to animals helps to increase the animals' efficiency of feed conversion, thereby saving the farmer money. Porcine somatotropin (PST) may help improve human health, as well as lower the farmer's cost of production. PST improves feed efficiency in hogs by 15 to 20 per cent. It also reduces fat deposition, allowing PST-treated hogs to provide consumers with leaner cuts of pork. The ability to produce lean pork has enormous implications for improving human health by reducing dietary fat and cholesterol. Another growth protein, bovine somatotropin (BST), is being administered to make dairy cows more efficient, thereby lowering the dairy farmer's costs. Regular administration of small amounts of BST increases a cow's milk yield by 10 to 25 per cent. Although feed intake also increases, this feed increase is proportionately less than the increase in milk production.

Studies indicate that BST can improve the milk-to-feed ratio by 5 to 15 per cent. Extensive laboratory testing has shown that the meat and milk of treated animals contain no more PST or BST than the meat or milk from untreated animals. Other tests have shown that these proteins do not act on the human body, and therefore, they are safe to eat. Nevertheless, the use of BST remains controversial.

Testing is under way to make sure that animals are not adversely affected by treatment with somatotropins. Another protein that can be administered to animals to increase feed efficiency is called growth hormone releasing factor (GHRF). While not a growth protein itself, GHRF causes the animal to increase production of growth proteins.

IMPROVED ANIMAL BREEDS

Long before anyone knew about genes or DNA, people were selectively breeding animals to produce desirable traits, such as strength, feed efficiency, and disease resistance. Selective breeding is a crude technology, because each cross between two animals mixes hundreds of thousands of genes, both desirable and undesirable. Recombinant DNA technology is now making it possible for scientists to breed animals with great precision. One or more new genes can be inserted into an animal embryo without disturbing the rest of the animal's genes. The resulting animal with the new gene or genes is referred to as a transgenic animal. Despite the differences in technique, the scientist using recombinant DNA has the same objectives as the selective breeder: improving efficiency, disease resistance, and other desirable traits. But recombinant DNA provides a wider array of possible new traits, since recombinant techniques allow animal scientists to insert genes not only from the same or closely related species, but also from distantly related animals, plants, and even microbes. One possible application of gene transfer in animals is to develop dairy cows that produce more nutritious milk. Bottle feeding human infants with cow's milk is not an ideal substitute for breast feeding, because cow's milk differs from human milk in some important components. Scientists are looking for ways to transfer genes into dairy cows that will enable the cows to produce milk that more closely resembles human milk.

Advances in Control of Parasitic Diseases Continue to Rely on Biotechnology Parasitic disease continues to be a major health problem worldwide in humans and domestic animals. Applying biotechnology has led to progress in prevention and control in many areas, particularly in regard to diagnosis. Parasites, like most organisms, carry repetitive sequences in their genomes. These are sometimes tandem repeats (back-to-back in a chromosome), or they may be located at random throughout the genome. Often, repetitive sequences are intergenic and noncoding, and are thus far more subject to evolution and change, increasing the likelihood that they will be genus or even species specific. Since detection of these sequences is not dependent upon gene expression, which may vary over time, repetitive sequence-based assays allow detection of the parasite at any life stage.

The effectiveness of repetitive sequence-based polymerase chain reaction (PCR) methods of identification has been demonstrated for parasites of several clinically important genera (1). A PCR assay has also been developed to differentiate two closely related nematode species that are virtually indistinguishable by morphologic examination (2), as well as for diagnosis of Toxoplasma and Cryptosporidium infections (3). Attempts to develop anti-parasite vaccines have met with multiple failures. The problems often result from lack of post-translational modification by currently used vector-host systems and the inability to regenerate the secondary and tertiary structure of the native immunogens. Antigen presentation in recombinant parasite vaccines has also been thought to be a problem resulting in failed efficacy. There is now some evidence to indicate that cloning of genes for antigens of interest to yield chimeric molecules expressed in the BCG strain of Mycobacterium bovis may provide improved antigen presentation to stimulate both humoural and cellular responses (4); however, it is unlikely that future applications of this technology will be simple. Nonetheless, work is proceeding on the application of biotechnology to the development of vaccines against parasites in most food animals. A particularly striking exception to the marginal successes reported in other systems has been a recombinant vaccine for a parasitic helminth of sheep (5). This product yielded greater than 94% protection in experimental work and is now available as a commercial product. A further possibility is the application of anti-idiotypic monoclonal antibodies. When monoclonal antibodies are prepared against a monoclonal antibody, it is possible to obtain, in a very small number of cases, an antibody which mimics the original antigen. Results of work with Cryptosporidium perfringens epsilon toxin suggest that animals immunized with anti-idiotype monoclonals can be protected against toxin challenge.

VETERINARY CARE

ATTENDING VETERINARIAN

The agricultural animal health care programme is the responsibility of the attending veterinarian. The Institute for Laboratory Animal Research (ILAR), National Research Council *Guide for the Care and Use of Laboratory Animals* defines the attending veterinarian as "a veterinarian who has direct or delegated authority" and who "should give research personnel advice that ensures that humane needs are met and are compatible with scientific requirements."

Animal Welfare Act regulations and the Public Health Service policy require that the attending veterinarian have the authority to oversee the adequacy of other aspects of animal care and use, including animal husbandry and nutrition, sanitation practices, zoonoses control, and hazard containment.

Research and teaching institutions must provide investigators and instructors with access to a veterinarian who has experience in the care of agricultural animals. The veterinarian can be full-time or part-time and must have authority to ensure that the provisions of the programme are met. The attending veterinarian must be provided access to all research and teaching animals and to any related documents including health care records.

The attending veterinarian also must be involved in the development and oversight of the veterinary care programme, as well as in other aspects of animal care and use such as protocol review, establishment of anesthetic and analgesic guidelines, study removal criteria, training of animal users, and responsible conduct of research activities. Veterinary involvement in these activities helps to ensure animal health and well-being. The attending veterinarian is not required to be the sole provider of veterinary care and can delegate authority to another qualified veterinarian. However, the attending veterinarian must communicate with, and oversee veterinary care provided by, other veterinarians. When necessary, the attending veterinarian should utilize the expertise of other professionals when making determinations about agricultural animal care. Trained nonveterinary staff may administer treatments according to standard operating procedures approved by the attending veterinarian.

PREVENTIVE MEDICINE

Adequate agricultural animal health care in research and teaching involves a written and implemented programme for disease prevention, surveillance, diagnosis, treatment, and endpoint resolution. The objectives of such a programme are to ensure animal health and wellbeing, minimize pain and distress, maintain animal production, prevent zoonoses, provide assistance to investigators on study-related animal health issues, and avoid contaminants or residues in animal products.

The programme should include training for animal users regarding animal behaviour, humane restraint, anesthesia, analgesia, surgical and postsurgical care, and euthanasia. A mechanism for direct, frequent, and regular communication must be established among personnel who are responsible for daily animal care and observation, animal users, and the attending veterinarian. This will help ensure that timely and accurate animal health information is effectively communicated.

SICK, INJURED, AND DEAD ANIMALS

Animal care personnel must be trained to recognize signs of illness and injury. In general, sick and injured animals should be segregated from the main group to protect them and the other animals, observed at least once daily, and provided with veterinary care as appropriate. When animals are separated, a mechanism should be in place to communicate to staff the status of the animals and to ensure proper daily, weekend, holiday, and emergency care.

In some circumstances, segregation is not feasible or may disrupt the social hierarchy, cause additional stress to the animal, or adversely affect research. The advantages of segregation should be weighed against its disadvantages, especially for mild illnesses or injuries that can be easily managed. Care should be taken to minimize spread of pathogens from ill animals to healthy animals by observing appropriate biocontainment measures. Incurably ill animals or ill or injured animals with unrelievable pain or distress should be humanely killed as soon as possible.

Unexpected deaths should be reported to the attending veterinarian. Dead animals are potential sources of infection and should be disposed of promptly by a commercial rendering service or other appropriate means following applicable state and local ordinances and regulations. Postmortem examination of fresh or well-preserved animals may provide important animal health information and aid in preventing further losses. When warranted, waste and bedding that have been removed from a site occupied by an animal that has died should be moved to an area that is inaccessible to other animals and the site appropriately disinfected.

MEDICAL RECORDS

An important component of an agricultural animal health programme is maintaining records that can be used to monitor animal health events, both physical and behavioural health events, as well as outcomes and levels of production. Medical records should comply with the American College of Laboratory Animal Medicine (ACLAM) statement on medical records.

Group health records may be appropriate for animals that are kept as cohorts, particularly when animals undergo periodic evaluation by means of examining several representative individuals of the group. The institution, under the guidance of the attending veterinarian, should determine the method(s) by which medical records are maintained. Oversight of medical records is the responsibility of the attending veterinarian and the IACUC.

When institutional representatives determine that a medical record should be created, the record typically contains the following information:

- Identification of the animal(s) or group(s);
- Clinical information, such as the animal's behaviour, results of physical examinations, and observed abnormalities, illnesses, and/or injuries;
- Immunizations and other prophylactic treatments and procedures;
- Documentation and interpretation of diagnostic tests;
- Documentation of research interventions;
- Treatments prescribed and administered;
- Clinical response and follow up information;
- Descriptions of surgical procedures, anesthesia, analgesia, and perioperative care;
- Methods used to control pain and distress;
- Documentation of resolution;
- Documentation of euthanasia or other disposition; and
- Necropsy findings if necropsy is indicated.

The record system must be structured so that information is easily collected, gathered, analysed, summarized, and available to the veterinarian, the principal scientist, and the IACUC.

The ACLAM statement on *Medical Records for Animals used in Research Teaching and Testing* suggests that:

- Notations in the medical record should be made by individuals who have administered treatments, or made direct observations or evaluations of the animal(s) or their diagnostic results, or their designee. Individuals typically responsible for making notations in the record include veterinary staff, animal husbandry staff, and research staff. All entries in the record should be dated, indicate the originator of the entry and be legible to someone other than the writer.

IMPROVING HEALTH THROUGH DEVELOPING VACCINES

Most biotechnologies related to health focus on the needs of the developed world, meaning that 90% of health research is devoted to the health problems of 10% of the world's population. Two main approaches are being used to develop vaccines using recombinant DNA technology. The first involves deleting genes that determine the virulence of the pathogen, thus producing attenuated organisms (non-pathogens) that can be used as live vaccines. Currently, this strategy is more effective against viral and bacterial diseases than against parasites. Attenuated live vaccines have been developed against the herpes viruses that cause pseudorabies in pigs and infectious bovine rhinotracheitis in cattle. A number of candidate *Salmonella* vaccines have also been produced. The second approach is to identify protein subunits of pathogens that can stimulate immunity.

The International Livestock Research Institute (ILRI) used this approach to develop a vaccine against *Theileria parva*, the parasite that causes East Coast fever in African cattle. A novel strategy for developing vaccines against bloodsucking parasites involves using components of the gut wall of the parasite that are not usually exposed to the immune system of the host. When the parasite feeds, it ingests antibodies induced by the vaccine, which destroy the gut wall and, consequently, kill the parasite. This strategy has been used successfully to develop a vaccine against the one-host tick *Boophilus microplus*.

Vaccination is one of the most effective and sustainable methods of controlling disease. Vaccines against parasitic diseases in Africa and viral diseases in Asia have been shown to control disease effectively and increase livestock productivity. A recent approach has been to use vaccines based on DNA. The use of DNA in vaccines is based on the discovery that injecting genes in the form of plasmid DNA can stimulate an immune response to the respective gene products. This immune response is a result of the genes being taken up and expressed by cells in the animal after injection. The live-vector and DNA vaccination systems could be manipulated further to enhance the immunity conferred by the gene products. Experimental studies have demonstrated that these vaccines can potentially induce appropriate and enduring immune responses.

This technology is, in principle, one of the simplest and yet most versatile methods of inducing both humeral and cellular immune responses, as well as protecting against a variety of infectious agents. However, although immune responses have been induced in a number of larger species, most of the information on the efficacy of DNA immunisation comes from studies of mice. An exhaustive review of the information available on the use of DNA vaccines in farm animals, including cattle, pigs and poultry, has identified the areas that need specific attention before this technology can be used routinely.

These areas include the delivery, safety and compatibility of plasmids in multivalent vaccines and the potential for using immune stimulants as part of a DNA vaccine. Korean scientists have developed a combined vaccine against pleuropneumonia, pneumonic pasteurellosis and enzootic pneumonia in swine. Molecular biology has been used to produce an improved vaccine against swine fever. In the Philippines, a vaccine has been developed that protects cattle and water buffalo against haemorrhagic septicaemia, which is the leading cause of death in these animals. The new vaccine provides improved protection at a very low cost.

DIAGNOSTICS AND EPIDEMIOLOGY

Advanced diagnostic tests that use biotechnology enable the agents causing disease to be identified and the impact of disease control programmes to be monitored more precisely than was previously possible. Molecular epidemiology characterises pathogens (viruses, bacteria, parasites and fungi) by nucleotide sequencing, enabling their origins to be traced.

This is particularly important for epidemic diseases, in which pinpointing the source of the infection can significantly improve disease control. For example, the molecular analysis of rinderpest viruses has been vital in determining the lineages circulating in the world and instrumental in aiding the Global Rinderpest Eradication Programme. Enzyme-linked immunosorbent assays have become the standard means of diagnosing and monitoring many animal and fish diseases worldwide, and the PCR technique is especially useful in diagnosing livestock disease.

Many diagnostic techniques currently used in developing countries are cumbersome and unsuitable for low-resource settings. Molecular diagnostic technologies that are either already in use or being tested in low-income regions include polymerase chain reaction (PCR), monoclonal antibodies and recombinant antigens. These technologies can be modified to facilitate their application in the developing world. Simple hand-held devices that rely on the binding specificity of monoclonal antibodies or recombinant antigens to diagnose infection may be easily adapted for use in settings without running water, refrigeration or electricity. Molecular characterisation of the virus serotypes causing foot and mouth disease has helped in the vaccination and control programmes in Asia.

In Japan and Taiwan, DNA testing is being used to diagnose hereditary weaknesses of livestock. One test looks for the presence of the gene responsible for porcine stress syndrome in pigs. Pigs with this gene tend to produce pale poor-quality meat because of their reaction to the stress of transport and slaughter. Pigs with this gene can now be excluded from breeding programmes, so the gene will become less common.

In addition, DNA testing is being used in Japan to check for the gene that causes leucocyte adhesion deficiency in Holstein cattle. Cattle with this condition suffer from gum disease, tooth loss and stunted growth. They usually die before they are one year old. By using DNA testing, carriers can be identified and eliminated from breeding herds. Bulls used for breeding can also be tested to make sure that they are not carriers. Another DNA test identifies a gene that leads to anaemia and retarded growth in Japanese Black cattle.

NUTRITION AND FEED UTILISATION

The shortage of feed in most developing countries and the increasing cost of feed ingredients mean that there is a need to improve feed utilisation. Aids to animal nutrition, such as enzymes, probiotics, single-cell proteins and antibiotics in feed, are already widely used in intensive production systems worldwide to improve the nutrient availability of feeds and the productivity of livestock. Gene-based technologies are being increasingly used to improve animal nutrition, either through modifying the feeds to make them more digestible or through modifying the digestive and metabolic systems of the animals to enable them to make better use of the available feeds. Feeds derived from GM plants (a quarter of which are now grown in developing countries), such as grain, silage and hay, have contributed to increases in growth rates and milk yield.

Genetically modified crops with improved amino acid profiles can be used to decrease nitrogen excretion in pigs and poultry. Increasing the levels of amino acids in grain means that the essential amino acid requirements of pigs and poultry can be met by diets that are lower in protein. Metabolic modifiers have also been used to increase production efficiency (weight gain or milk yield per feed unit), improve carcass composition (meat-fat ratio), increase milk yield and decrease animal fat.

The use of recombinant bovine somatotropin (rBST) in dairy cows increases both milk yield and production efficiency and decreases animal fat. In the USA, the use of rBST typically increases milk yield by 10% to 15%. Although trials conducted in developing countries have reported a similar percentage increase, this increase is not significant because of the low milk yields and the high cost-benefit ratio. However, rBST is being used commercially in 19 countries where the economic returns make its use worthwhile. A porcine somatotropin has been developed that increases muscle growth and reduces body-fat deposition, resulting in pigs that are leaner and of greater market value.

CONSTRAINTS ON APPLYING THE TECHNOLOGY

The application of new molecular biotechnologies and new breeding strategies to the livestock breeds used in smallholder production systems in developing countries is constrained by a number of factors. In the developing world, poverty, malnutrition, disease, poor hygiene and unemployment are widespread, and biotechnologies must be able to be applied in this context.

Over the last few decades, the green revolution has brought comparative prosperity to farmers with land, but the majority of farmers, who are landless or marginal farmers and subsist only on livestock, have been neglected and remain poor.

The major constraints on applying biotechnologies have been enumerated by Madan and include:

- The absence of an accurate and complete database on livestock and animal owners so that programmes can be implemented
- The biodiversity present within species and breeds in agro-ecological systems
- The fact that models of biotechnological intervention differ distinctly between developed and developing economies
- The fact that many animal species and breeds are unique to the developing world; each has its own distinct developmental, production, disease resistance and nutrient utilisation characteristics
- The lack of trained scientists, technicians and fieldworkers to develop and apply the technologies, both in the government and in the private sectors
- The absence of an interface between industry, universities and institutions, which is necessary to translate technologies into products
- The inability to access technologies from the developed world at an affordable price in order to make a rightful, positive and sustainable contribution to livestock production and the economic welfare of farmers
- The high cost of technological inputs such as materials, biologicals and equipment
- The failure to address issues of biosafety and to conduct risk analyses of new biologicals, gene products, transgenics and modified food items, and, above all
- The negligible investment in animal biotechnology.

The critical issues affecting livestock productivity have recently been re-examined. Research that aims to enhance productivity and sustainability should focus on improving livestock feeds and nutrition, improving animal health, managing natural resources relating to the livestock sector, assesing the impact of technological interventions, and strengthening the capacity of the national agricultural research systems of developing countries.

Furthermore, the potential production capacity and contribution of livestock to the economy are still not being achieved in developing countries because the transfer, adaptation and adoption of technology is hampered by the lack of a clear policy for livestock development that is conducive to the introduction of new proven technology and by the lack of information flow from and to decision makers.

In developing countries, there is a wealth gap between urban and rural areas, which persists and may even be widening; the rural-urban divide also tends to be reflected in education and health indicators. In addition, women in rural (and urban) areas who are predominantly involved in animal husbandry have higher illiteracy rates than men. A survey of 21 African countries recently highlighted the substantial disparities in primary schooling between urban and rural areas, in favour of urban dwellers.

Special attention must be given to the knowledge and information needed to enable rural people to apply biotechnology. There is a need to identify alternative delivery systems for animal healthcare and to propose new roles for the state and the private sector in service delivery.

BUILDING CAPACITY

Owing to the constraints outlined above, the economic benefits of animal biotechnology cannot be realised without a conscious, sustained, holistic, multi-stakeholder, participatory approach. There is a great need to ensure that capacity is not just created but also is retained and enhanced. Capacity-building activities must be carried out at all levels: the awareness of policy and decision makers must be raised, the necessary legal and regulatory frameworks must be initiated, the technical and regulatory capacities must be enhanced and institutions may need to be overhauled. More importantly, it is necessary to assess and deploy competent operators and institutional capacity continuously so that, as biotechnology advances, the procedures required for its safe use can be constantly evaluated, upgraded and applied. This is a daunting task, but it can be achieved through firm commitment and partnerships.

FUNDING TO IMPLEMENT TECHNOLOGY

Developing and commercialising improved technologies in most developing countries has been the responsibility of the public sector, and technology has been disseminated freely. This situation will have to continue if superior genetics, diagnostics and vaccines are to be delivered. However, research and almost all commercial development of biotechnology in the developed world are being driven largely by the private sector. The global trends in funding for research and development and production do not address the concerns, needs and opportunities of the developing world. Developing countries are finding it increasingly expensive to access and use new technologies. There is limited private-and public-sector investment in animal health and production, particularly in relation to modern biotechnologies that are 'resource hungry'. Although several discoveries have been made in laboratories in the developing world, in most cases these have not been converted into useful technologies or products. The key potential users–resource-poor often illiterate farmers with a limited knowledge base–do not feel that applying these technologies is worth the effort, cost and risk involved. This is mainly because there is no agency or industry that can scale up and package the technology. Also, in the developed world, there is an economic incentive to market biotechnological services and products; this is lacking in the developing world because of the limited purchasing power of resource-poor stakeholders.

Research in biotechnology in recent years has also been motivated by economic considerations, and little research is conducted in the developing world because of the probable lack of returns on the investment. For understandable reasons, current funding policies in developing countries focus on areas that will yield practical benefit in the short term. In determining future policy, policy-makers and funding bodies must not lose sight of the substantial benefits that can be gained in the longer term by investing in strategic research into vaccine development. Adequate multi-institutional (national and international) support through an international donor consortium is needed to develop cost-effective, cheap and easily adaptable biotechnological products.

nt spent by international agencies on animal biotechnology in developing countries is currently very low and constitutes only a small percentage of the total spending on agriculture. The World Bank, the Food and Agriculture Organization, the Consultative Group on International Agricultural Research, the United Nations Development Programme, the United States Agency for International Development, the Swedish International Development Cooperation Agency, the International Development Research Centre, the Asian Development Bank and other donor and funding agencies have to designate a higher percentage of funds to the livestock sector. It has been convincingly shown that investing in livestock has a dramatic and far-reaching impact on the human development index. This is a strong argument in favour of investing heavily in animal production and health biotechnologies in order to bring economic prosperity, nutritional security, rural development and health improvements to poor populations in the developing world.

Although animal production is being changed significantly by advances made in thousands of biotechnology laboratories around the world, benefits are reaching the developing world in only a few areas of conservation, animal improvement, healthcare (including diagnosis and control of disease) and the augmentation of feed resources. Adopting biotechnology has resulted in distinct benefits in terms of animal improvement and economic returns to the farmers. Over the past decade, the ILRI has focused on biotechnological applications, especially in Africa, and several developing countries now have multi-institutional programmes to develop and apply biotechnology. The developing world will have to respond to the many genebased technologies now being developed with a sense of commitment, trained manpower, infrastructure and funding.

ANIMAL HEALTH: DISEASE DIAGNOSIS

Successful control of a disease requires accurate diagnosis. This has been greatly improved in recent years through developments in biotechnology. The most recent major development, the finding that it is possible to immortalise individual antibody-producing cells by hybridisation to produce antibodies of a given class, specificity and affinity (*i.e.* monoclonal antibodies) has provided a tool that permits the analysis of virtually any antigenic molecule. The use of monoclonal antibodies has revealed that the failure of vaccines (*e.g.* of rabies) to provide protection in all parts of the world was due to the diversity in the antigenic composition of the causative virus. The (monoclonal antibody) technology is relatively simple and can readily be applied in developing countries. Monoclonal antibodies are currently supplied to developing countries directly or in the form of kits and simple reagents for completion of the tests. For example, kits for rinderpest virus diagnosis used in African countries come in this form.

The ability to generate highly specific antigens by recombinant DNA techniques has made it possible for an increasing number of enzyme-linked immunosorbent assays (ELISA) to have the capacity to differentiate between immune responses generated by vaccination from those due to infection.

This has made it possible to overcome one of the major drawbacks of antibody detection tests: the fact that, because antibodies can persist in animals for long periods, their presence may not indicate current infection. ILRAD has developed a technique to overcome this problem in diagnosis of trypanosomiasis. The parasite antigen detection test uses monoclonal antibodies raised in laboratory mice to capture the parasite antigens which are then revealed by their reaction with a second layer antibody to which is conjugated an easily detectable enzyme. This test reveals current infections and facilitates differentiation between the major trypanosome species. This has important implications for disease control, especially because of the association of different parasite species with different epidemiological and disease circumstances.

The advent of PCR has enhanced the sensitivity of DNA detection tests considerably. For example, PCR used in combination with hybridisation analysis, has been shown to provide a sensitive diagnostic assay to detect bovine leukosis virus.

Other diagnostic techniques include nucleic acid hybridisation (NAD) and restriction endonuclease mapping (REM). As has been indicated above, one of the most valuable features of these molecular techniques is their specificity and sensitivity. A good example of the specificity of NAD is its application in distinguishing infections caused by *peste des petite ruminants* (PPR) virus from rinderpest, diseases whose symptoms are clinically identical and which cannot be distinguished antigenically with available serological reagents. This technique also allows comparison of virus isolates from different geographical locations. Detailed overview of biotechnological tools for diagnosis of livestock diseases has been provided by Bourne and Bostock and Robinson and McEvoy.

Other available diagnostic techniques which may have application in small ruminants and/or cattle include: nucleic acid probes (NAP) for heartwater *(Cowdria ruminantium), chlamydia psitacci,* Paratuberculosis and Bluetongue; restriction endonuclease reaction (RENR) for diagnosis of *Corynebacterium pseudotuberculosis*; PCR for characterising subtypes of Bluetongue from different geographical regions; monoclonal antibodies (MAB) for differentiating false positive anti-Brucella titres caused by *Yersinia enterocolitica* and true positive anti-Brucella titres in latent infected animals; and MAB for diagnosis of Toxoplamosis, Pasteurellosis, *Mycoplasma* spp, PPR and Boder Disease. Examples of tropical diseases for which diagnostic tests are available are presented in Table. Teale reviewed the diagnostic techniques that are currently available or which could be developed in the near future.

Table. Tropical Diseases for which Probes and Monoclonal Antibodies (MAB) are Available.

Viral diseases	Bacterial diseases
Rinderpest	Cowdriosis
Peste des petite ruminants	Contagious bovine pleuropneumonia
Bluetongue	Contagious agalactia
African horse sickness	Contagious caprine pleuropneumonia
Foot-and-mouth disease	Anaplasmosis (A. centrale and A. marginale)
Vesicular stomatitis	Haemorrhagic septicemia (only MAB available)

VACCINES

Conventional means of controlling major livestock diseases include chemotherapy, vector control, vaccination, slaughter of infected stock, and other management practices (including grazing management and controlled stock movements).

Vector control requires continuous application of pesticides. These are often unaffordable to farmers in developing country. Moreover, where these drugs or pesticides are used, resistance by parasites is often encountered and reinfection following administration of drugs against parasitic diseases usually occurs. Additionally, in many cases drugs are not readily available locally. In some cases where they are available, they are ineffective, either because they have been partly preserved or they are not genuine.

Immunisation remains one of the most economical means of preventing specific diseases. An effective vaccine can produce long-lasting immunity. In some cases, vaccination can provide lifetime immunity. Moreover a small number of doses is usually required for protection. Level of infrastructure and logistical support required for a large-scale vaccination programme is such that a successful vaccination campaign can be implemented in remote rural areas. In general, vaccines offer a substantial benefit for comparatively low cost, a primary consideration for developing countries.

Vaccines have conventionally been produced by several methods some of which have become rather static with regard to efficacy, safety, stability and cost. Very effective vaccines against animal diseases such as rinderpest and pig cholera have been in use for more than 20 years and have helped to significantly reduce the incidence of these diseases world-wide. However, vaccines of questionable efficacy also exist. Impotency, instability, adverse side effects, and reversion of attenuated organisms to wild (disease-causing) forms represent some of the problems. However, research strategies for the development of better, cheaper and safer vaccines are constantly being sought. Through the use of monoclonal antibodies and recombinant DNA technologies, it is now possible to define and produce immunogenic components much more rapidly. These technologies are increasingly being used to clarify the pathogenetic mechanisms and immune response to microbial diseases.

This should lead to the production of more effective vaccines in the future. To date many candidate vaccines have been produced by these techniques. However, only few of these are being produced commercially.

Table summarises some vaccines developed by recombinant DNA technology. There are other vaccines under various stages of development.

Table. Example of Some Novel Animal Vaccines

Species	Disease	Producer
Avian	Coccidiosis,	Genex and A.H. Robins
	Newcastle virus	Codon and Salsbury labs
Bovine	Papilloma virus	Molecular genetics
	Viral diarrhoea	California Biotechnology
	Brucellosis	Ribi ImmunoChem
	Rinderpest	USDA and University of California, Davis
Swine	Parvovirus	Applied Biotechnology
	Dysentery	Codon
Equine	Influenza, Herpes	California Biotechnology
		Applied Biotechnology
Companion	Canine parvovirus	Applied Biotechnology

The following are of particular relevance to small ruminants: recombinant vaccines against Bluetongue and Rift Valley Fever may soon be available for field use; a hybrid virus vaccine against Orf, a thermostable recombinant vaccine against PPR and recombinant vaccines against *Taenia ovis* and Echinococcus; tick gut antigens and *Haemonchus* gut antigens produced by molecular techniques offer ways for prophylaxis against these parasites.

Progress is also being made towards the development of vaccines against Babesiosis and Theileriosis in cattle which may indicate prospects for similar immune-prophylaxis becoming available for sheep and goats as well.

Physiology of Lactation and Growth

Recombinant bovine somatotropin (BST) is a genetically engineered synthetic analog of the natural growth hormone. Since the 1970s there have been a number of studies on the effects of BST on milk yield, reproductive performance and health as well as its likely effect on humans who consume such milk. Under good management and feeding, regular BST administration to lactating dairy cows increases milk yield by 15-30% and also increases efficiency of milk production.

BST is now readily available and is already in commercial use in the United States. However, the appropriateness of BST use to increase milk production in the USA is doubtful as the country already has a milk surplus and the public is also concerned about its effects on health. Indeed, the benefit of BST in the USA is perceived to be towards reducing the production costs of large dairy farms, but this could push smaller farmers out of the market. The appropriateness of BST use for developing countries is still a matter of debate. Those supporting its introduction argue that its use in commercial dairy farms could increase the national milk output. Those opposed point out the fact that BST does not improve milk yield in indigenous non-dairy breeds and that its use on crossbred and exotic dairy cattle will require more feeding, and that provision of adequate nutrition is already a problem for most dairy operations in developing countries. Whether or not BST affects reproductive performance has not yet been conclusively established. Although most reports have indicated nonsignificant effects of BST on reproductive performance, most such studies have been on single lactations; the effect of BST on lifetime productivity needs to be investigated. Some studies have shown that feed conversion efficiency declines in subsequent lactations.

The study by Phipps *et al.* also indicated a decline in the incremental amount of milk yield from BST application in subsequent lactations. Additionally, the effect of BST on reproductive performance is likely to be more adverse in the presence of higher biotic and abiotic stresses, including nutritional stresses. There is also need to examine the economics of BST application in view of the known association of its use with mastitis and other infections. BST use is thus bound to be associated with increased use of antibiotics and other veterinary drugs. Thus, in evaluating the potential role of BST in developing countries, one needs to consider not only the possible response levels of the cattle in these countries to BST treatment, but also the cost of BST, the amount and cost of other incremental inputs required for effective use of BST, and the milk prices. Ultimately, the main technical constraint to BST use in developing countries will not only be its cost, but the absence of an efficient delivery system; current use of the technology requires regular injections.

Porcine somatotropin (PST) and recombinant growth hormone stimulatory peptides (*e.g.* growth hormone releasing factor, GRF) along with BST have been shown to increase growth rates by 8-38% in cattle, sheep and pigs. In almost all cases, administration of exogenous growth hormones have been associated with increased carcass protein and reduced carcass fat. Other growth-promoting agents (*e.g.* anabolic steroids and beta agonists) have been shown to have even larger effects but public concerns over the possible residual effects in meats have led to their being banned in most developing countries. The use of anabolic implants is, however, permitted in some countries such as the USA.

SAFETY AND ANIMAL HANDLING

Stockmen working in animal housing and associated handling facilities are exposed to a number of potential risks and health hazards. Many stockmen work alone, handling big, sometimes dangerous animals as well as large groups of animals.

Several work tasks involve daily lifting, carrying of heavy loads and awkward working postures, which may cause overload injuries on the worker's musculo-skeletal system. Other potential risk factors contributing to accidents as well as to physical health problems and physiological stress for the worker is the fact that they operate in an environment where they are exposed to airborne particles, gases, noise. The work environment is partly the same environment, which is provided for the animals, and it should be noted that there is a strong relationship between the animal welfare issues and the handling of the work environment problems in cattle houses.

HEALTH AND INJURIES AMONG LIVESTOCK FARMERS

Accident Agents

In an accident situation there is a complex interaction between man and injury agents. In animal husbandry the human is subjected to a high degree of exposure to a large number of factors that singly or in combination can be accident provoking, such as animals, equipment and building structures. Risky situations are often due to wet and slippery floors as well as systems for material and animal handling.

Prevention of Accidents

With reference to accident statistics some accident preventing facilities and measures can be listed:

- Good cattle handling and handling facilities
- Pens from which the herdsman can easily escape if necessary in case of for instance frightened cattle and aggressive bulls
- Personnel passages through yard fences, properly placed
- Wide feeding passage or possibility to shut out the cattle from the feeding area
- Non slip flooring
- Satisfactory lighting considering quality and quantity
- Well designed ladders and stairways
- Protective railings around the fodder hay-chute in two storey houses
- Handles and bolts that do not stand out in alleys and pathways
- Protectors for the knuckles on barrows
- Emergency stops on machinery

Table. Recommended Illumination Level for Different Work Areas in Buildings Used for Beef Production. Lux Values Refer to the Work Plane. In Addition to the Quantity, the Quality of Light must be Considered

Building or Work Area/Task	Intensity of Illumination, Lux
Animal housing area	150
Animal handling area	250
Veterinary treatment	1,000
Feed processing, feed alley	250
Silo, feed storage	100
Ladders and stairs	250

Ergonomics

Any normally fit adult person, regardless of sex, physical strength and body dimensions should be able to work comfortably in livestock buildings. Despite an increasing degree of mechanisation and automation many jobs in animal husbandry are still associated with lifting heavy loads, moving and carrying equipment, feed and other materials with manual methods. An important step towards improved human welfare within cattle housing is to eliminate unnecessary loads on the human musculoskeletal system.

Health Aspects Related to Air Quality

It has been known for decades that farm workers suffer health risks due to inhalation of gases and airborne particles commonly referred to as dust. The particles may be liquid droplets or solid. Several respiratory disorders have been reported, which are associated with dust and work in confined livestock buildings such as hypersensitivity, pneumonia, acute inflammation, chronic bronchitis, occupational asthma and toxin fever. The respiratory problems connected to gases range from mild irritations of the respiratory tract to lethal effects. For the most frequent components in the air environment in livestock buildings there are statutory threshold limit values, which differ from country to country. It should be noted that threshold limit values in connection with animal welfare in many countries often are lower than the work environment legislation.

Dust and Dust Reducing

Dust in animal houses is generated inside the building and arises primarily from the animals and bedding and feed material. The airborne particles are often carriers of biologically active material like endotoxins, pathogens and allergens that can highly be disease provoking. Of special interest is the concentration (*i.e.* number of particles per air volume) of respirable particles with a diameter less than 5 μm that can be inhaled and deposited in the lung.

The smaller particle the higher possibility that the particle reaches the lower respiratory tract and the deeper the particle is deposited the higher is the risk for illness. Even if "common dust" can be seen by human eyes (larger diameter than 100 ìm), the very small respirable particles cannot be, so we are indeed dealing with a danger our senses are not able to detect.

There are basically four approaches that can be used for airborne dust reducing:

1. Prevent particle formation
2. Prevent particle release
3. Remove suspended particles from enclosed workspaces
4. Isolate workers from dust clouds in workspace.

Of course the best approach in this matter is to have the cattle and the working sites outdoors or in open barns. In the cattle barn the most effective way to obtain a low dust concentration level is to prevent formation by always using hygienically perfect forage and bedding material. Providing an effective ventilation system in a barn is a good practice to remove suspended respirable particles as well as gases.

Gases and Reduction of Gas Levels

In animal buildings over 150 different gases and volatile compounds have been identified. Most of the volatile compounds originate from biodegradation of animal excreta or are produced by the animals themselves such as carbon dioxide (by the respiration of the animal). The gases that are found in the highest elevated concen-trations are ammonia, carbon dioxide, hydrogen sulphide and methane.

It must be emphasised that it is dangerous to enter any manure tank without either using a self-contained air supply or combining several measures: testing the air, constant and adequate ventilation of fresh air and using an harness and lifeline on the person entering the tank. As well, human and animals positioned over a slatted floor with manure storage underneath are at risk during agitation, mixing or pumping of manure due to the gases, especially hydrogen sulphide. A good strategy is to evacuate dangerous zones during these manure-handling actions.

Ammonia

Ammonia is the most common polluting gas in the atmosphere of the cattle barn. Animal manure is the main source of ammonia. The gas together with hydrogen sulphide a major component in what is termed odour, noticeable at ammonia concentrations of 5 ppm or more. Higher concentrations than that cause irritation to the respiratory organs and can aggravate the negative health effects of high dust concentrations. Many factors affect the emission and concentration of ammonia such as air flow rate, manure and air temperature, manure surface area, density of animals, the degree of mixing of urine and faeces, time intervals between manuring, the pH-value, carbon/nitrogen ratio, mois-ture content and type of bedding. Consequently, many measures can be taken such as: proper ventilation and sufficient number of air changes, low air temperature in the barn and good urine drainage characteristics of the floor.

Hydrogen Sulphide

Hydrogen sulphide is the most toxic gas in animal confinement houses and 200 ppm and more are lethal for humans. The presence of the gas is of concern in buildings with liquid manure systems particularly. During slurry clearing operations concentrations in the range of 1,500 to 2,000 ppm have been reported. Since hydrogen sulphide is lethal the aim should be to keep it below detectable concentration at all working sites.

Special care has to be taken when pumping or agitating the manure. During emptying a slurry cellar the workers and the animal should be evacuated from the barn. In mechanical ventilated, closed animal houses continuously evacuation of the gases in an indoor slurry pit through a perforated duct under the floor and along the pit can reduce the gas concentrations above the slatted floor. By removing the slurry daily from the cattle barn to an outdoor storage it is possible to hold the concentration at an acceptable level.

Methane

Methane is generated when manure is stored under anaerobic conditions. This occurs in all non-aerated manure storages. Primarily, methane is dangerous because of its flammability. If methane is mixed in a proper proportion with oxygen a spark will be able to set off an explosion.

Carbon Dioxide

Carbon dioxide is present in all air, indoor as well as outdoors. Only extremely high concentrations (70,000 ppm or 7% and over) could have severe health effects. Mostly, incidents and accidents involving carbon dioxide are connected with asphyxiation due to oxygen deficiency rather than the direct effect of carbon dioxide itself. For instance, it is dangerous to enter a haylage silo during storing. In animal buildings the gas is more an indicator of ventilation efficiency than a health risk.

Noise

Noise can be defined as "unwanted sound, rapid, annoying pressure vibration in the surrounding air". Ventilation fans, grinders, vacuum pumps and mechanical feeding and manure systems generate noise. High sound levels are a real stress factor. Exposure to noise will be able to cause temporary deafness and permanent hearing loss after a variable period if the level is over 85 dB(A). The unit, dB(A), is exponential and the human sense of hearing will perceive the sound level as halving if a reduction of 10 dB(A) occurs.

It is fairly easy to reduce the sound level of ventilation fans and other mechanical noise generators. Maximum noise level recommendations and regulations differ from country to country. Furthermore, it is important to reduce noise when handling cattle in order to improve animal movement and to make cattle handling safer. Cattle are more sensitive to high frequency noise than human. Animals will be calmer and easier to handle, if noise level is reduced. Clanging and banging metal parts should be silenced with rubber pads. Equipment operated with hydraulics should be engineered for quietness.

CATTLE HANDLING AND CATTLE HANDLING FACILITIES

Handling facilities are an essential part of a safe, easy and rapid handling of cattle. Appropriate handling and handling facilities remove much of the stress and frustration of the workmen, which inevitably occurs with excited, stubborn or aggressive animals. Properly constructed facilities confine cattle safely and efficiently with minimal animal stress and risk of injury to both cattle and workers.

Animal Behaviour and Improving Animal Handling

Understanding cattle behaviour can help farm workers to avoid dangerous situations and minimise accidents to handlers, as well as to design the handling facilities appropriately. Animals have natural boundary called the Flight Zone.

Deep penetration into the Flight Zone can cause panic and escape attempts. Handlers should have the possibility to remove, and should remove themselves from the Flight Zone, if the animal becomes aggressive. To move an animal forward: stand in the shaded area position B, behind the Point of Balance at the shoulder.

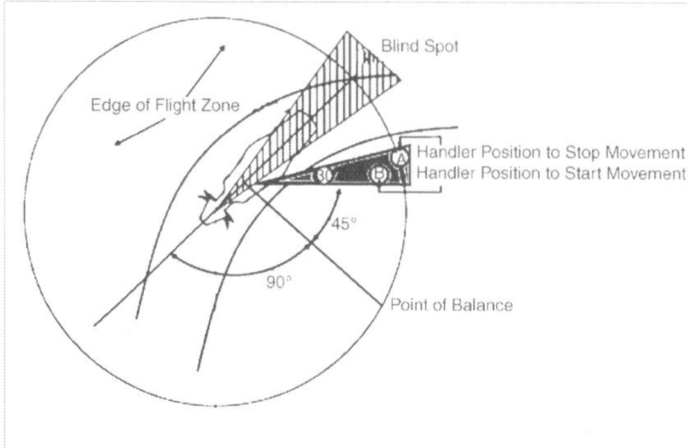

Fig. The Flight Zone of Animals

Keep out of the Blind Spot at the rear of the animal. To stop movement: back off to position A. To make an animal back up: stand in front of the Point of Balance. To make an animal turn left or right: approach the animal head on. Factors reducing an animal's Flight Zone size: frequent contact with people; history of gentle handling; calm environment. Factors enlarging the Flight zone: infrequent contact with people; history of rough or abusive handling; excitement.

To improve animal handling and animal movement, many technical measures can be taken such as:

- Appropriate handling facility design.
- Animals tend to move from a dark to a more brightly lighted area. The light should illuminate the chute up ahead. Eliminate shadows and patches of light and dark, which may confuse animals. An approach is to illuminate the entire working area. Lamps should not shine into the eyes of approaching animals because glaring and blinding light impedes movement. Illumination should be uniform and diffuse.
- Prevent distractions, such as a chain hanging down in an entrance. Avoid sparkling reflection in a puddle, a moving reflection on a sheet of metal or bars of shadow across an otherwise sunlit alleyway. Dark colours can create shadow effects. Bright colours such as white and light yellow have been proven satisfactory.
- Reduce noise
- Animals might refuse to move, if they can see people ahead. Install shields to prevent animals from seeing farther ahead. Gates can be rigged with motor controls so a handler standing behind the cattle can open them.
- Solid sides that prevent the cattle from seeing outside the fences should be provided in the races and crowding pens. The crowding gate on the crowding pen should be solid as well, preventing animals from attempting to turn back to were they come from.
- Cattle are sensitive to changes in type and texture of floors and fences.

Changes in type of flooring can cause balking. Use the same type of flooring throughout a facility, if possible. Use non-slip flooring. Drains should be located outside main drive alleys, chutes and crowd pens.

BIOTECHNOLOGY AND ANIMAL HEALTH

The potential areas of impact for biotechnology on the control of infectious diseases and improved animal health are manifold. They include the identification of susceptible stock, tests for the rapid and early diagnosis of disease, an understanding of the mechanisms of pathogenesis and the development of safe and effective vaccines or prophylactics. The principal methodologies which will enable these targets to be approached are those of monoclonal-antibody production and recombinant-DNA technology. The benefits of selecting stock on the basis of resistance to specific infectious agents would appear at present to be restricted to those areas of the world where the importance of a single disease predominates over all others (*e.g.* trypanosomiasis). Whilst the possibility of selecting on an antigen non-specific basis is likely to be of more general relevance, at present the development of vaccination strategies that can effectively control infectious agents is likely to be of greatest significance for improved animal health.

To this end monoclonal antibodies can be used in the development of diagnostic sera and reagents, and the characterization of virulence and adhesion determinants. Once identified such determinants can be isolated in a pure form from crude mixtures with monoclonal immunosorbents. Infectious diseases may be controlled by the prophylactic use of monoclonal antibodies, but this direct effect will be of limited use in animal production, by comparison with their potential use as 'educators' of the immune system in anti-idiotype vaccines. Recombinant-DNA technology can provide a range of potential immunogens including attenuated or non-pathogenic strains and subunit vaccines. Since the majority of infectious agents are met at mucosal surfaces, the problem that remains is how these vaccines may be delivered so as to provide effective presentation and so stimulate a protective response.

MUCOSAL DEFENCE

The level of challenge to, and the degree of sterility maintained differs markedly between the various mucosal surfaces; from the gastrointestinal and upper respiratory tracts to the lower respiratory tract, reproductive tract and mammary gland which at least in health remain sterile. They are presented with a vast array of different antigens ranging from potential pathogens to relatively harmless dietary components. Together with the skin they provide the interface between the host and its environment. However, unlike the skin there are other physiological roles (*e.g.* gaseous exchange, absorption of nutrients, etc.) which restrict their ability to keep antigens out. Since the mucosal surfaces provide such a primary point of contact it is at these sites that the products of biotechnology might be expected to provide the most relevant and beneficial effects. For convenience the major part of the present review will be restricted to antigens presented to the gut. The immune system of the gastrointestinal tract is presented with two types of antigens, those capable of causing infection and those, such as dietary antigens, which in themselves impose no direct threat to the animal's health. The type of response stimulated is critical, for whilst failure to respond to potential pathogens may be fatal, an over-reaction to dietary antigens may lead to damaging hypersensitivity responses. Latter may in turn lead to altered handling of potential pathogens. Therefore, although the most obvious area of impact of biotechnology will be directly on infectious agents, regulating responses to food antigens may also be of significance.

THE RESPONSE TO FOOD ANTIGENS

Three responses have been described that may protect an individual from damaging hypersensitivity reactions to dietary antigens. Immune exclusion mediated by locally produced IgA may reduce the absorption of intact proteins across the gut mucosae and in certain species it has been shown that any antigen that is absorbed may be eliminated via bile. These processes which restrict the access of antigenic material to the immune system are not absolute and the development of an acquired specific hyporesponsive state is probably of greater significance. Thus orally induced tolerance may actively down-regulate the response to dietary proteins. Before the induction of this 'protected state' all animals pass through a transient phase of hypersensitivity, during which time damaging allergic reactions can be demonstrated. In farm species it has been shown that following weaning, piglets are sensitive to antigens in the post-weaning diet and since these changes occur before the proliferation of *Escherichia coli* it has been postulated that response to food proteins may be an important predisposing factor in the aetiology of post-weaning diarrhoea.

The importance of gut flora and bacterial lipopolysaccharide in regulating the response to dietary antigens has been clearly shown in rodents, thus by the use of engineered strains of bacteria it may be possible to regulate the response to dietary proteins and so prevent the transient allergic responses that reduce the gut's ability to handle potential pathogens. This brief description of the regulation of the response to food proteins attempts to serve two purposes.

First, it may indicate a role for engineered strains in the controls of hypersensitivity reactions and the consequences thereof. Second, the type of response generated to food proteins differs markedly from that required to eliminate infectious agents. It is then perhaps not surprising, as will be discussed later, that dead mucosal microbial vaccines are generally not very effective.

THE RESPONSE TO INFECTION

The implications of biotechnology on animal health directly related to infections are numerous; it is therefore proposed to discuss them sequentially from diagnosis to vaccine development. The conventional methods for the identification of the causal agent have in the past depended heavily on an ability to isolate and grow the organism in pure culture. The use of monoclonal antibodies should aid both the identification of infectious agents and their virulence factors. Unlike conventional anti-serums, which by virtue of the nature of the immune response are very heterogenous, monoclonal proteins have a unique specificity which allows such discrimination to be made.

For example, by use of the enzyme-linked immunosorbent assay (ELISA) it is possible to detect subtle antigen differences between strains of an organism as well as to characterize the immune response of the infected animal. Once having identified a particular virus, for example as being important in the pathogenesis of diarrhoea, how then might one proceed towards an effective vaccine? Consider for instance that virus X has been shown to be an important pathogen but is closely related to similar commensals. It can produce disease, and animals that recover from infection show solid protection. In neutralization tests it has been possible to show the specificity of X, but it is very difficult to grow! Further progress may then involve the development of a sensitive immunoassay for research and sero-epidemiology. Production of a monoclonal antibody (either conventionally from rodents or preferably from infected farm species) may allow isolation of a strain-specific antigen. Subsequently the antigen should be produced by cloning and expressing the gene in, for example E. *coli,* by recombinant-DNA technology. Such a procedure may form the basis for the development of a vaccine.

CONTROL OF INFECTION

Without attempting to consider ways of eradicating an organism from the environment there are two types of approach to the preventionxontrol of infection within an animal. The first is essentially passive and involves little or no requirement of the host's immune system, whilst the second requires the stimulation of an active vigourous immune response.

PASSIVE PROTECTION

Monoclonal antibodies have been used therapeutically to passively protect neonatal pigs and calves against challenge with a lethal dose of a K99 positive pathogenic strain of E. *coli.* It has been pointed out that such an approach has a theoretical advantage over conventional vaccination of the dam, in that it does not require anticipation of the problem. Further, in this case where the disease is restricted to the first 24 h of life and the effects of the bacteria are contained within the intestine, it is directly treated by oral administration of the monoclonal antibody. A similar approach has been used to protect mice from intranasal infection with Respiratory Syncital Virus, but generally it would appear that the therapeutic use of monoclonal antibodies will be of benefit in specific, defined individual areas, notably in human medicine, rather than on a herd basis in veterinary medicine. The rationale for this argument is the amount of antibody required, particularly if it is necessary that it is given orally. Monoclonal antibodies to infectious agents may, however, be of benefit in the 'education' of the immune system and this active process will be discussed later. Another passive form of treatment involves the use of anti-microbials, and in this context biotechnology has already made a significant impact. Human leucocyte interferon has been isolated by monoclonal anti-immunosorbents and on a commercial level recombinant interferon and interleukin 2 are now generally available.

ACTIVE PROTECTION

Before considering the potential impact of biotechnology on the development of safe, effective vaccines, it is necessary to consider what is required of such products in order that they may stimulate a protective response. As has already been mentioned, the vast majority of infectious agents are met at a mucosal surface and therefore vaccines should be directed towards stimulating protection at these sites. To this end it is clear that by far the most effective method of eliciting such a response involves presentation of the antigen at the mucosal surface. In general parenteral vaccines have not been effective, and where exceptions exist they almost always involve some mucosal component (*e.g.* oral + parenteral).

The other exception to this principle involves immunization of the dam to achieve passive protection of the sucking neonate, by virtue of the presence of antibodies in colostrum. In the case of domestic species, antibodies in colostrum are derived from serum and these may be effectively stimulated by parenteral immunization. Unlike colostrum, milk antibodies are locally produced and in this case, like other true mucosal secretions, mucosal immunization is required to stimulate production of antibodies.

Having established that mucosal vaccines are those of choice for both theoretical and practical reasons, it is, however, disappointing that it is surprisingly difficult to stimulate a vigourous, protective response via this route. As has already been pointed out, this is of benefit with regard to the 'harmless antigens' (*e.g.* dietary antigens) also presented to the mucosae. To date the majority of effective mucosal vaccines are living attenuated organisms reflecting the now generally held axiom that either replication or at least adherence are essential prerequisites of mucosal immunogenicity. Whilst there is now clear evidence of memory in the secretory immune system, vital if any vaccination regimen is to be effective, it has also been shown that the response is highly dose-dependent. Thus where dead oral vaccines have been used, very large numbers of bacteria are required and even then the level of immunity is sub-optimal by comparison with that following the use of live organisms. Another problem associated with mucosal immunization, unique to the gastrointestinal tract, relates to an ability to pass through the stomach in a form that remains antigenic. In this context the use of capsules or liposomes may be of benefit. Alternatively the observed immunological link between mucosal sites, such that oral immunization leads to the appearance of antibodies in milk, may be of benefit.

The New Vaccines

Vaccines derived as products of biotechnology may take a variety of forms and rather than discussing them on a specific disease basis it is proposed to briefly describe the general feature of each type. The genes coding for protein or peptide antigens derived from a number of bacteria and viruses have been cloned and expressed in appropriate bacteria. For example, with the foot and mouth disease virus one of the structural capsid proteins, WI, has been identified as carrying the antigenic determinants necessary to raise neutralizing antibodies in experimental animals.

The viral RNA encoding for WI has been copied into double-stranded cDNA, inserted in a bacterial vector and expressed in *E. coli*. Subsequently a range of oligopeptides from VPI have been isolated or synthesized and a major antigenic determinant comprising nineteen amino acids identified. The question that remains is how these proteins or peptides may be presented to the immune system in order to stimulate a protective response. The isolated VPI protein, for instance, is not as antigenic as the virus and a 100 to 1000-fold excess on a molar basis is required in order to obtain satisfactory protection. With the oligopeptide it is necessary to couple to a carrier protein (*e.g.*, KLH) and although a single injection of 200 pg has been shown to protect guinea-pigs from virus challenge, to date no satisfactory method has been found to stimulate a protective response in cattle. Similar results have been obtained with *E. coli* antigens (K88ab, ac, ad and K99 have all been cloned and expressed) but it still remains to be determined how a protective mucosal response may be stimulated.

Whilst to date this synthetic peptide vaccine approach has been disappointing, its theoretical attraction over conventional dead vaccines remains since it avoids the requirement for handling large volumes of infectious materials, or vaccinal strains that grow poorly in bulk-suspension cultures, and the risk associated with failure to completely inactivate the virus before incorporation into a vaccine. An alternative approach involves the use of attenuated strains and these may take a variety of forms, and have been used against a number of agents. Deletion mutants of Aujeszky's disease virus have been prepared and although shown to have greatly reduced virulence are still highly immunogenic in pigs.

Similarly an attenuated strain of *Salmonella typhi,* Ty2la, which is unable to utilize galactose, has been derived. The organism is able to survive for a limited number of divisions, but antigenicity remains unaltered. For another enteric pathogen, cholera, an appropriate mutant is also available. In this case it produces only the B subunit of the toxin and thus, whilst being able to bind to the epithelial cell and give good protection, does not generally produce diarrhoea. Where attenuated strains of a particular organism are not available or appropriate the use of live recombinant vaccine may be of benefit. Intranasal inoculation with recombinant *Vaccinia* virus has been shown to protect mice against upper and lower respiratory tract infections with influenza. Alternative vectors to *Vaccinia* are also available, for example yeasts or engineered non-pathogenic strains of E. *coli* that produce only the B subunit of the labile toxin, and so bind to enterocytes but do not produce diarrhoea.

The general acceptability of live attenuated vaccines remains to be determined and the possibility of developing techniques whereby peptide vaccines can be delivered in a way that stimulates a protective mucosal response remains attractive. In this context the use of peptide conjugates coupled to molecules such as cholera toxin or lectins that bind avidly to epithelial cells may be of benefit. Alternatively the potential role of antiidiotypic antibodies (monoclonal) may be of interest, it having been recently shown that such antibodies may prime for protection against E. *coli*. The capacity of biotechnology to produce a range of antigenic moieties derived from infectious agents would appear to be unlimited. Further, it should not be difficult to identify those antigens that are important in the pathogenesis of disease.

The major problem that remains is how best these products might be presented to the immune system in order to stimulate a protective immune response. It is therefore to this target of understanding how antigens are presented to and processed by the mucosal immune system that intense future research should be directed.

3

Agricultural Development in India

INDIAN ADVANTAGE IN AGRICULTURE

India, today, holds a small share of the global biotech market, but has all the capabilities to become a dominant player. The consumption of biotech products in India is expected to quadruple in the next decade. The human and animal segment of the industry alone is growing by at least 20%.

India has a rich human capital, which is the strongest asset for this knowledge-based industry. India has a large English speaking base and, according to Confederation of Indian Industry estimates, produces roughly 2.5 million graduates in IT, engineering and life sciences, about 650,000 postgraduates and nearly 1500 PhDs qualified in biosciences and engineering each year. India has proved its competency in selected areas of biotechnology such as, to name a few: capacity in bioprocess engineering, skills in gene manipulation of microbes and animal cells, capacity in downstream processing and isolation methods, and its competence in recombinant DNA technology of plants and animals. India has also allowed assisted stem cell research that permits researchers to use embryos from fertility clinics upon informed consent of the donors, thus giving it a clear head start in this new and promising field in Biotech. Clearly, India has the strength and capabilities in this industry, and a definite advantage to forge ahead and become the chosen location for many biotech companies looking for large markets and low cost qualified manpower to work in their R&D division.

AGRICULTURE IN INDIA

India's record of progress in agriculture over the past four decades has been quite impressive. The agriculture sector has been successful in keeping pace with rising demand for food. The contribution of increased land area under agricultural production has declined over time and increases in production in the past two decades have been almost entirely due to increased productivity.

Contribution of agricultural growth to overall progress has been widespread. Increased productivity has helped to feed the poor, enhanced farm income and provided opportunities for both direct and indirect employment. The success of India's agriculture is attributed to a series of steps that led to availability of farm technologies which brought about dramatic increases in productivity in 70s and 80s often described as the Green Revolution era. The major sources of agricultural growth during this period were the spread of modern crop varieties, intensification of input use and investments leading to expansion in the irrigated area. In areas where 'Green Revolution' technologies had major impact, growth has now slowed.

New technologies are needed to push out yield frontiers, utilize inputs more efficiently and diversify to more sustainable and higher value cropping patterns. At the same time there is urgency to better exploit potential of rainfed and other less endowed areas if we are to meet targets of agricultural growth and poverty alleviation. Given the wide range of agroecological setting and producers, Indian agriculture is faced with a great diversity of needs, opportunities and prospects. Future growth needs to be more rapid, more widely distributed and better targeted. These challenges have profound implications for the way farmers' problems are conceived, researched and transferred to the farmers. On the one hand agricultural research will increasingly be required to address location specific problems facing the communities on the other the systems will have to position themselves in an increasingly competitive environment to generate and adopt cutting edge technologies to bear upon the solutions facing a vast majority of resource poor farmers.

In the past agriculture has played and will continue to play a dominant role in the growth of Indian economy in the foreseeable future. It represents the largest sector producing around 28 percent of the GDP, is the largest employer providing more than 60 percent of the jobs and is the prime arbiter of living standards for seventy percent of India's population living in the rural areas.

These factors together with a strong determination to achieve self-sufficiency in food grains production have ensured a high priority for agriculture sector in the successive development plans of the country.

An important facet of progress in agriculture is its success in eradication of its critical dependence on imported foodgrains. In the 1950's nearly 5 percent of the total foodgrains available in the country were imported. This dependence worsened during the 1960's when two severe drought years led to a sharp increase in import of foodgrains. During 1966 India had to import more than 10 million tonnes of foodgrains as against a domestic production of 72 million tonnes. In the following year again, nearly twelve million tonnes had to be imported. On the average well over seven percent of the total availability of foodgrains during the 1960s had to be imported.

Indian agriculture has progressed a long way from an era of frequent droughts and vulnerability to food shortages to becoming a significant exporter of agricultural commodities. This has been possible due to persistent efforts at harnessing the potential of land and water resources for agricultural purposes. Indian agriculture, which grew at the rate of about 1 percent per annum during the fifty years before independence, has grown at the rate of about 3 percent per annum in the post independence era.

AGRICULTURE – SUB-SECTORS

Indian agriculture broadly consists of four sub-sectors. Agriculture proper including all food-crops oilseeds, fiber, plantation crops, fruits and vegetables is the largest accounting for nearly 70 percent of the agriculture sector as a whole.

The rapid growth in this sub-sector through exploitation of wastelands and fallows, spread of irrigation and adoption of production enhancing technologies was critical in transforming India from a country vulnerable to food shortages to one of exportable surplus. Although this sub-sector has made impressive progress its share in the sector as a whole has declined from 78 percent in 1960-61 to less than 70 percent by early 90s.

Table. Agriculture sub-sectors – value of output (%)

Sub-sector	1960-61	1970-71	1980-81	1993-94
Agriculture	78.0	79.3	75.3	68.5
Livestock	16.6	15.1	17.2	24.5
Forestry	4.4	4.4	5.9	3.8
Fisheries	1.0	1.2	1.6	3.2
	100.0	100.0	100.0	100.0
Total value at Current price (Rs. Billion)	89	218	615	2740

Correspondingly the share of livestock sector has increased considerably. The livestock industry has grown from Rs. 15 billion in early 1960s to Rs. 100 billion by 1980-81 and Rs. 672 billion by 1993-94. In nominal terms the sector grew at almost 15 percent per annum during 1980s. Milk production, which was almost stagnant for two decades ending 1970, grew by over 5 percent per annum in the 80s. Similarly, production of eggs increased at the rate of about 6.5 percent during the same period. As a result the share of livestock increased from about 17 percent till early 80s to 25 percent by 1993-94.

Though it plays relatively a minor role within the sector as a whole, fishing sub-sector activities have been on the rise. The sub-sector has grown from only Rs. 3 billion in 1970-71 to nearly Rs. 90 billion in 1993-94. The growth was particularly rapid in 70s and 80s. Value added increased at over 5 percent per annum during this period.

In real terms forestry and logging activities have been on the decline since mid seventies. As of 1993-94, the size of the industry in terms of value of output was 103 billion.

Table. Foodgrains Production

Year	Area (million ha)	Production (million tonnes)	Yield kg/ha
1950-51	97.32	50.82	522
1960-61	115.58	82.02	709
1970-71	124.32	108.42	872
1980-81	126.67	129.59	1023
1990-91	127.84	176.39	1379
1998-99	124.0	195.25	1574

Over the past three decades, the country has successfully transformed itself from a food deficit economy to one which is essentially self sufficient in availability of foodgrains and other essential commodities, albeit only at the prevailing level of effective demand. Annual aggregate foodgrains production, which averaged about 82 million tonnes in 1960-61 increased to 123.7 and 172.5 million tonnes for the trienniums ending 1980-81 and 1990-91 respectively. Current (1998-99) production level is 195 million tonnes and the country has been able to accumulate substantial, (35 million tonnes) stocks of foodgrains to cope up with any sudden difficulties arising from drought or a similar situation in any part of the country.

Increased outputs, have been achieved chiefly by adopting, since mid sixties, a strategy aimed at increasing foodgrains production by concentrating public sector efforts and resources in regions with a high potential for quick and substantial productivity gains through increased cropping intensity and average yields. These were the areas favoured by agroclimatic resource conditions and where irrigation facilities already existed or could be developed relatively rapidly. The main elements of this strategy were:

- Expansion of irrigation coverage,
- Increased provision and utilization of key inputs – mainly high yielding varieties (HYVs) of crops, mainly of wheat and rice and chemical fertilizers and plant protection chemicals,
- Expansion and improvement of institutional support services such as research and extension and
- Price policies favourable to producers of major foodgrains.

The success of this strategy was made possible by development and availability of replicable production technology packages, so called 'Green Revolution' technologies. Irrigation facilitated double cropping and widespread adoption of HYVs. The HYVs performed particularly well under irrigated conditions, were highly responsive to fertilizers and their short duration permitted increases in cropping intensities.

Irrigation development was the cornerstone of the strategy. Undivided India was amongst the largest irrigated areas in the world. With partition nearly one-third of the irrigated area went to Pakistan. At the time of independence the net irrigated area was 20.9 million ha (gross irrigated area 22.6 million ha).

Recognizing large-scale development of irrigation facilities as critical to rapid agricultural growth, the country has spent about Rs. 45,000 crores on irrigation development in the first four decades after independence. During the period 1950-51 to 1965-66 development of irrigation through government canals grew from 7.2 million ha to 9.8 million ha – a growth rate of 2.1 percent per annum. During 1970s this pace dropped slightly to 1.9 percent. In 1980s the rate of increase dropped significantly to 1.1 percent per annum.

The growth of tube-well irrigation, however, increased rapidly from 4.5 million ha in 1970-71 to 9.5 million ha in 1980-81 and then to 14.3 million ha by 1990-91. The net irrigated area increased from 31 million ha in 1970-71 to 53.5 million ha in 1995-96 which corresponds to 22 percent of the net sown area in 1970-71 and 37.63 percent in 1995-96 *(Tables 3, 4).* With improvements in irrigation efficiency the gross irrigated areas has increased to 71.51 million ha. The percentage of gross cropped area service by irrigation increased from 18.3 percent in 1960-61 to 23.0 percent in 1970-71 and to over 38 percent at present.

Table. India – irrigated area

Year	Net	Gross	Net as % of net sown	Gross as % of gross cropped
	——— million ha	———	————— %	—————
1970-71	31.10	38.19	22.17	23.04
1980-81	38.72	49.78	27.66	28.83
1990-91	48.02	63.20	33.61	34.03
1995-96	53.51	71.51	37.63	38.33

Table. Sources of irrigation

Year	Net Irrigated Area (million ha)	Sources: Canals	Tanks	tube-wells	Other wells	Other Sources
		(% net irrigated area)				
1960-61	24.80	10.4	4.6	0.2	7.2	2.4
1970-71	31.10	41.28	13.22	14.34	23.88	7.29
1980-81	38.72	39.49	8.22	24.62	21.08	6.59
1990-91	48.02	36.34	6.13	29.69	21.73	6.11
1995-96	53.01	32.04	5.81	33.52	22.16	6.47

Fertilizers have constituted yet another key input in addition to expanded irrigation and spread of HYVs in achieving goals of high production and productivity. India currently occupies third position in the world, after China and USA, in terms of fertilizer production and consumption. Consumption of fertilizers has increased from 1.54 million tonnes in 1967-68, representing the pre green revolution era, to 17.31 million tonnes currently (1997-98).

The average per hectare use of fertilizers currently around 85 kg per hectares is the lowest among several Asian countries.

However, rice and wheat account for a major fraction, around 65 percent of the total fertilizer consumed in the country, with very little fertilizers going to the rainfed areas. According to some current projections, fertilizer's use will need to increase to 30-35 million tonnes to meet the foodgrains need of 2020. The demand for nutrients will stretch by almost another 15 million tonnes if requirements for horticulture, vegetables and plantation and commercial crops are included. At present domestic production of N and P fertilizers (13.42 million tonnes) falls short of consumption by over 20 percent. In addition the entire requirement of K fertilizer is imported.

Most agricultural development programmes initiated in 1960s were concentrated in regions of high potential. Thus five states, Punjab, Haryana, Uttar Pradesh, Andhra Pradesh and Tamil Nadu account for 50 percent of the country's net irrigated and 53 percent of the gross irrigated area. The combination of expanding irrigation coverage and widespread adoption of short duration HYVs led to significant increases in cropping intensities. Acreage cropped more than once per year increased from 13 million ha in 1950-51 to about 44 million ha at present.

Average cropping intensity for the country as a whole rose from 115 percent in 1960-61 to 131 in 1993-94. By 1993-94 cropping intensity has risen to 187 percent in Punjab, 167 percent in Haryana and 142 percent in Uttar Pradesh.

Table. Cropping intensity

Years	Net sown area	Gross cropped area	Area cropped more than once	Cropping intensity
	million ha			(%)
1950-51	118.8	131.9	13.1	111.1
1960-61	133.2	152.8	19.6	114.3
1970-71	140.3	165.8	25.5	118.2
1980-81	140.0	172.6	32.6	123.3
1990-91	143.0	185.7	42.7	129.9
1993-94	142.0	186.4	44.4	131.2

An important consequence of the strategies adopted since sixties has been to boost production of, chiefly, two crops rice and wheat. Their share in total foodgrains production went up from 57 percent in 1970-71 to more than 75 percent in 1990-91. Production of foodgrains other than rice and wheat did not increase significantly and in the eastern region even the yield of rice did not increase.

Agricultural production and income rose substantially in the north-western states of Punjab, Haryana, Western Uttar Pradesh, parts of Rajesthan, Tamil Nadu and Andhra Pradesh. By contrast productivity and output growth have been modest in eastern and central India and in deccan plateau. Progress was particularly slow in rainfed areas, which account for over 60 percent of the cropped area and where a great majority of rural poor are concentrated. An important impact of the strategies pursued in the 'Green Revolution' period has been intensification of regional disparities and imbalances in agricultural development and food availability and hence levels of food security. For the country as a whole while per capita availability of cereals has increased substantially, that of pulses has decreased significantly.

In summary an annual increase in foodgrains production of 3.22 percent during fifties was mainly because of expansion in area. Sixties recorded a low annual growth rate of 1.72 percent necessitating large-scale imports of foodgrains.

Table. Rice and Wheat Production (million tonnes)

Year	Rice	Wheat	Foodgrains	R+W as % of Foodgrains
	million tonnes			
1970-71	40.80	20.86	108.4	57.4
1980-81	49.91	34.55	129.6	65.4
1990-91	72.78	53.03	176.4	75.0
1996-97	81.74	69.35	199.4	75.0

Annual growth of 2.08 percent was recorded during seventies. This decade was the turning point in India's foodgrains economy leading to self-sufficiency through significant productivity increase first in wheat and later in rice in the eighties. An annual growth of 3.5 percent in foodgrains in eighties was the hallmark of green revolution that enabled India to become self-sufficient and even a marginal exporter. The pace of growth slowed in nineties barely making or even slower than the population growth rate. This is a matter of concern.

Table. Annual compound growth rate 3of foodgrains production

Crop	1950-51 to 1959-60	1960-61 to 1969-70	1970-71 to 1979-80	1980-81 to 1989-90	1990-91 to 1997-98
Rice	3.28	- 8.05	1.91	4.29	1.53
Wheat	4.51	5.90	4.69	4.24	3.67
Coarse cereals	2.75	1.48	0.74	0.74	- 0.49
Total cereals	3.00	2.51	2.37	3.63	1.84
Pulses	2.72	1.35	- 0.54	2.78	0.76
Total foodgrains	3.22	1.72	2.08	3.54	1.66

Table. Per capita net availability of cereals and pulses per day (g)

Year	Cereals	Pulses	Total
1951	334	61	395
1961	400	69	469
1971	418	51	469
1981	417	38	455
1991	468	42	510
1998	451	33	484

HORTICULTURE

The diversity of physiographic, climate and soil characteristics enables India to grow a large variety of horticultural crops – fruits, vegetables, flowers, spices, aromatic and medicinal plants, plantation crops etc. India is the largest producer of fruits in the world and second largest producer of vegetables. The area under fruits estimated at 1.45 million ha in 1970-71 grew to 2.8 million ha in 1991-92 and then more rapidly to 5 to 6 million ha by 1994-95. This sector is likely to grow rapidly in the future both on account of internal demands and export opportunities.

ANIMAL HUSBANDRY AND FISHERIES

Animal husbandry and dairying sub-sector plays an important role in overall economy and in social development. The contribution of the sub-sector is estimated to be about 25 percent of the total value of output of agricultural sector.

Table. Production of milk, eggs and fish in India

Year	Milk production / availability million tonnes/annum	g./day	Fish (000 tonnes)	Eggs (million)
1950-51	17.0	124	752	1832
1960-61	20.0	124	1160	2881
1970-71	22.0	112	1756	6172
1980-81	31.6	128	2442	10060
1990-91	53.9	176	3836	21101
1998-99	75.0	210	5388	28400

The sector also plays a significant role in supplementing family incomes and generating employment in the rural sector particularly among the land-less, small and marginal farmers and women besides providing nutritious food. Production of suitable cross breeds and their wider adoptions has contributed to increasing country's milk production, which has now reached 75 million tonnes annually. Similarly, genetic improvement and better management practices have markedly pushed up production of poultry and eggs.

Table. Number of holdings (million)

Category of Farmers	1970-71	1980-81	1990-91
Marginal	35.7 (50.6)*	50.1 (56.4)	62.1 (59.0)
Small	13.4 (19.0)	16.1 (18.1)	20.0 (19.0)
Semi-medium	10.7 (15.2)	12.5 (14.0)	13.9 (13.2)
Medium	7.9 (11.2)	8.1 (9.1)	7.6 (7.3)
Large	2.8 (3.9)	2.2 (2.4)	1.7 (1.6)
All groups	70.5 (100)	89.0 (100)	105.3 (100)

Through the overall contribution of fisheries sub-sector is small, development of prolific and fast growing forms of several common fish species coupled with breakthrough in breeding under captivity, fish seed production and multi-layer fish culture has resulted in registering a very high annual growth of 11 percent in aquaculture production during the past decade.

Table. Area operated - % of total area

Category of Farmers	1970-71	1980-81	1990-91
Marginal	9.0	12.1	14.9
Small	11.9	14.2	17.3
Semi-medium	18.5	21.1	23.1
Medium	29.7	29.6	27.2
Large	30.9	23.0	17.4
All groups (million ha)	162.1 (100)	163.8 (100)	165.6 (100)

SUSTAINABLE DEVELOPMENT POLICY

An example of a holistic and sustainable development policy based on human ecology can be demonstrated by the situation on the Andaman and Nicobar Islands. The suggestions developed below represent a step in the direction of policy-orientated counselling; as such they target application and implementation.

DATABASE

Information on the Andaman and Nicobar Islands has been sketchy and obscure, and the database limited. The first information recorded on these islands came from Claudius Ptolemaeus, the great Roman geographer, who lived in Alexandria in the second century. Ptolemaeus, who had never seen Andaman or Nicobar, described the islands as "Islands of the Cannibals".

The next reference to these islands is found in the writings of I'Tsing, a Chinese traveller who arrived in India in AD 673. He referred to the islands as "Andaban" and also described their inhabitants, evidently on the basis of hearsay, as cannibals. Marco Polo, the famous Venetian traveller, journeyed near the islands around 1260. He described the inhabitants like this: "All the men of this island have heads like dogs, and teeth and eyes like dogs; for I assure you that the whole aspect of their faces is that of big mastiffs".

Despite these obvious historical inaccuracies, there is nevertheless abundant written information on these islands. There is also much expertise to be drawn upon in Port Blair itself, and there is in fact a small, but accurate, firsthand empirical database to supplement the previous ones, that is observational impressions.

Port Blair was founded by the British as a penal colony for their Indian colonial empire. During the Second World War, the islands were occupied by the Japanese. They passed to independent India in 1947 as a centrally administered union territory with limited sovereignty.

The Andaman and Nicobar Islands are located 1200km east of the Indian subcontinent and are the remains of a former land bridge that extended from

Burma to Sumatra. They extend between 6° and 13° on the northern parallel, thus falling in the tropical rain zone with 200 days of precipitation per annum. Dense tropical rain forests are the dominant form of vegetation on these almost unspoiled islands. Along the coast, they are encircled by mangrove forests. Coral banks in shallow sea stretch out in front of white, sandy beaches. The entire area encompasses $8300km^2$.

Most of the approximately 550 islands are uninhabited. The Indian census estimates the number of inhabitants in 1981 at 190000, up from 50000 in 1971. Of these, 22000 are negroid descendants of the island's original inhabitants, and Asian immigrants of a tribal character, some of whom continue to live in complete isolation from civilization.

The rest of the population consists of the so-called locals, descendants of the Indian prisoners in and around Port Blair, as well as Indian immigrants and administrators from the mainland who have formed the political and economic elite since 1947.

Twenty per cent of the population is engaged in agricultural production. The chief crops grown on the 48 000 ha of cultivated land are rice and vegetables, followed by coconut and fruit trees. Tropical wood is processed in saw mills: in 1984 this amounted to $182000m^3$. Despite the existence of a plenitude of sea life, the fishing industry has hardly developed. Except for the saw mills, there is no industrial production.

Some Political Proposals and their Supposed Social and Ecological Impacts

Before I visited Port Blair, the capital of the Andaman and Nicobar Islands, in 1987, I tried to sort out those political proposals that could have major impacts on the future development of the islands. The basis of most of my information consisted of rumours, newspaper articles, and information gathered from consultations with experts.

According to these sources, three major development proposals could be identified:

- A proposal for a defence area to test missiles,
- A proposal for a tourist centre,
- A proposal for a free-trade zone including a duty-free port.

Missile Testing Site

According to *India Today*, the Defence Ministry had been trying for the past four years to set up a National Test Range (NTR) in Orissa, Balasore District (estimated cost: Rs30000 million). Because of extremely strong opposition from local inhabitants it was suggested that the NTR be located in the Andaman Islands where fewer individuals would be affected — and possibly where opposition would not be so strong. There are defence and budgetary arguments both for and against the Andaman Islands as a testing site, but more relevant to my purposes is a consideration of why the Orissa population resisted so strongly, because the reasons may prove valid for the Andamans as well.

For the Orissa farmers, "the area, spread over 400km^2 along the seashore and comprising 132 villages, is a virtual gold-mine with cash crops like betel-vines, coconuts and cashew besides profitable fishing in the shallow waters...". Local farmers claimed that the NTR project would destroy vast paddy fields, 30000 betel-vines, thousands of coconut palms and cashew plants at a value of over Rs1750 million.

Thus, the planned resettlement of farmers and their families could not only bring about the destruction of an ecosystem, it could also destroy a human ecological system where societal groups live in harmony with nature. This same argument could, of course, be applied to the Andaman or the Nicobar Islands should the NTR be located there.

Tourist Centre

What is tourism? A tourist was previously understood to be a person who was curious about the nature and lifestyles of other persons in other parts of the world. It was this curiosity that made the tourist travel. Because of this simple definition, we sometimes still believe that tourism promotes understanding between peoples.

Today, however, the tourist has been degraded into a foreign-exchange-carrying animal from whom local inhabitants or governments attempt to extract as much money as possible. Because tourists are willing to surrender to this exploitation, they are permitted to retaliate by exploiting the local environment and disrupting local social structures.

For example:

- Trekking in Nepal has had severe environmental consequences that threaten traditional Nepalese society.
- Coastal tourism in East Africa led to rising expectations among the local population. Children began to earn much more than their fathers as hard-working fishermen had ever earned; social hierarchies were upturned.
- Sex tourism in Thailand has disrupted local cultural values and beliefs.

What do these examples show? Again, although the impacts may be more subtle than those of the proposed NTR, human ecological systems will be affected. The impacts may be for the better or the worse, so the question should be raised as to how these systems can be affected for the better, that is, how can the greatest damage be avoided?

Ghetto tourism, where tourists living in a camp are kept apart from the indigenous population outside, was a partial answer to this, but ghetto tourism seems to be on the decline. According to Willibald Pahr, Secretary General of the World Tourism Organisation (WTO), experiences with ghetto tourism have not been favourable. Another answer to this problem was given by countries like China, Burma and Tanzania. To cite one example, Gertrude Mongella, Minister of Tourism in Tanzania, stated that her country was opposed to mass tourism and, consequently, sex tourism does not play a role. Tanzania's tourist policy is to make the country attractive for educated persons who show a specific interest for the local countryside and local culture.

Interestingly, the Tanzanian tourist policy led back to the original definition of a tourist, which was thought to be obsolete — a person who is curious about other people. It is also worth noting that this concept of tourism had an elitist element to it. Moreover, this elitist way of thinking was being propagated by socialist countries — an amazing switch of values! My question is, therefore: is elitist behaviour necessary to protect human ecological systems, that is, to protect social systems and their natural environments?

Free-trade Zone and Port

For years, there has been talk of making Great Nicobar a duty-free port. Hong Kong's future remains uncertain, Singapore is crowded, and Great Nicobar is situated on the trade routes between India, Burma, Southeast Asia and the Far East. What role does Hong Kong play in this? It could either serve as a model for Great Nicobar or it could provide the needed capital for development as a result of its own political and economic uncertainty.

It is not likely that Hong Kong will provide the source of capital for Great Nicobar. First, those Chinese who were able and willing to leave the crown colony have already done so and made new homesteads in the Caribbean. Secondly, Hong Kong is flowering economically at present.

Thirdly, and most importantly, there is no ethnic Chinese minority on the Andaman or the Nicobar Islands. There is a strong Hindustani-speaking tradition on these islands, so the chances are better that they will attract floating Indian capital, including that of Hong Kong Indians who will lose their national identity in 1997 when the crown colony becomes part of the People's Republic of China. There are many possible objections to the proposed free-trade zone, but one is particularly significant. On the Nicobar Islands, there is still a growing aboriginal population. These islands are one of the few areas in the world that are almost untouched, with a functioning human ecological system virtually free from outside influences.

For this reason, the Indian Anthropological Survey protested against the institution of a free-trade port. Whether, as Iqbal Singh suggests, it would be possible to make Great Nicobar a duty-free port but leave the other islands untouched, is still an open question. It is not my opinion that all efforts to develop the Andaman and Nicobar Islands are a *priori* wrong. However, I do believe that those responsible must be extremely careful. There is no government in the world that can easily afford to destroy human ecological systems. Any recommendation should take this into account. The costs and benefits of development must also be calculated, but not in purely monetary terms. What would be the price, for example, of the Nicobarese human ecological systems, which, once they are destroyed, cannot be rehabilitated. Finally, it will be necessary to differentiate between who will share the benefits and who will bear the costs of development.

Three Models for Future Development

The following recommendations represent an attempt to present some ideas — following the concept of holistic human ecology — concerning the future prospects of the islands, albeit based on limited observational data. This includes possible benefits to their inhabitants. I shall therefore proceed with a good deal of caution. As a result of the information collected, three basic models of future development would seem to apply to the Andaman and the Nicobar Islands:

- An ecological preservation/social stagnation model
- A macropolis model, and
- An ecodevelopment model.

Ecological Preservation and Social Stagnation

These islands represent one of the very few, perhaps the only, specimens of truly natural ecosystems that have hardly been influenced anthropogenically. It is argued, therefore, that these islands must be preserved as they are. This has consequences for the three population groups on the islands, the tribals, including an aboriginal population, the so-called "locals" and recent immigrants, including government and administrative officials.

Some islands are populated with aborigines who live in harmony with their island environment. In view of the present slaughtering of Amazonian peoples, the aboriginal inhabitants of the Andamans and Nicobars may soon become the last undisturbed autochthonous communities on Earth and, as such, their lives, lifestyles, culture and environment ought to be preserved. To achieve this, outside influence should be kept to a minimum, but this can only be realized if "civilized" humans are not permitted to disturb the aborigines. The "locals" should be confined to Port Blair and a few "civilized" peripheral spots. This implies that no further colonization of the islands will be permitted and that immigration is ruled out. It further implies social stagnation in order to preserve an original human culture. It is obvious that this model cannot tolerate development that would include military ventures, capital- and technology-intensive endeavours like a free-trade zone, or any significant form of tourism.

Macropolis

This model would make a new Hong Kong of Great Nicobar. A possible five million people living on some 860km^2 would result in a population density of

5800 per km^2. In view of this figure, it would not be too far-fetched to say that more islands would be needed, be they for reasons of pleasure or for agricultural supply; such a development creates its own dynamics. According to the proponents of this model, the development would be desirable because of its expected pull effect on the mainland economy. The anticipated economic benefits would outweigh the social costs of development that — in "extrapolating" the present scenario — could include the following: the "locals" will lose their cultural identity; the Shompen, the tribals on Great Nicobar, will disappear; the Nicobarese will decay and their numbers will dwindle. The economic benefits, it is argued, would also outweigh the ecological destruction suffered by Great Nicobar and the other islands. To the contrary, it is argued that human welfare is more important than nature protection, and that the economic and social benefits of the macropolis model will reach the locals and Nicobarese through the trickle-down mechanism, once these peoples have been "civilized".

Ecodevelopment

The ecodevelopment model is located between the extremes. Ecodevelopment has two roots: ecology and development. Therefore the contradiction between development or growth and environment becomes the dialectics of development plus environment. It implies that accelerated economic development be, at the same time, ecologically and socially sound or at least benign. Ecodevelopment policy begins by considering the needs of the people. Thus, the elimination of poverty is a precondition of economic development, not its result. Similarly, the priority no longer lies in export earnings, but in creating employment in the domestic economy and in enhancing a broad income distribution. Employment includes selfemployment at the monetary as well as at the subsistence level. The planning of ecodevelopment projects is shaped by several factors. First, the sociocultural values and behavioural patterns of the peasant family, and the politico-administrative scope of the village community are major determinants. Secondly, the economic limitations, in terms of low income earners who may be partially embedded in a subsistence economy, are also important. Finally, the biophysical environment may impose natural constraints. Selecting an appropriate location for a project is governed by the ability of those responsible for the planning process to harmonize sociocultural, economic and ecological factors, which may require democratic resolution of social conflicts. How does this concept translate into action for the Andaman and the Nicobar Islands? Let us consider the answer to this query in terms of the three elements, needs, self-reliance and environmental compatibility.

THE APPLICATION OF AIM TO THE INSTITUTIONAL SUSTAINABILITY INITIATIVE

AIM emerged as a product of a USAID-funded applied research effort, the Institutional Sustainability Initiative (ISI).

The Action Stream of ISI

During the 1980s, the ISI emerged from the recognition that sustaining the benefits of donor financed development efforts was a major policy problem. At USAID's 1987 conference on the future of agricultural and rural development in Asia and the Near East, sectoral specialists from the agency's field offices (missions) emphasized institutional sustainability as a priority. In response USAID pledged to develop practical field guidance on this complex policy issue. Thus, the preparatory phase of the action-inquiry stream for the sustainability policy issue began; with a field level call for action and a USAID-Washington response. USAID's Bureau for Asia and the Near East took a leadership role in the ISI, and provided some initial funding. The ISI stressed "high quality and practical field guidance"; minimal funds were available for research. What research would be done had to be low cost and closely tied to practical results.

The next step in the preparatory phase was to obtain field level commitment to the ISI. Field missions were encouraged to request a sustainability assessment and design for their new and ongoing projects in the agriculture and rural development sectors; few projects had been designed with serious attention to sustainability.

In the investigation phase of AIM, the conceptual model devised in step one is systematically applied in the field to identify sustainability and conclusions. Informal contacts with field missions facilitated the selection process, and invitations for applying the model were plentiful.

The planning for the initial ISI applications was somewhat ragged. The first field teams went out while AIM was still being developed, and their instructions were minimal. Experienced consultants and USAID personnel were left to their own devices more than hindsight indicates was optimal. Some misunderstandings occurred. Changes were instituted later to assure that field teams were better oriented prior to their departure for the field. The SCOPE framework was used during orientation in two recent field cases, Thailand and CARDI, to orient the ISI team.

For each field site, a scope of work for the sustainability consultancy was negotiated, which melded the goals of the donor, the mission, and the assessment/design team. The team met prior to departure, and planned their activities, adapting issues and questions to the field setting. When the ISI team arrived in the field, they established an in-country team, which then reviewed and modified the interview and analysis formats. Finally, the sustainability questionnaire was used to gather relevant data and carry out the assessment.

Several difficulties were encountered in the investigation phase. In many instances the scope of work required time-consuming adaptation to add the distinctive action-inquiry perspective, plus extra negotiations once the teams were in the field. Consultants had to be trained in the use of the questionnaire. However, it is noteworthy that field responses to the assessments were very positive. Clients were pleased, willing to cover local costs, and interested in arranging future field work. The third phase of the action stream involves the design of alternatives: recommending policy, structural and management changes to improve sustainability. The team used the SCOPE framework, the assessment data, and suggestions from local stakeholders to recommend changes. The ISI team members tried to act as resource persons and facilitators in this process. From the interim assessment and dialogue with interested parties, the team developed a final assessment and made recommendations to USAID and developing country officials.

In step four of AIM, sustainability recommendations are accepted by the organization and are implemented with monitoring of results built into the process. Follow-up visits to several countries indicate that many recommendations have been agreed to and some implemented. However, to date, no formal mechanism exists to track this dimension. One of the current ISI activities is the design of such a mechanism to follow up the work of the first three action stream phases.

Finally, in knowledge consolidation and dissemination, evaluations will continue to be conducted in the field and assessments will be fed back into the research stream. For ISI this would mean that additional sustainability evaluations will be done in the field, and assessments of the utility of ISI action guidelines will be made. In fact, time is needed to develop the guidelines and assess their utility from a sustainability perspective. Currently there are neither mechanisms nor resources for this type of systematic evaluation.

The Inquiry Stream

The preparatory phase in the research component of AIM began with a review and synthesis of previous findings on sustainability, leading to the development of the SCOPE framework and the selection of indicators for its variables. It is important to point out that converting a conceptual framework into indicators and research hypotheses normally leads to some revision of the framework. Moving from theory to research design requires further clarification of concepts and relationships.

Phase two in the inquiry stream consisted of planning the investigation, the field testing of the SCOPE framework, and its application to an existing data set. The major activity in planning the research was the construction of an information questionnaire for the field team, incorporating the SCOPE variables. The logistics and politics of the seven field applications led to the decision that only the Washington-based social scientist members of the ISI teams filled out the questionnaire in six of these cases.

Team members conducting field inquiries were experienced researchers and interviewers; they determined their own research tactics and methods. Invariably they interviewed key project personnel and stakeholders and reviewed previous studies and documents on the project. Their knowledge was converted into a form that was comparable across cases when they completed the information questionnaire on their project or institution. They also reported on the aspects of SCOPE that were most useful for analyzing sustainability issues in the particular case, and what was not useful. Furthermore, they discussed errors and weaknesses in the application of the SCOPE hypotheses to the case.

The SCOPE framework has also been applied to a set of 25 agricultural education institutions. Twenty-five expert questionnaires have been statistically analyzed, and the results support all of the SCOPE propositions, though not always at statistically significant levels. In the design of the inquiry stream, the team compiled and analyzed the data from the field studies and data set studies, and revised the SCOPE framework accordingly. The field cases have not been selected randomly but are the total set of projects for which a request was made for a sustainability analysis. It may be assumed that the on-site interest in action surpassed the research interest in the selection of cases.

This means that the sample parameters cannot be used as estimates of the population parameters. It also means that correlation findings cannot be generalized uncritically. However, unless the cases are rather peculiar, the correlation analysis can be cautiously used as a guide until more representative cases can be analyzed. The main precaution to take in using the findings is to note the unique features of the set of cases. Two such features stand out. Since each project requested a sustainability analysis, these cases probably have sustainability problems and also have project personnel who are concerned about them. Second, six of the seven cases studied to date involve agricultural education or research institutions. (The seventh involves the institutionalization of a credit programme.) The institutional homogeneity of the case limits the generalizability of the correlation results.

Preliminary findings from the data set study of 25 agricultural education institutions indicate that most management factors have surprisingly weak correlations with prognoses of sustainability. Relations with the minister of agriculture seem to override most other contributing factors. It is hoped that further analysis will ferret out the contingency factors that determine which aspects of the SCOPE framework are critical in different conditions.

The implementation phase of the research component of AIM involves research on the implementation of changes in the studied projects, and the development of guidelines for designing and implementing projects for sustainability for donor and country use. ISI has not reached this phase or the final phase — when the team returns to the field and analyzes the results of the implemented sustainability changes, revises the SCOPE framework and the guidelines, and disseminates the findings.

INDIA'S ECONOMIC REFORMS IN AGRICULTURE

Although India's economic reforms were initiated in June 1991, the process of liberalization was implemented gradually and thus it is difficult to assess the full impact of the liberalization measures. Nevertheless, an attempt is made to discuss what is observable in terms of agricultural growth.

One observation is that the expected increase in exports due to liberalization simply did not occur. India's share in world exports was 0.6 per cent in 1997; India has to aim for at least 4 per cent by 2005 in order to meet the growing import demands for capital goods, raw materials and crude oil as well as to meet her external financial commitments. For the last decade or so, India's share in world exports of agriculture has been between 2 per cent and 3 per cent. India is not as competitive as the other countries and calculations show that India's crop yields have increased at a slower rate over the 1990s.

In addition, the agricultural sector's output growth decreased to 2.9 per cent during 1992-1993 to 1998-1999. Kalirajan and others (2001) explain that two important reasons for the slowdown are that there was no major breakthrough in developing new high-yielding varieties during the 1990s and there was a decline in the environmental quality of land which reduced the marginal productivity of the modern inputs. What could this mean in terms of the effectiveness of the policies of reduced protection to industry, a market determined exchange rate and the opening of the agricultural sector to foreign trade?

First, although the reduction to protection of industry is substantial, there is reason to believe that the reduction was not necessarily sufficient to benefit the agricultural sector whose tariffs were also drastically reduced. Hence, the expected shift in resources to agriculture did not occur. Second, is the apparent ineffectiveness of the market determined exchange rate in boosting exports. This is however not surprising as the exchange rate may not be a key factor determining agricultural export demand for India. In general, unlike manufacturing industries, agriculture did not benefit much from these two policies because the share of imported inputs in the value of agricultural production is small. It is likely that a change in the mindset and attitude of farmers has yet to take place and there are delays or hesitation in embracing India's openness.

Third, in opening up the agricultural sector to foreign trade, India has taken major steps towards trade liberalization since 1991, partly on its own initiative and partly from its commitments to WTO. Kalirajan and others (2001) provide a detailed review of these reform procedures. But why have the benefits from trade liberalization been slow to come?

One reason is that prospects for growth in agricultural exports depend partly on domestic policies and partly on the removal of protectionist policies pursued by developed countries such as Japan and members of the European

Union (EU). An OECD report (1998) estimated that the producer equivalent subsidy in the OECD countries increased by US$ 9.3 billion from 1988 to 1993 and this subsidy as a percentage of the value of production in 1997 was 9 per cent in Australia, 20 per cent in Canada, 47 per cent in EU and 70 per cent in Japan. These protectionist practices do not seem likely to come to an early end. An UNCTAD report (1999) noted that 29 member countries of the OECD spent an average of US$ 350 billion a year in agricultural support between 1996-98. Schumacher (2000) further reports that the EU provides product-specific trade distorting domestic support to at least 50 different agricultural products. The implication of these reports is that food exports from India may not show a large increase given the international environment and the still-existing restrictions on exports in the major importing markets based on the self-sufficiency argument and food security. Other macroeconomic factors, such as the recession in developed countries in 1996-98 as well as the 1997 South-East Asian financial crisis, have clouded the possibilities of increasing Indian exports.

Another problem faced by Indian agricultural exporters is the protectionist measures in the form of non-trade barriers that developed countries use to restrict market access. This is by tightening requirements of quality, testing and labeling, and anti-dumping and countervailing measures. For example, in May 1997, the EU banned marine products from India citing unhygienic processing conditions. The extra costs of meeting the standards required in export markets as well as costs associated with changes in the production mix and transactions associated with exports may well be discouraging Indian exporters.

One existing problem of India's agricultural protection is the use of input subsidies. The general argument favouring this has been that it is necessary to encourage the use of particular inputs for production for various benefits. For India, Gulati and Sharma (1995) show that the input subsidy in per cent of GDP increased from 2.13 in the triennium ending 1982-1983 to 2.73 in the triennium ending 1992-1993.

But the benefits of these subsidies have accrued to only certain classes of farmers in some regions cultivating irrigated crops. Furthermore, highly subsidized prices of inputs such as irrigation water and electricity for pump sets have encouraged cultivation of water-intensive crops, over-use of water, ground water depletion/salinity and water logging in many areas. Subsidy for nitrogen fertilizer on the other hand has resulted in nitrogen phosphorous potassium imbalance and acted as a disincentive for use of the environmentally friendly organic manure. As a result, the linkage between food crops and non-food crops, which include fodder, has been reduced.

These adverse consequences are a drain on the fiscal burden of central and state Governments. Thus, if not properly monitored, input subsidies can be counterproductive and, in this context, protection to lower costs of production should be done selectively in the course of liberalization.

In fact, Agenda 21 of the United Nations Conference on Environment and Development in 1992 stressed that there is a need for integration of environmental considerations in the pricing of natural and other resources in such a way that prices reflect social costs.

Such a pricing policy will not only lead to a more efficient use of scarce resources but also result in subsidy reductions and improvements in environmental quality. The money saved from the reduction of subsidies can be spent in the development of rural infrastructures, agricultural research, farmers' education and other forms of support for agriculture.

AGRICULTURAL GROWTH AND PERFORMANCE

While the above analysis has provided a general view of the impact of economic reforms, this chaper examines agricultural growth and performance in the states of Bihar, Karnataka, Tamil Nadu and Punjab with their attendant policy implications.

Thc yiclds for various crops in these states differ greatly. While Tamil Nadu had the highest yield in rice, oil seeds and sugarcane, Punjab enjoyed the highest yields in wheat, coarse cereals, pulses and food grains. Karnataka on the other hand is seen to do well in cotton and Bihar performed quite well in pulses and coarse cereals. Further analysis and findings by Kalirajan and others (2001) show that Punjab had made remarkable achievements on the agricultural front while Bihar had remained stagnant in the last two decades, with Karnataka and Tamil Nadu showing moderate achievement. Clearly, differences in physical endowments, climatic conditions and institutional characteristics are some of the reasons for the varying productivity performance. Thus, having across the board economic reforms is likely to work less effectively than state-specific policy measures that enable each state's agricultural yields to reach their full potential. The comparative advantage of each state's agricultural production should be determined and with inter-state restrictions removed, total agricultural output would see a very significant increase.

For example, Karnataka with less favourable soil and water resources should be given incentives to concentrate on agro-processed products and corporate agriculture in horticulture, floriculture and animal husbandry, or to undertake watershed development to help with dry land agriculture. Many studies have indicated that with watershed areas, productivity growth has been mainly due to seed and fertilizer use. Thus, this state has to be given input subsidies for high yielding seed varieties but at the same time, the farmers need to be educated on the over use of chemical fertilizers.

With Bihar, agricultural performance is problematic on many fronts. First, although demographic pressure has increased and agricultural technology has improved, most of the uncultivated land is concentrated in southern Bihar, where irrigation facilities have not kept pace and the soil is of poor quality. Given the physiography of southern Bihar, wells are also unsuitable and thus the dominant mode of irrigation has been through tanks whose expansion and maintenance has been neglected. Second, the infrastructural facilities of Bihar have been lagging as seen by the infrastructure development. Due to infrastructural bottlenecks, availability of modern goods and services has not increased or their supply remains costly or unreliable.

Third, agriculture in Bihar is dominated by small and marginal farmers and the prevalence of mass poverty is largely related to the backwardness of agriculture. Fourth and importantly, the state agricultural policies in Bihar are in dire need of review. The semi-feudal production condition still exists in rural areas and the ineffective protection of tenancy rights has hindered agricultural growth. The slow pace of land consolidation reflects inadequate financial outlays and a shortage of manpower. Kalirajan and others (2001) note that marketing and extension services in Bihar are also rather weak compared to the other states.

Punjab on the other hand, was one of the few states which enjoyed the success of land reforms and the high priority of investment in rural infrastructure. Also, the irrigation base of the small and medium sized farms was comparable to that of large farms. In addition, the Punjab Agricultural University at Ludhiana contributed to the development of new seed varieties.

However, there are clear signs of a decline in crop yields since the 1990s and this has been associated with the increasing use of fertilizers and excessive water use which have increased the unit cost of production as a result of declining soil quality. Hence, care is needed when providing further input subsidies in fertilizer and water use. Another related fact is the steep increase in wages in Punjab and in the absence of productivity increases, the cost increase has affected the profitability of farmers. With Tamil Nadu, the main crop has been rice as this state is blessed with two monsoons. But from 1992-1997, there has been a steady decline in the areas irrigated by canals and an increase in well-irrigated areas while the use of tanks remains an unreliable source of irrigation. However, major improvements in about 10 rice varieties released in the early 1990s can be expected to improve productivity growth in rice production although pests and diseases as well as imbalance in the use of fertilizers are major constraints. Thus Tamil Nadu could do with subsidies of pesticides and farmers should be educated on the more effective use of fertilizers to obtain high yields. Interestingly, the cropping pattern of late has shown increasing substitution of food crops by commercial crops but there is concern that the benefits will reach farmers only with the development of adequate infrastructure such as roads and markets. However, that Tamil Nadu has a higher index than the all India average of infrastructure.

Challenges of Globalization

It is important to realize that globalization poses many challenges to a developing country like India, which had relied on a state directed and regulated policy regime for more than four decades. In moving to a more open, market-based economy there are many transitional problems that the country has to manage. The Government must play a pro-active role in facilitating the globalization process so that the opportunity sets for the economic agents are widened and the adverse effects of globalization are minimized. The Indian Government must also prepare the necessary information base and develop its capacity to articulate India's concerns and policy trade-offs in the international forums for multilateral trade and environmental negotiations.

In addition, the Government should embark on an extensive programme to educate farmers on the need to meet the standards required in the export markets. In fact, India needs to seek technical assistance in creating the capacity for meeting such standards and to consider watershed developments for environmental considerations. Equally important is the need to disseminate information about possible export markets to farmers, so that market access is achieved at minimum cost. Given the requisite information about markets and profitability, the likelihood of farmers investing in post-harvest and processing technologies and storage and efficient transportation arrangements as well as developing supporting infrastructure is very high.

Although the brave and bold move by India to reduce the tariff rate for agricultural products from 113 per cent in 1990-1991 to 26 per cent in 1997-1998 deserves to be applauded, the question of whether India is ready to compete in world markets remains to be seen. The infant industry argument may still hold for India to shield itself from external competition but one can easily question the length of time that is required to that end. Also, a delay in opening up to foreign trade has the danger that local producers may become too complacent and never be ready for competition.

As India opens up externally, it is also expected to face vulnerability in the wider international price fluctuations and thus Acharya (1998) claims that a minimum price support scheme is important. These prices can also act as a signal to adopt modern inputs and invest in yield-raising infrastructure for increasing production. For instance, keeping basic staple food grains at reasonable prices would induce farmers to switch over to high value crops. However, during the 1990s, Kalirajan and others (2001) shows that procurement prices especially for rice and wheat have been increasing faster than the general price level. Such high prices along with guaranteed purchases by the Food Corporation of India have pushed up market prices. These higher prices are partly responsible for the large buffer stocks with the Food Corporation. If this trend continues, India's comparative advantage will be eroded. With openness and high price instability, unstable export revenue can also be expected. One way of reducing such risk is for India to diversify her agricultural exports. For example, since 1990, even in commodities such as tea, coffee, cocoa and spices, where India is supposed to have a comparative advantage international prices have been unstable. Besides increasing the type of exports to obtain more export revenue, India should also seriously consider exporting more value added agricultural products through agro-processing such as processed vegetables, fruits, fish and meat products given that export or even local demand for basic agricultural products would decline as incomes rise. The move to higher value added activities within the agricultural sector also spells greater opportunities for industrialization and vice versa as borne by Kalirajan and Shand's (1997) findings of a bi-directional relationship between agriculture and industry for most Indian states. On the other hand, Sivakumar and others (1999) establish empirical evidence of high forward linkages of agriculture due to the presence of agro-industries while Satyasai and Viswanathan (1999) show the significance of the spillover effects to the industrial sector via the intensive use of purchased inputs in the agricultural sector.

The lack or slow pace of internal or domestic liberalization is also seen to hinder the possible gains from external or trade liberalization. For example, although central zoning restrictions have been abolished, state government restrictions on inter-state and even inter-district restrictions on marketing and movement of goods still exist in many cases. This interferes with the benefits from crop specialization and economies of scale arising from comparative advantage. The land market is another example of distortion whereby land ceilings exist preventing the operation of large-sized farms. This has led to the emergence of a large number of small economically unviable land holdings. The easy leasing of land should be permitted with assurance of resumption. Yet another problem lies with the insufficiency of credit to agriculture. From 1995-1996, the Rural Infrastructure Development Fund was set up to allocate funds for the completion of projects and the government has committed itself to strengthening the cooperative credit structure through substantial refinancing and restructuring of the Regional Rural Banks. However, as mentioned earlier, due to varying institutional factors in the Indian states, these domestic reforms can be expected to yield quite different results.

Although India missed the opportunity to open up two decades ago, its attempts to do so now must be regarded as better late than never. Others such as Desai (1999) observe that, "the logic of the global economy as well as India's interests dictate that India become proactive in its liberalization policies. India must liberalize not because it has no choice but because it is the best choice".

His lament that India has adopted a 'victim mentality' when it really needs to adopt a 'winner mentality' has become less of a concern as over time, India has shown commitment to stay on the bandwagon of globalization. Having realized that globalization is a necessary but not a sufficient condition for high growth production, India has undertaken economic reforms, both internal and external. However, it must be ensured that these reforms are synchronized so that the pace of both reforms is set right in order to work hand in hand to promote agricultural productivity growth.

Thus, training the farmers and educating them appropriately to change their mindset and reorienting them to take up new activities or adopt foreign technology is of utmost importance. In this context, it is necessary to involve non-governmental organizations in training and mobilizing the rural poor to face the challenge of liberalization.

Also, with domestic economic reforms, more care needs to be exercised to draw up state-specific liberalization measures to maximize their benefits. Lastly, in the implementation of these reforms for successful globalization, one crucial element, not entirely within control is the need for good governance and stability in the political and economic environment. Political leaders who are the ultimate decision makers in these matters need to examine their own role dispassionately.

It is quite apparent that at this relatively early stage, there is little observable evidence of gains to India's agricultural performance after opening up. However, there could easily be benefits that have not yet surfaced, or are yet to be identified and perhaps too difficult or intangible to measure.

Whatever the case, it is highly likely that it is too soon to assess the full impact of globalization and economic reforms. Furthermore, the process of liberalization has been gradual and remains incomplete. For example, the complete removal of quantitative restrictions after March 2001 will have provided an opportunity for Indian farmers to tap world markets and, if they are successful, results should start to become evident soon. Export promotion via the development of export and trading houses as well as effective liberalizing export promotion zone schemes for agriculture are fairly recent measures and only time will tell as to how effective these measures are. Other possibilities such as agro-industry parks for promoting exports are also in the pipeline.

In conclusion, India has successfully set sail on the waters of globalization and economic reforms and even in the wake of economic and political instability, she has to carefully steer her course in order to reap the benefits of increased productivity growth in the agricultural sector.

INDIA'S AGRICULTURAL ECONOMY

In the early 1950s, half of India's GDP came from the agricultural sector. By 1995, that contribution was halved again to about 25 per cent. As would be expected of virtually all countries in the process of development, India's agricultural sector's share has declined consistently over time as seen in the table below.

Table. Share of agricultural output in India's GDP

Year	1950/51	1965	1976	1985	1991	1999
Percentage share	52.2	43.6	37.4	32.8	28.3	24.4

In the last five decades, the Government's objectives in agricultural policy and the instruments used to realize the objectives have changed from time to time, depending on both internal and external factors. Agricultural policies at the sectoral level can be further divided into supply side and demand side policies. The former include those relating to land reform and land use, development and diffusion of new technologies, public investment in irrigation and rural infrastructure and agricultural price supports. The demand side policies on the other hand, include state interventions in agricultural markets as well as operation of public distribution systems. Such policies also have macro effects in terms of their impact on government budgets. Macro level policies include policies to strengthen agricultural and non-agricultural sector linkages and industrial policies that affect input supplies to agriculture and the supply of agricultural materials.

During the pre-green revolution period, from independence to 1964-1965, the agricultural sector grew at annual average of 2.7 per cent. This period saw a major policy thrust towards land reform and the development of irrigation. With the green revolution period from the mid-1960s to 1991, the agricultural sector grew at 3.2 per cent during 1965-1966 to 1975-1976, and at 3.1 per cent during 1976-1977 to 1991-1992. Acharya (1998) explains that the policy package for this period was substantial and consisted of:

- Introduction of high-yielding varieties of wheat and rice by strengthening agricultural research and extension services,
- Measures to increase the supply of agricultural inputs such as chemical fertilizers and pesticides,
- Expansion of major and minor irrigation facilities,
- Announcement of minimum support prices for major crops, government procurement of cereals for building buffer stocks and to meet public distribution needs, and
- The provision of agricultural credit on a priority basis. This period also witnessed a number of market intervention measures by the central and state Governments.

The promotional measures relate to the development and regulation of primary markets in the nature of physical and institutional infrastructure at the first contact point for farmers to sell their surplus products.

Acharya (1998) also notes that the rate of growth of productivity per hectare of all crops taken together increased from 2.07 per cent in the decade ending 1985-1986 to 2.51 per cent per annum during the decade ending 1994-1995. Similar evidence of an increase in yields, a partial measure of productivity gains given by output per unit of land area is seen below for various crops.

Although productivity gains were sustained in the 1990s after the liberalization process began, the yield rates for most of the agricultural products in India are far below comparable rates in a number of other countries. Except for sugarcane, tea, coffee and jute, India's yields are lower than the world average. It should be noted that India is ranked second both in area and output for sugarcane production and is the largest producer of tea and jute in the world. Although India is doing quite well in wheat production, the average yields in the Netherlands and Ireland are more than three times India's yield rates. In all other major crops, India's productivity performance seems to lag behind others.

Table. Yield for various crops (kg/ha)

	1950/51	1960/61	1970/71	1980/81	1990/91	1995/96	1998/99
Rice	668	1 013	1 123	1 336	1 740	1 855	1 905
Wheat	663	851	1 307	1 630	2 281	2 483	2 596
Coarse cereals	408	528	665	695	900	941	1 035
Pulses	441	539	524	473	578	552	661
Food grains	522	710	872	1 023	1 380	1 499	1 611
Oil seeds	481	507	579	532	771	851	948
Cotton	88	125	106	152	225	246	240
Sugarcane	33 422	45 549	48 322	57 844	65 395	68 369	69 288

Why Globalize?

Globalization in the context of agriculture can be best discussed in the context of three components – improvement of productive efficiency by ensuring the convergence of potential and realized output, increase in agricultural exports and value added activities using agricultural produce, and finally, improved access to domestic and international markets that are either tightly regulated or are overly protected.

These components are linked in various ways. For example, productive efficiency would enhance value added activities in agriculture through agro-processing and exports of agricultural and agro-based products. These activities in turn would increase income and employment in the industrial processing sector. Thus globalizing agriculture has the potential to transform subsistence agriculture to commercialized agriculture and to improve the living conditions of the rural community. Ahluwalia (1996) explains that this indirectly requires an improvement in agricultural growth from between 2 and 3 per cent in the past to about 4 per cent per year. Although initially, with respect to agriculture, there was no major policy reform package in the 1990s, it was however anticipated that the opening up of the agricultural sector to foreign trade, the move to a market determined exchange rate and reduction of protection for industry would, over time, benefit the agricultural sector.

Manmohan Singh (1995), the then Finance Minister, in his inaugural address at the 54th Annual Conference of the Indian Society of Agricultural

Economics, brought to notice that a policy of heavy protection of the industrial sector operated to the disadvantage of the agricultural sector when industrial prices were raised relative to world prices and thus the profitability of investing in industry was raised relative to agriculture. This would lead to a shift of resources from agriculture to industry. A policy of heavy industrial protection also led to an appreciation of the exchange rate. Ahluwalia (1996) noted that over-valuation of the exchange rate (before the Indian rupee was devalued by 18 per cent in two phases starting in July 1991) discouraged agricultural exports more than industrial exports because Indian industrial policy had sought to offset the constraints faced by industries via a system of export incentives for market support. Agricultural exports on the other hand were denied any such incentives as they did not use imported inputs.

Ahluwalia (1996) argued that in the past, the agricultural sector was negatively protected because of the above two reasons and the fact that farmers were denied access to the world markets due to trade barriers. Exports of plantation crops and a few commercial crops were free from export restriction but exports of essential commodities, particularly food products, were subject to bans, quotas and other restrictions. Interestingly, Kruger and others (1991) showed that while many developed countries continue to protect agriculture, developing countries do not do so. However, no formal attempt or theoretical framework has yet been used to assess the extent of negative protection in Indian agriculture. The implementation of economic reform in the Indian agricultural sector has been a gradual process. These include an 87 per cent cut in tariff on agricultural products, sustenance of high-yield crop varieties, removal of minimum export price on selected agricultural products, a lift on quantity restrictions on the export of some crops and various land reforms related to tenancy rights and land ceilings.

PRODUCTIVITY GAINS FROM GLOBALIZATION AND ECONOMIC REFORMS

In the wake of India's efforts towards globalization and economic reforms, the expected benefits of total factor productivity (TFP) growth can be represented using the production frontier. The production frontier traces out the maximum output obtainable from the use of inputs.

Opportunities from globalization and economic reforms can lead to:

- shift from A to B due to technical efficiency
- shift from B to C on existing frontier due to input growth
- upward shift from C to D due to technological progress

Which constitute various sources of TFP growth, can be linked with trade gains. The movement from A to B led by technical efficiency allows increases in output when inputs and technology are used to their fullest potential to obtain the greatest yield. Given that India has been involved in agricultural production for so long, there would be learning-by-doing gains that can help boost production given the expected increase in demand as India opens up. The increased production would enable a better utilization of inputs, especially that of advanced capital technology. The reduction in the tariff rate for agricultural products from 113 per cent in 1990-1991 to 26 per cent in 1997-1998 is also expected to motivate local producers into rethinking their production techniques and efficiently utilizing the inputs and technology to keep costs of production down in order to remain competitive. The optimum or efficient use of land and water resources would then allow agriculture to respond to the demand for other products such as horticulture and livestock which is expected to increase following a rising trend in the per capita incomes of both rural and urban groups.

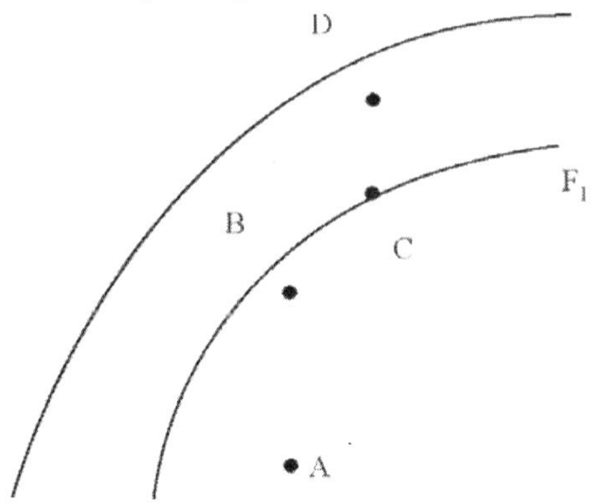

Fig. Total Factor Productivity Gains from Globalization and Economic Reforms

The move from an overvalued exchange rate to that of a market determined rate would also make agricultural exports cheaper and hence boost exports. The new trading opportunities would necessitate an increased use in the quantity of inputs to boost output and this allows for the movement from B to C along the existing production possibility frontier. Increased exports would bring about economies of scale and as Verdoon's law states, output growth would lead to productivity growth.

The scale of output under increased exports would justify the huge fixed costs underlying technologically advanced equipment and hence increase incentives to adopt high quality inputs. The use of such inputs would result in technological progress and this is represented by the shift from C to D. The reduction in tariff rates in industry from 1990-1991 to 1997-1998 range from 153 per cent to 25 per cent for consumer goods, 77 per cent to 18 per cent for intermediate goods and 97 per cent to 24 per cent for capital goods.

This means that farmers now have relatively cheaper access to imported new technology and better capital equipment as well as the option of adopting better farming techniques and this should lead to technological progress. In particular, the development of agro-processing as an instrument for agricultural and rural modernization will bring benefits, given its capital-intensive and technology-intensive nature. Lower duty rates on plastics and metals also lower costs of packaging. These forms of cost efficiency should allow competitive pricing of products. In addition, external competition can be expected to motivate local producers into the production of improved quality intermediate inputs for agriculture.

The importance of technology in agricultural development was first demonstrated in the 1970s with impressive growth in yields following the introduction of new wheat and rice varieties. But this technology was limited to areas of assured irrigation as the new seeds also required heavy inputs of fertilizers and pesticides for optimal results. However, the potential for further extending this technology is not yet exhausted as there is scope for expanding irrigation further and improving the quality of irrigation in many areas. For further technological progress, genetic engineering and the biotechnology revolution provides a prospect of developing new varieties that can flourish with less dependence on water and chemical inputs. Such reduced dependence upon chemical fertilizers and pesticides is also desirable because of environmental considerations, which are an increasing concern.

It must however be acknowledged that the link between trade liberalization and productivity growth is two-way as they both feed on each other. The productivity gains can be obtained from openness but to benefit from openness via increased demand for exports, agricultural products need to be priced competitively. In other words, productivity growth is necessary to lower the costs of production.

TECHNOLOGICAL NEEDS AND FUTURE AGRICULTURE

It is apparent that the tasks of meeting the consumption needs of the projected population are going to be more difficult given the higher productivity base than in 1960s. There is also a growing realization that previous strategies of generating and promoting technologies have contributed to serious and widespread problems of environmental and natural resource degradation. This implies that in future the technologies that are developed and promoted must result not only in increased productivity level but also ensure that the quality of natural resource base is preserved and enhanced. In short, they lead to sustainable improvements in agricultural production.

Productivity gains during the 'Green Revolution' era were largely confined to relatively well endowed areas. Given the wide range of agroecological setting and producers, Indian agriculture is faced with a great diversity of needs, opportunities and prospects. Future growth needs to be more rapid, more widely distributed and better targeted. Responding to these challenges will call for more efficient and sustainable use of increasingly scarce land water and germplasm resources.

Technical solutions required to solve problems will be increasingly location-specific and matched to the huge agroecological/climatic diversity. Detailed indigenous knowledge and greater skills in blending modern and traditional technologies to enhance productive efficiency will be more than ever before, key to the farming success and sectoral growth. Most technological solutions will have to be generated and adapted locally to make them compatible with socio-economic conditions of farming community.

New technologies are needed to push the yield frontiers further, utilize inputs more efficiently and diversify to more sustainable and higher value cropping patterns. These are all knowledge intensive technologies that require both a strong research and extension system and skilled farmers but also a reinvigorated interface where the emphasis is on mutual exchange of information bringing advantages to all. At the same time potential of less favoured areas must be better exploited to meet the targets of growth and poverty alleviation.These challenges have profound implications for products of agricultural research.

The way they are transferred to the farmers and indeed the way research is organized and conducted. One thing is, however, clear – the new generation of technologies will have to be much more site specific, based on high quality science and a heightened opportunity for end user participation in the identification of targets. These must be not only aimed at increasing farmers' technical knowledge and understanding of science based agriculture but also taking advantage of opportunities for full integration with indigenous knowledge. It will also need to take on the challenges of incorporating the socio-economic context and role of markets.

With the passage of time and accelerated by macro-economic reforms undertaken in recent years, the Institutional arrangements as well as the mode of functions of bodies responsible for providing technical underpinning to agricultural growth are proving increasingly inadequate. Changes are needed urgently to respond to new demands for agricultural technologies from several directions. Increasing pressure to maintain and enhance the integrity of degrading natural resources, changes in demands and opportunities arising from economic liberalization, unprecedented opportunities arising from advances in biotechnology, information revolution and most importantly the need and urgency to reach the poor and disadvantaged who have been by passed by the green revolution technologies.

Another important implication of increasing globalization relates to the need for greater attention to the quality of produce and products both for the domestic and the foreign markets. This would imply that production must be tuned to actual rapidly changing product demand. Such adaptation to global markets would require state of the art research, which can be achieved only by setting global standards of research, focus on well defined priorities and mechanisms which permit close interaction of farmers with researchers, the private sector and markets.

CHALLENGE OF INDIAN AGRICULTURE

There are several concerns about Indian agriculture that we need to address. Even in the green revolution crops, rice and wheat, the Indian average yield ranks around 50th in the world. A weak agro-based industry, the problems of weak marketing, storage, transportation, credit support, etc, haunt us. Only 0.3% of agricultural GDP is spent on agricultural R &D. The population is growing at an alarming rate of 1.8% per annum. It will be around 1.3 billion by 2020 and would require 50% additional foodgrains. A total of 200 million people are still below the poverty line and have insufficient access to food. Declining resource base, degeneration of natural resources, including soil fertility, water environment, etc, are some other concerns.

Food production systems centre around a few states only. Out of 17 big states, only 7 are self-sufficient and surplus in foodgrains. There was a significant decline in per capita intake of calories and protein between 1975 and 1990.

Inspite of all these problems, there is plenty of good news. India harbours a vast diversity of basic natural resources. Tremendous bio-diversity of agriculturally important micro-organisms exists, which are not yet been harnessed. We have a variety of crops, including quality fruits, vegetables, foodgrains, and a number of plantation and commercial crops. Indian fisheries resource has spread over 8,100 km of coastline on the east and west, and provides access to marine resources and transportation.

Very large proportion of area of various food crops is under low productivity category. It can be increased with inputs of high class science. There is a scope to cultivate sizeable portion of waste lands through soil amendment and introduction of agroforestry. Several million hectares of saturated soil can be potentially exploited from rainfed lowlands of eastern India. The India Exclusive Economic Zone (EEZ) of the marine sector, of about 2 million sq.km., has high potential harvestable yields. Only about 20% of the cultivated area is covered under tractor cultivation, great scope of making quantism quantum jump in production and productivity exists, changing consumption patterns offer scope for diversified agriculture. Trade liberalization provides opportunities to reach for outside markets, which were not accessible otherwise. Great gains are possible through agro-based industrialization and overall rural development.

The cultivators and farmers of India have a rich heritage of traditional agricultural wisdom on crop and livestock husbandry and fisheries; based on the principles of the conservation of natural resources and environment. India has an extensive network of extension channels for dissemination of technologies generated by research institutes. India has a strong manufacturing capability and a large network of about 20,000 manufacturers and nearly one million village artisans for indigeneous production of all types of agricultural machinery. Therefore, inspite of several disadvantages, we have a great chance to leap forward, if we enforce the right policy initiatives and implement them.

BIOTECHNOLOGY AND INDIA AGRICULTURE

We need to recongnise that, just as the food reequirements of today's population of nearly 6 billion people could not have been met by the technologies of the 1940's, we cannot assume that current practices will feed the population of 8 billion expected by 2020. New approaches are needed in addition to the continued improvement of existing methods of crop and animal husbandry and food processing.

The strategic integration of biotechnology tools into India agricultural systems can revolutionise Indian farming and usher in a new era. Genetic engineering is clearly the most revolutionary tool to impact agricultural research since the discovery of genetics by Mendel. Prior to genetic engineering, the exchange of DNA material was possible only between individual organisms of the same species. With the advent of genetic engineering in 1972, scientists have been able to identify specific genes associated with desirable traits in one organism. For example, a gene from bacteria, virus or animals may be transferred into plants to produce genetically modified plants having changed characteristics. This method therefore allows mixing of the genetic material from species that cannot otherwise breed naturally.

Genetic engineering can be vital for an agrarian country like India. It can help in minimising the crop damage through disease and pest resistant varieties, reducing the use of chemicals, enhancing stress tolerance in crop plants, thus permitting productive farming on unproductive lands, etc. One could even extend the growing season of crops and minimise losses due to environmental factors on the one hand and increase the shelf life of fruits and vegetables, on the other, thus minimising losses due to food spoilage. We can expand the market vista and improve food quality.

Biotechnology can also produce plants that possess healthy fats and oils, possess increased nutritive value, and create a whole range of higher value feeds. Biotechnology has even the power of producing biodegradable plastics, edible vaccines, etc.

It is only through the blending of the 'gene revolution' with our experience in the 'geen revolution' that we can reach our goal of 'evergreen revolution' and also 'nutritional revolution' The advantage of the gene revolution is that it is relatively scale neutral, benefiting big and small farmers alike, It is also environment friendly. Thus, it can be of great help to the smallest farmer with limited resources in increasing farm productivity through the availability of improved but powerful seed. It can also reduce a farmer's dependency on chemical inputs such as pesticides and fertilisers.

Modern biotechnology offers unlimited opportunity for enhancing genetic potential of crops and other commodities, management of biotic and abiotic stresses, bio-remediation and organic recycling. India is one of the main centres of agricultural biodiversity and its gene richness can greatly complement the developments in modern biotechnology. Likewise, new developments in GIS, remote sensing, and crop modelling, provide new opportunities for integrated management of natural resources. The revolution in informatics provides opportunities for sharing latest information for research and planning in a highly organised and efficient manner.

With the globalisation of economy, the opportunities for value addition and post harvest management are also immense. With clever blending of technologies, the India farmer can usher in new era of confidence and performance. Sir Francis Bacon had once said: *'It would be an unsound fancy to expect that things which have never yet been done, can be done except by methods which have never been tried'* In the new context, we have to review the new role of biotechnology. Also, it is in this context that we have to view the rapid advances that modern biotechnology is making, including the area of transgenics. Let us focus on some of the developments in this area, since considerable interest and controversy has been created around the world today in this exciting area.

MOVEMENT IN TRANSGENICS

The advent of genetic modification over the last two decades has enabled plant breeders to develop new varieties of crops at a faster rate than was possible using traditional methods, with huge potential for further beneficial development.

Tobacco was the first plant to be genetically transformed\, in 1983, with cereals beginning in 1990, but it is only recently that products such as GM soya have reached the market place in Europe, and, for a variety of reasons, given rise to concern & controversy. The concern must be addressed seriously if society is to exploit the new technologies appropriately. We will address this issue rather extensively a little later. Whereas in India, we have been discussing and debating the issues, the rest of the world is galloping ahead.

Table. Ag Biotech Product Sales

$ Millions	Sales			Annual Growth Rate	
	1994	1997	2002	1994-97	1997-02
Transgenic seeds and plants	20	405	2,030	173%	38%
Animal growth horomes	80	225	405	31	12
Biopesticides	35	65	110	23	11
Other	107	180	340	19	14
Total U.S. Sales	$242	$875	$2,885	53%	27%

Development, field testing, and commercialisation of transgenic crops, is now recongnized by the international scientific community to be an essential strategy for food security. Fortyfive countries, including India, have conducted transgenic crop field trials up to 1997. The activity in the developing countries has been shown in Table below.

Table. Releases of Transgenic Plants in Developing Countries (by Species and Introduced Trait)

Species	Introduced Trait					
	Total Fied releases	Herbicide Resistance	Insect Resistance	Virus Resistance	Product quality	Others
Maize	46	29	16	1	3	5
Soya bean	27	25	—	1	1	—
Cotton	24	16	15	—	—	—
Tomato	19	—	2	1	16	—
Potato	13	—	1	6	—	6
Subtotal	129	70	34	9	20	11
Other species	30					
Total	159					

The emphasis has been mainly on traits for herbicide, insect, virus & fungal resistance, and on product quality with a limited number of genes. Out of a total 159 releases. Argentina leads the way with 43, followed by chile, Mexico, Puerto Rico and the Republic of South Africa, with 17-20 each. India is far behind with just a few releases. There are fears relating to transgenic plants in society. These need to be addressed, not by brushing them aside, but by careful scientific analysis and educational programmes. All biosafety protocols must be followed rigidly. Let us deal with this issue more exhaustively now.

PUBLIC CONCERNS

The potential of biotechnology as a method to enhance agriculture productivity in the future has been accepted globally. However, because of its revolutionary nature, risk and uncertainty may be created by the process of genetic engineering and by the resulting genetically modified products. This may result from the introduction of a new unrelated DNA sequence into a recipient organism. The introduced DNA may have some unexpected effects on the cellular processes of the recipient organism. In addition to this, introduction of antibiotic resistance genes, as selection markers, pose serious implications in public health especially in genetically modified plants that are directly used for food purposes.

Risks are also associated with the genetically modified plants that are released into the environment. The nature of interactions with other organisms of the natural ecosystems cannot be anticipated without proper scientific testing. For example, modified plants with enhanced resistance to pests or disease threaten to transfer resistance to the wild relatives, which will have implications for biodiversity and ecosystem integrity. These, and many more, doubts plague the minds of common people. Therefore, it is very clear that if the biotechnology potential has to be completely realized, it has to be done in an extremely responsible manner. At each step, proper testing and scientific data have to provided. Not only this, proper regulatory and policy mechanisms have to be put into place.

It is therefore, absolutely essential, that conscious efforts be made especially by the developing countries to initiate well-defined programmes for the development and regulations of genetically modified plants. To make this possible and to benefically, ethically and sustainably reap the benefits of biotechnology, there is a pressing need for scientists, researchers, policy makers, NGOS, progressive farmers, industrialists, and representatives of the government, to come together on a common forum in order to discuss the common concerns and find solutions to them.

Will modified plants transfer their introduced genes into wild relatives growing nearby? Will modified plants that produce new compounds disrupt the 'balance of nature' in some way?

Could the planting of a restricted number of cultivators leave crop plants more susceptible to disease? Could the planting of a restricted number of cultivators lead to a reduction in biodiversity (of crop plants, weed species, insects and microflora in the fields in question)?

Will genetically modified plants be able to avoid the factors that regulate natural populations and thereby change the usual ' balance' between populations?

Similarly, there are concerns about the social and economic effects. For example, some of these concerns are:

- How will the structure of farming (particularly in developing countries) be affected by biotechnology?
- How will patent laws affect traditional breeders 'rights (*e.g.* the right to save seed from one year to the next)?
- Will plant breeding be left increasingly in hands of a few companies, and if so, what effects might this have?
- Will some countries be 'plundered' for their genetic resources?
- We also have a set of ethical and moral issues that need addressing. In particular,
- The consumer has the right. What will consumers be told about the new food products?
- Is it acceptable to 'interfere' with nature through genetic engineering?
- Do we have the right not to use all means available to improve crop plants, especially when so many people are under or malnourished?

Then there are several regulatory issues. Among these, the three key issues are:

- Do current regulations give sufficient protection to farmers, consumers, those who have invested in, and those engaged in research?

- Is there sufficient international legislation to ensure environmental protection?
- Do current regulations compromise the competitiveness of biotechnology companies by being excessively restrictive?

The Royal society of London had appointed a group of experts to examine various aspects including the scientific evidence concerning the risk of transfer of genes from Genetically Modified (GM) crop plants to wild species and non-GM crops, the uptake of genes from GM food by the digestive system, and the current state of the regulatory system. The experts concluded that the chances of gene transfer happening are slight, provided the regulatory processes are followed, but that this must be kept under consideration.

The experts also concluded that the uptake of genes via the food chain is not a new issue because genes (*i.e.* DNA) are normal constituents of the human diet. Many products from GM Plants, such as sugar prepared from GM tomato paste, are so similar that they are regarded as 'substantially equivalent' others, for example flour from GM soya, may contain a new gene or its product, although many of the purification processes involved in food production will destroy and DNA present in the raw material. Some GM foods have been produced using an antibiotic resistance 'market' gene, which is a laboratory device designed to identify genetically transformed plants.

The Royal Society has shown concern about the use of such genes in food products and it suggested that any further increase in the use of such markers in the human or animal food chain would be undersirable. Gm crop plants have been produced to improve insect tolerance and virus resistance, and to include herbicide tolerance, so that transferred to non-target species of plants, and that the development of resistance by target pests is minimised, the regulatory authorities must be assured that any negative effects would be no greater than those resulting from conventional procedures; and that any long term effects on the environment and ecology would be closely monitored, with statutory restrictions in place to control marketing; and also that best practice advice was adopted by growers.

The reliance on a case by case approach to the legislation, may result in lack of analysis of the overall impact of the technology on agriculture and the environment, and of the long term effects of gm. Although mechanisms are already in place to regulate many individual aspects of GM technology, there is no means for looking at GM technology as a whole. As independent, overarching, regulatory body is needed to span departmental responsibilities, monitor the enforcement of existing or future regulations, and strengthen the guidelines to growers of GM crops, such as those specifying the isolation distances between Gm and non-GM crops. The body should also review and monitor the membership of advisory committees and regulatory bodies.

4

Agricultural Practices in Poultry Farming

For handling birds and for all practices under this heading, experienced and skilled persons should carry out or train and supervise those who carry out these procedures.

BEAK TRIMMING

Trimming of the tip of the beak is done to minimize injury and death due to aggressive and cannibalistic behaviour. Outbreaks of cannibalism among egg-laying strains of chickens, turkeys, and ducks can occur with any housing system, resulting in a serious welfare problem. If the trimmed beak grows back, a second trim may be needed. An alternative to beak trimming is use of low light intensity in housing systems where light control is feasible. Genetic stock that shows little tendency towards cannibalistic behaviour and feather pecking should be used when possible.

EGG-STRAIN CHICKENS

Production, behaviour, and physiological measurements of stress and pain as indicated by neural transmission in the trimmed beak are used as criteria to determine well-being in beaktrimmed birds. In addition, the welfare of those hens that are pecked by beak-intact hens has been evaluated. Disadvantages of beak trimming include shortterm stress as well as short-term, and perhaps long-term, pain following the trimming of the beak. Because feeding behaviour must adapt to a new beak shape, a bird's efficiency in eating is impaired following a trim.

Welfare advantages include decreased mortality; reduced feather pulling, pecking, and cannibalism; better feather condition; less chronic stress; and less fearfulness and nervousness. Welfare advantages are more applicable to the interactive flock, whereas welfare disadvantages are applicable to individual birds whose beaks are trimmed. Genetic lines differ in their aggressiveness and beak-trimming requirements. Genetic selection is effective in reducing or eliminating most feather-pecking and beak-inflicted injuries and heritability estimates for survival suggest that the prospects for improving livability through genetic selection are good. Therefore, when feasible, stocks should be used that require either minimal or no beak trimming. Nevertheless, beak trimming is justified in stocks that otherwise are likely to suffer extensive feather-pecking and cannibalistic losses.

Management guides, available from most breeders, indicate methods for beak trimming to reduce these vices. Beak trimming should be carried out when birds are 10 d of age or younger. The amount of beak removed should be 50 per cent or less to avoid neuroma formation and to allow the keratinized tissue to regenerate. The length of the upper beak distal from the nostrils that remains following trimming should be 2 to 3 mm. The lower beak should be slightly longer than the upper beak. If a second trim is needed due to regrowth of the beak, it is recommended that it be done before the pullets are 8 wk of age to avoid a decrease in egg production.

BROILER-TYPE CHICKENS

Beak trimming is generally not required in young broilers raised for meat production. For broiler breeders, early beak trim before 10 d of age is generally sufficient to control feather-pecking and cannibalism in breeder stocks.

TURKEYS

Beak trimming of turkeys is a standard management practice. Strains of turkeys and sexes differ in their requirement for and their response to beak trimming. In strains of turkeys that exhibit a high incidence of beak-inflicted injuries, arc-type beak trimming at hatching is effective in reducing such injuries. Severe arc-type beak trimming increased mortality relative to hot-blade trimming of the upper beak at 11 d of age.

There was no evidence that arc-type beak trimming 1.5 mm from the nostrils at hatching or hot-blade trimming of the upper beak at 11 d of age increased mortality relative to leaving beaks intact. Beak trimming completed shortly after hatch did not modify performance or behaviour in commercial market toms compared with non-trimmed controls and also reduced pecking damage when beak regrowth did not occur. Arc-type beak trimming 1.5 mm anterior to the nostrils or hot-blade trimming of the upper beak at 11 d of age is recommended to prevent cannibalism in strains of turkeys that exhibit a high incidence of beak-inflicted injuries.

DUCKS

Feather pecking is a behaviour that sometimes occurs in ducks and may be controlled either by partial removal of the nail of the upper bill or inhibition of the growth of the nail by heat treatment. If not controlled, feather pecking injures the feather follicles of the tail, wings, and back, and the protective feather and down covering breaks down. Tip searing using cautery only may be a preferred method of bill trimming in Pekin ducks because of better weight gains following a trim and fewer changes in the morphology of the bill. For all species of poultry it is critical that the equipment used to trim beaks is maintained in good working condition. Personnel involved in beak trimming should receive species-specific training on proper procedures to use during beak trimming.

TOE TRIMMING

Because of the size and weight of the birds involved and the sharpness of their toenails, broiler breeder males and market turkeys generally have certain toes trimmed to prevent them from inflicting serious injuries to the hens during natural matings or to their penmates. Toe trimming should be done at 1 d of age using an electrical device that removes and cauterizes the third phalanx of the toes involved.

Microwave energy application to the tip of the toe is also used to restrict toenail growth and is conducted using specialized equipment at the hatchery. In chickens, the microwave method did not result in increased stress or fearfulness. Provision of abrasive strips or hard surfaces in the facility may help to control excessive claw growth and reduce the need for declawing. Trimming toes for the purpose of identification is unjustified and should not be performed.

EGG-LAYING STRAINS OF CHICKENS

Leghorn hatchlings whose claws were trimmed through use of microwave energy experienced increased mortality and reduced feed consumption and BW during the pullet grow-out period. Removal of the claws resulted in a reduced foot spread allowing the toe of some pullets to slip into the wired mesh of the cage floor.

The pressure on the web between the toes led to a splitting of the foot epidermis in 24 of the 1,200 pullets whose claws were trimmed. Compton *et al.* reported similar results when using a hot blade to reduce claw length and suggested that chick movement about the wired cage was difficult until the toe grew long enough to allow the foot to spread across the wired cage floor. These results suggest that trimming the claws of egg-laying strains of chickens is not recommended.

BROILER BREEDER MALES

When meat-type males of certain genetic lines are to be used in natural matings, the practice of trimming certain toes at 1 d of age can be considered; toe trimming of breeding males may prevent injury to the female during natural mating. However, there is also evidence that toe trimming may impair the mating ability of males. The removal of one nail does not appear to cause chronic pain. For those genetic lines with long spurs, the spur bud on the back of the cockerel's leg may be removed at 1 d of age using a heated wire. Use of genetic lines with short, blunt spurs is preferable over spur removal. Most commercially available broiler breeder lines do not need to have their spurs removed.

TURKEYS

Toe trimming is a widespread management practice in turkey production. The number of toes trimmed per foot varies from 1 to 3 plus the dewclaw. Carcass grade of turkeys may or may not be improved by toe trimming, although rate of early mortality may be increased. Toe trimming may be justified when excessive injuries are likely to occur, but alternative methods should be considered to prevent bird injury.

SNOOD REMOVAL

Turkeys have a frontal process called a snood, which is an ornamental appendage for the adult male. The snood can be grasped by other turkeys during fighting and can be torn or damaged.

Breaks in the snood skin can be a health concern among older turkeys or those housed on pasture or on ranges. Data collected from industry showed that snood removal in tom poults reduced the odds of mortality. To avoid injury and possible infection, the snood can be removed from the newly hatched male poult by clipping or pinching the snood from its base on the head. If removed, the process should occur as soon as possible after hatching and no later than 3 wk of age. Snood removal after 3 wk of age is possible by clipping but not recommended without veterinary advice as the snood will continue to increase in size and vascularization especially in the males.

PARTIAL COMB AND WATTLE REMOVAL

Removal of part of the comb and wattles of chickens may be needed if birds are kept in cages. Combs and wattles can get caught in wire openings or feeders after significant comb and wattle growth has occurred. Comb and wattle removal is more commonly performed on cockerels because these structures are larger in males. Dubbing or removal of part of the wattles should only be used as a last resort when equipment or housing conditions cannot be modified to prevent torn or damaged combs or wattles. To perform successful comb and wattle removal with minimal bleeding and excellent long-term results, surgical scissors, scalpel blade, or electrocautery/radiosurgery electrode should be used to remove part of the comb and wattle during the first few days after hatching. To reduce risk of infection between birds, the scissor blades can be disinfected.

PINIONING

Surgical pinioning, which involves amputation of the wing tip from which primary feathers grow, or tendonotomy is used mainly in exhibit birds to render them permanently incapable of flight. Pinioning is not recommended as a means of reducing bird flightiness in chickens broilers, and ducks used for research and teaching. If flightiness is problematic, the primary feathers of one wing may be clipped.

INDUCED MOLTING

In birds, plumage is normally replaced before sexual maturity through a natural molt. Molting also occurs naturally after sexual maturity and is associated with a pause in egg production, which can be lengthy and take place out of synchrony with others in the flock. Inducing synchronized molting is used to rejuvenate laying flocks to extend the productive life of hens for 2 or 3 cycles of production.

Molting has become a common procedure for commercial table-egg layers and sometimes for broiler breeders and turkey breeders. In recycled egg-laying strains of chickens, molting decreases the demand for chicks by 47 per cent and thereby reduces the need to process, render, or bury the same percentage of spent hens. Rejuvenation of flocks also prevents the annual euthanasia of one hundred million additional male chicks. Additional advantages of molting include feather rejuvenation, thus improving thermoregulation. After a molt, livability and egg quality are improved during the second cycle of egg production compared with a non-molt control group.

EGG-STRAIN CHICKENS

Several procedures used to induce a molt have included short-term and long-term feed withdrawal; manipulation of dietary energy, protein levels, and dietary ingredients such as calcium, iodine, sodium, or zinc; and addition of feed additives that influence the neuroendocrine system such as iodinated casein. These procedures have been used coupled with a reduction in the daily photoperiod. These methods cause a cessation of egg production along with decreased BW and feather loss.

To allow for a return to egg laying, feather regrowth and BW gain are accomp-lished by feeding a diet designed to meet the nutritional requirements for a non-ovulating, feather-growing hen. Until 2000, the most common procedure used to induce a molt was to withdraw feed for 4 to 14 d without water restriction. Feed withdrawal for inducement of ovarian arrest is stressful leading to increased mortality during the first 2 wk of the molt. Hens are more fearful during a fasted molt compared with before and after a molt.

Temporary frustration as indicated by a moderate increase in aggression on the first day of feed removal has been noted in molted hens compared with non-molted full-fed controls. Aggression dissipated by the end of the first day, and molting hens showed elevated activity on the second day of fasting as indicated by increased non-nutritive pecking, standing, and head movement.

Resting behaviour increased by d 3 of fasting, and although non- nutritive pecking decreased from d 2, this pecking, interpreted as a redirection of foraging activity, remained higher than in control hens. Resting behaviour persisted for the remaining part of the fast. Similar changes in behaviour of hens subjected to a fasting molting regimen have been reported by Simonsen and Aggrey *et al.* with the notation of an additional behavioural repertoire of increased pre-ening on d 8 to 10 post-feed removal, most likely coinciding with the dropping of feathers. Hens subjected to a fasting molt compared with non-molted controls demonstrated decreased skeletal integrity, immunity, helper T cells and heterophil phagocytic activity. In addition, hens subjected to a fasting molt showed an increase in *Salmonella enteriditis* (SE) fecal shedding, the prevalence of SE in organs, inflammation of the intestines, the recurrence of a previous SE infection, and susceptibility to SE infection compared with non-molting controls.

Salmonella enteriditis was readily transmitted horizontally among molting birds under simulated field conditions, whereas in actual field settings, increased environmental *Salmonella* was observed in molted versus non-molted hens. As an alternative to fasting, hens subjected to non-feed-removal molting regimens show post-molt performance not unlike the hens of the fasting molting regimen. Examples of successful non-feed-removal molting methods include the ad libitum feeding of diets high in corn gluten, wheat middlings, corn, or a combination of 71 per cent wheat middlings and 23 per cent corn. *Salmonella* shedding, intestinal inflammation, and internal organ contamination of SE-challenged hens were reduced and bone mineral density improved through the use of non-feed-withdrawal molting programmes compared with hens of a fasted molt.

Environmental presence of *Salmonella* increases during the molt in rooms containing fasting hens, but not in rooms of hens molted through wheat middlings. *Salmonella* fecal populations did not increase during a non-feed-removal molting programme compared with the pre-molt and post-molt periods, with *Salmonella* prevalence being the lowest during the molting period. Biggs *et al.* reported no differences in social behaviour between fasted hens and hens subjected to a non-feed-removal molting programme.

These results on increased resistance to *Salmonella* and improved skeletal integrity suggest that non-feed-withdrawal methods of molting should be used rather than the more conventional feed-withdrawal molting regimens. During the non-fast molt, hens should be monitored for health, mortality, and body weight. Water withdrawal or restriction, which can lead to increase mortality especially during hot weather, is not recommended.

BROILER BREEDERS, TURKEY BREEDERS, AND DUCK BREEDERS

Induced molt is occasionally done on parent breeding stock using feed withdrawal methods. Molting methods for breeder ducks are similar to those used for broiler breeders. Non-fasting methods of inducing a molt have not been reported in breeder stock.

SPECIAL CONSIDERATIONS

Genetically Modified Birds

To date, there are no special animal care requirements for transgenic or cloned poultry. Transgenic birds are cared for in the same manner as conventionally domesticated birds unless the genetic manipulation affects basic bird needs. Future transgenic animals may have special requirements and they should be cared for based on their genotype and phenotype rather than based on the technology that was used to create them.

SURGERIES

All intrathoracic and intraabdominal invasive surgeries require anesthesia. Caponization, or removal of the testes, is an invasive surgical procedure that requires anesthesia.

OTHER BIRD SPECIES

Gaunt and Oring and the Canadian Council on Animal Care offer recommendations on the care and use of wild birds, pigeons, doves, non-domesticated waterfowl, budgerigars, and quail. Parkhurst and Mountney provide animal care recommendations for geese, Coturnix quail, Bobwhite quail, chukar partridge, pheasants, guinea fowl, peafowl, pigeons, and swan. The Standing Committee of the European Convention for the Protection of Animals Kept for Farming Purposes provides recommendations and minimum standards for the welfare of ostrich and emu. Recommendations from New Zealand provide animal care guidelines for ratites. These references are given not as an endorsement but as referral material only.

AGRICULTURAL ANIMAL HEALTH CARE AND EUTHANASIA

Agricultural Animal Health Care and by the American Veterinary Medical Association (AVMA) *Guidelines on Euthanasia.* For the purpose of euthanasia, the AVMA accepts administration of barbiturates, inhalant anesthetics, carbon dioxide, carbon monoxide, gunshot and stunning followed by exsanguination, and conditionally accepts nitrogen and argon gases, cervical dislocation, decapitation, and maceration. Methods of euthanasia should ensure death and be selected to take into account any special requirements of experimental protocols so that useful data are not lost.

Anesthetic agents are generally acceptable, and most avian species can be quickly and humanely killed with an overdose of a barbiturate administered intravenously. When relatively large numbers are involved, exposure to gas euthanasia agents such as carbon dioxide in enclosed containers may be used.

Atmospheres containing a significant amount of carbon dioxide, with or without the presence of oxygen, cause birds to head shake and breathe deeply, but scientific evidence indicates that these behaviours are not associated with distress. These behavioural changes are not caused by irritation of mucosal epithelia in the nares or throat because they occur at carbon dioxide levels considerably below the threshold of trigeminal nerve nociception; that is, 40 to 50 per cent carbon dioxide based on lab study of nerve fibre activity in chickens.

Furthermore, although poultry can detect atmospheres containing significant concentrations of carbon dioxide and may show responses indicative of some degree of aversion, several studies have demonstrated that most chickens and turkeys will voluntarily enter carbon dioxide concentrations as high as 60 to 80 per cent. Because poultry can be rendered unconscious with 30 per cent carbon dioxide in air, or less if enough time is allowed, and concentrations of carbon dioxide above 50 per cent quickly kill adult birds, it is not necessary to measure the carbon dioxide concentration closely when performing euthanasia.

However, it is important that the process be observed and carbon dioxide added, if necessary, to ensure that death is attained without undue delay. Although euthanasia of poultry in high concentrations of carbon dioxide is relatively rapid, it also tends to promote vigourous convulsive wing flapping after loss of posture. Although the birds are not conscious when this occurs, the sight can be disagreeable to human observers.

Slower induction of unconsciousness using lower concentrations of carbon dioxide appears to sedate birds and greatly reduces convulsions after loss of posture. Newly hatched chicks and poults have a greater tolerance to carbon dioxide so concentrations of 60 to 70 per cent should be used to kill these birds. Anoxia using argon or nitrogen, or mixtures of these gases with carbon dioxide, has been found to be effective and to produce minimal distress, but residual oxygen should be kept below 2 per cent.

Anoxia causes strong convulsive wing flapping after loss of posture. When employing anoxia, the final gas concentration should be achieved quickly to avoid development of ataxia in conscious birds. It is acceptable for an individual who has been properly trained to use cervical dislocation without stunning or anesthesia when small numbers of birds that are small in size require euthanasia.

When enough experienced personnel are available for a given period of time, large numbers of birds can be euthanized via cervical dislocation, as long as operator fatigue is avoided. Cervical dislocation is not recommended with larger poultry such as turkeys and adult ducks or when one individual is required to kill a large number of birds. Following cervical dislocation, the necks of small birds should be checked for dislocation of vertebrae to ensure that the procedure was done correctly. Use of a captive bolt device for euthanizing large birds such as adult ducks and turkeys can be used by a skilled operator provided bolt diameter, mass and velocity, and angle of bolt impact are appropriate. Restraint of the head without compromising the handler is a major concern with use of captive bolt, so safety and restraint issues need to be considered. Both cervical dislocation and captive bolt killing are followed by severe convulsive wing flapping. A Burdizzo, a flatedged clamp used for crushing tissue, may be used by trained individuals for the euthanasia of large poultry, particularly turkeys older than 10 wk of age. Birds must be rendered insensible before crushing the cervical vertebral column with a Burdizzo. Embryonated eggs may be destroyed by chilling or freezing at a temperature of 4°C for 4 h. Decapitation or anesthetic overdose are suitable methods for embryos that have been exposed for experimental purposes.

Maceration in a purposedesigned macerator, a mechanical apparatus with rotating blades, is also considered a humane method for killing embryos and surplus neo-natal chicks. Chicks are rapidly fragmented by maceration, which results in immediate death.

SLAUGHTER

Slaughter of animals entering the human food chain must comply with regulations as outlined in the *Federal Humane Slaughter Act*. The processing area for poultry slaughter should be designed and managed to minimize bird discomfort and distress. The manager or person in charge of the processing area should be competently trained in animal slaughter and is responsible for training all staff to carry out their duties responsibly and humanely. The holding area for birds to be processed should be adequately ventilated and protected from temperature extremes and adverse weather such as wind, rain, sleet, snow, and hail.

Upon arrival, birds should be inspected to ensure that none are injured or suffering from heat or cold stress. Injured birds with signs of severe stress should be humanely killed or slaughtered immediately. If numbers are in excess, the farm manager should be contacted immediately. Birds should be processed as soon as possible once they arrive at the slaughter facility. All birds should be slaughtered within 12 h of feed and water withdrawal.

Feed withdrawal minimizes microbial contamination of the carcass by preventing breakage of the gastrointestinal tract during processing. All transport crates and trucks should be inspected to make sure that all of the birds have been removed for processing. Birds should be handled carefully when removed from crates or, in the case of large turkeys, from livestock trailers.

In plants with automated lines, birds should be shackled with a line running at a speed that permits the proper positioning of the birds to prevent injuries such as broken bones or bruising and to minimize discomfort and stress. Shackles should be of proper size to prevent bird escape and discomfort. Both legs should be hung on the shackles. To keep birds in the proper position for stunning, the height of the line should be adequate. Measures should be taken to minimize wing flapping such as use of funnels, breast bars, curtains, low light intensity or blue lights, reduction in noise, running a hand down birds after shackling, and avoiding bends in the line between the shackling area and the stunner. In non-automated systems, cones should be of appropriate size. Birds should not be suspended upside down in cones or shackles for more than 90 s before they are stunned. Poultry killed using exsanguination should first be stunned using electrical or gas methods.

Stunned birds may recover consciousness quickly; therefore, exsanguinations should be accomplished immediately after stunning to avoid recovery from consciousness. Exsanguination itself results in a rapid loss of consciousness if both carotid arteries are completely severed. Considerations involved in electrical stunning are discussed by Gregory and Wilkins, Bilgili, and Raj and Tserveni-Gousi. Electrocution is acceptable if the current travels through the brain and through the heart.

Occasionally some birds may not develop ventricular fibrillation after electrocution, so any birds showing signs of recovery should be immediately killed by other means such as by cervical dislocation, decapitation, or gas. Electrical stunners adjusted for sufficient current should render birds immediately insensible before neck cutting, and they should remain insensible during exsanguination. Acceptable stunners include a hand-operated stunner, stunning knife, a dry stunner incorporated into a metal bar or grid that is electrically live, or an electrical water bath.

Hand-held electrical stunners may be used for shackled birds or for those birds that are restrained in a cone. The electrodes are applied to either side of the head between the ear and eye. The stunner should be applied to shackled birds until wing flapping stops or until the legs become rigid and extended when using the cone. With respect to use of a water bath for stunning, the water level in the bath should be set so that the heads of all birds make effective contact with the water.

Use of an ammeter is recommended to monitor current flow through the water bath while it is loaded with birds. The water bath should be deep enough to prevent water overflow and the electrodes should extend the length of the water bath. Birds exiting the water bath should be regularly checked to ensure that stunning is effective. Characteristics of adequate stunning include rigidly extended legs, rapid and constant body tremors, wings held close to the body, open eyes, and an arched neck with the head directed vertically.

If cardiac arrest is induced during stunning, birds become limp with no breathing or reflex of the nictitating membrane. Pupils are dilated and the birds do not respond to a comb pinch. Stunning equipment should be maintained properly to ensure an adequate stun. Gassing birds before exsanguination may be a humane method of rendering birds insensible, but further research is needed to determine if it is a superior method. If birds are gassed before or immediately after removal from transport crates or vehicles, they avoid the stress of shackling and the potential of pre-stun electrical shock. Post-stun exsanguination should be initiated by making a ventral cut in the neck, wherein at least both carotid arteries or the carotid artery and the jugular vein on one side are severed. Properly stunned birds will not show voluntary behaviour such as eye blinking, coordinated head or limb movements, or attempt to escape the shackle or cone during exsanguination.

Some involuntary convulsive movement, such as a wing flap, is not unusual as the blood supply to the brain becomes depleted. In some cases there may be a need for kosher or halal slaughter of birds, which does not allow stunning. For this purpose a very sharp knife with a straight surface that is at least twice the length of the head should be used to cut the arteries, veins, trachea, and esophagus. A poultry scalpel can also be used effectively.

An aggressive single stroke cut is most effective. Birds must be permitted to bleed out before further work is conducted. This process should only be performed on birds that are adequately restrained such as by the use of a cone. Birds must be rendered insensitive in less than 30 s. Following exsanguination, birds must not be breathing when they enter the scalder. Birds must be monitored to make sure they are dead before entering the scalding tank. If any bird shows signs of consciousness, they must be removed from the processing line and promptly stunned.

POULTRY PRODUCTION AND FARMING SYSTEM

Energy-Efficient Lighting for the Farm from the US National Sustainable Agriculture Information Service (ATTRA) provides an overview of energy-efficient lighting technology and explains how to select lighting options that are appropriate for the farm, including alternative poultry production. Energy-efficient lighting options present farmers with new opportunities to reduce electricity costs and help manage farms sustainably. Cost-effective energy-efficient lighting can be used to improve productivity and safety, and reduce operating costs.

Supplemental lighting is normally used by alternative egg producers to maintain productivity, and sometimes for alternative broiler production in northern climates. Small layer flocks housed during late spring through mid-summer with daily access to the outdoors do not require supplemental light. Supplemental lighting is necessary for pullets to maintain production during late fall and winter as days shorten. Poultry are very sensitive to three aspects of light: intensity of light (measured in footcandles), wavelength (measured in colour temperature), and day length (duration of light period). Research by Michael Darre and others has found that blue light wavelengths help calm birds; red wavelengths may be used to help reduce feather picking; blue-green wavelengths help maintain growth; and orange-red wavelength helps maintain reproduction. The light intensity for layers should be enough to read a newspaper by and will vary with the poultry breed. Generally, 'warm' wavelength lamps of less than 3,000K in the red-orange spectrum are best for small flocks with outdoor access. The day length should never be extended past 16 hours or the longest day of the year. Solar photovoltaic lighting provides a simple solution to maintaining egg production during shorter days. Solar lighting systems basically consist of a solar module, a deep-cycle battery, a charge controller, a 12V programmable timer, and an efficient DC lighting fixture with lamp. Energy-efficient LED lamps work very well with solar modules. All of the components to build a basic low-voltage solar lighting system can be purchased online for less than $300 or as a kit.

To conserve energy and keep poultry healthy, use timers to switch lights on and off. Programmable timers must be 12V when used in conjunction with a 12V solar lighting system. Th ere are 12V timers available online as well as schematics to convert a household programmable thermostat to a 12V timer. Timers also ensure that birds receive a uniform number of light hours each day. Set timers to light in the morning instead of the evening to give birds a natural dusk and allow them to roost. Check timers at least once a week, and clean lamps if dust builds up. Lamps should be free of obstructions that cause shadows on the floor. Baby chicks require additional light in their first 72 hours to help them find food and water. A low watt 'warm' lamp is recommended for every 200 square feet of floor space. (Hawes) The high heat from incandescent lamps may double as a brood light and heat source, although it may be more energy-efficient (and cost-effective) to use a separate heat source and a solar lighting system.

ALTERNATE POULTRY PRODUCTION IN INDIA

The alternate poultry production in India is gaining momentum and attention from the farmers, entrepreneurs, professionals and researchers. It is used for food, game, fancy, pet, and also for research purposes. Here, the importance, prospects and potentials of alternate poultry production in India has been reviewed.

QUAIL PRODUCTION

Quail is the most efficient biological machine for converting feed into animal protein of high biological value and hence is the cheapest source of animal protein for human diet. Consumption of quail began before domestication of poultry. In China, Taiwan, UK quail has been reared through centuries. During fifties its commercial production started in Italy. The CARI is the pioneer in introducing quail farming in 1974 by importing germplasm from U.S.A. It is introduced in India as an alternate avian species for growing poultry industry.

M/s. AVM Hatchery and Poultry Breeding Research Centre, Coimbatore has played a major role for production on extensive scale in private sector. Owing to the wild life act and ban on captivation of domestic quail, no worthy development and popularisation of quail production took place for last ten years. But now the ban has been lifted, it has tremendous scope to occupy a considerable portion of poultry meat market by 2010 A.D. as a profitable enterprise. Quail has unique qualities of hardiness and adaptability to diversified agro-climatic conditions. Its consumption may be referred by all, as it has no religious taboo. It has nutritive value, amasing taste, and game favour, tender, very delicious with low calorific value and high dry matter.

It is rich in protein, vitamins, essential amino acid, unsaturated fatty acids and saturated fatty acids and phospholipids. It may be recommended for children and pregnant women, for speedy recovery. It is a tiny inter breeding avian species quite different from wild quail. Because of twin capacity of prolific egg production and meat yield, it attains the status of viable commercial poultry enterprises as well as rural poultry production.

At present quail production is being done at Agricultural Universities or few private sectors only, but if we want to achieve our target we will have to extend our quail farming activities to rural areas where our bulk of population lives. Quail production requires less investment and gives quick returns, higher profits and hence can be adopted by rural masses quickly. A very important point is that so far benefits of quail farming have not fully reached rural masses, though quail production can generate additional income. Agriculture production can be boosted through the use of high value quail manure, as it is four times nutritive than cow dung manure.

DUCK PRODUCTION

Ducks are the second largest source of table eggs and there are about 23 million ducks in India. West Bengal has the highest duck population followed by Assam, Tamil Nadu, Andhra Pradesh, Bihar, Kerala, Orissa, Jammu and Kashmir and Tripura. Ducks predominantly are of indigenous type kept for egg production on natural foraging and have a production potential of about 130-140 eggs/bird/year. Duck eggs have a preference over chicken eggs in certain states and areas.Considering the importance and scope of duck rearing in our rural economy, improved strains and technologies are available in the country and the farmers are much benefited from the same.

Duck farming is primarily popular among small farmers, marginal farmers and agricultural labourers as well as poor section of the community. Ducks lays 40-50 eggs more than desi chicken. The duck egg is heavier than hen by 15-20g. Ducks require lesser attention and supplement their feed by foraging, eating fallen grains in harvested paddy fields, insects, snails, earthworms, small fishes and other aquatic materials in lakes and ponds hence incur reasonable feed cost. Ducks have a longer profitable life and lay well even in second year. Ducks do not require elaborate houses like chicken, reducing the capital investment. Ducks are quite hardly, more easily brooded and more resistant to common avian diseases. Ducks flourish well in marshy riverside, wetland and barren moors where other types of stocks do not.

Ducks have no problems of cannibalism or pugnacious behaviour. Ducks lay 95-98 per cent of their eggs early in the morning, before 9 AM, thus saves a lot of time and labour. Ducks are useful in controlling unwanted plants in ponds, lakes, and streams like green algae, duckweed, pond weed, musk grass, arrow head wild celery, etc. (5-10 ducks/ 0.405ha of water table). Ducks are good exterminators of potato beetles, grasshoppers, snails, and slugs. In areas plagued by liver flukes ducks can help correct the problem (2-6 ducks/ 0.405 ha of land). Ducks can be used to free the bodies of water form mosquito pupae and larvae (6-10 ducks/ 0.405 ha of water space).

Ducks are suitable for integrated farming systems such as duck-cum-fish farming, duck farming with rice culture, etc. In duck-cum-fish farming the droppings of ducks serve as feed for the fishes and no other feed manuring of the pond is necessary for fishes (200-300 ducks/ha of water area). Under integrated duck farming with rice culture, the ducks perform 4 essential functions, *viz.*, intertillage, weeding, insect control and manuring. The down and small body feathers of ducks are valuable as filler for pillows and as lining for comforters and winter clothing (one Pekin duck produce about 100g of down and small feathers).

The important Indian breeds are Sythet mete and Maheswari, of eastern region, Chara and Chamballi of Kerala, desi breeds of West Bengal and other states. Among improved varieties distinct breeds for egg and meat production are available. Khaki Campbell and Indian Runner are the most popular breeds for egg laying. White Pekin, Muscovy and Aylesbury are known for meat production. High-laying strains of ducks are available at the Central Duck Breeding Farm, Hessarghatta, and Bangalore. Duck units are functioning in some of the states like Assam, Tripura, Orissa, Madhya Pradesh, Andhra Pradesh, Kerala, Haryana and J & K.

TURKEY PRODUCTION

Turkey is mainly reared for meat and is quite popular in many western countries. It is one of the white meat choice and famous for its leanness and deliciously cooked whole turkey forms the traditional dish in Thanks giving and Christmas dinners. Turkeys are sold as whole or cut-up parts and further processed products such as sausages, patties, and nuggets.

Turkey farming is still at the stage of infancy and need to be popularised among farmers to provide diversified food and employment especially in India. Rearing turkeys can be an excellent family or youth project. They are quite suitable for upliftment of small and marginal farmers as the birds can be easily reared in free range or semi-intensive system with minimal investment for housing, equipment and management.

Even though turkey farming is practically non-existent in our country, few backyard birds are distributed in few areas. Also organised farms of turkeys have been maintained at Central Avian Research Institute, Izatnagar, CCS Haryana Agricultural University, Hissar, University of Agricultural Sciences, Bangalore, Rajasthan College of Agriculture, Udaipur, Kerala Agricultural University, Mannuthy and Tamil Nadu Veterinary and Animal Sciences University, Kattupakkam.

Turkeys are not difficult to raise. They require a little special care to get them off to a good start. Sometimes they are little slow in learning to eat and drink. If the farming is started with a good stock and provided optimum feed, housing and management, one can raise turkeys successfully.

There are seven standard breeds of turkeys, *viz.* Large White (Broad Breasted White), Bronze (Broad Breasted Bronze), Beltsville Small White, Bourban Red, Black, Slate and Narrgansett. The turkeys can be marketed as broiler at approximately 16 weeks of age when hens usually reach a live weight of about 8kg. Toms weigh approximately 12kg. Smaller, fryer roasters can be produced by slaughtering at an earlier age as per the market demand. Adequate research in the field of management, nutrition and disease control under tropical conditions as well as education are imperative before private entrepreneurs can be convinced for the viability of turkey enterprise in India.

OSTRICH AND EMU PRODUCTION

Ostrich and Emu farming is now-a-days considered to be one of the most profitable agricultural projects. They are referred as 'farms of the future' due to the large variety of products and hence the high profits potential. Ostrich meat is highly nutritive and earns enormous foreign exchange. Ostriches are raised commercially for their egg, meat, feathers and hide. The increased consumer awareness of the problems of cholesterol in the blood and their possible association of heart attacks and arteriosclerosis, the demand for ostrich meat in international market began to increase. A number of beef producers in Europe, America and Canada have most recently switched to commercial ostrich farming due to higher and faster financial returns on ostrich projects. Ostrich meat is in great demand in European and American countries and is already extremely popular and available in most of the restaurants and supermarkets.

Recently Tamil Nadu Veterinary and Animal Sciences University imported 100 chicks from Malaysia. Their adaptability to climatic conditions prevailing in the state would be tested and if successful it will be recommended to the interested farmers and entrepreneurs. Considering the above aspects of alternate poultry production, our country has a great potential to develop further in this area of farm production. The NABARD and associated banks are also ready with model projects and financing of such projects through reimbursement scheme which is also a boost for future developments.

GUINEA FOWL PRODUCTION

An important source of traditionally raised poultry species, guinea fowl was introduced into Indian sub-continent by African ethnic migrants a few centuries back. Definite population figures are not available, it is estimated that few million birds are raised every year. Guinea fowl ranks third among the poultry birds; it is referred by different local names in different regions, *viz.*, chittra in west, china murghi in south and titari in the northern plains. Earlier attempts to raise guinea fowl intensively were made at Christian Missionary Poultry Farm, Etah (UP), Allahabad Agriculture Institute, Naini (UP), and Government Poultry Farm, Gurdaspur (Punjab) and subsequently at Military Poultry Farm.

However preliminary survey of guinea fowl production in northern regions and the perusal of earlier published literature established its distinct popularity with marginal farmers, tribes and pastoralists may be attributed to its inherent hardiness and excellent foraging potentials to go as animal component of biomass-based polycultural systems.

In recent years this alternate poultry species witnessed increasing emphasis for low-input grain-saving aviculture (LISA).

The primary function of guinea fowl production is to produce low-cost game flavoured meat. There are two quite distinct production systems for rising of guinea fowl. One is the modern intensive system and other is the traditional extensive system. The former is characterised by high inputs, high costs, good husbandry, hygiene and a complete supply of feed. The intensive production is largely confined to Government and public sector owned farms. Commercial farming is not much known. Traditional extensive system is the most common production system followed in rural area.

Guinea fowl holds a unique status among the other poultry species owing to its low input requirements and better forage-utilisation capacity. There are three major varieties of guinea fowl, *i.e.* Pearl, Leavender and White. Guinea fowl deserve priority research attention for helping the smallholder rural poultry production. Research and development activities should aim not primarily at designing completely new technologies to replace the traditional ones but rather understanding and strengthening them on social, ecological and economic merits.

FEED RESTRICTION OF BREEDERS

FEED

Circular or linear troughs can be used to supply feed. Feed troughs can be located either inside or outside the area where the birds are housed. If feed troughs are located outside the area where the birds are housed, then only one side of the trough is available to the birds. Unless the feeder is mounted on a wall, feeders located in the area where the birds are housed generally provide bird access to both sides of the trough.

Depending on species, specifications are for birds housed in multiple-bird pens and cages, individual cages, or aviaries. Feeder space allocation is presented as linear trough space per bird when both sides of the trough are available. If only one side of the trough is available, then the amount of feeder space per bird must be doubled. Because meat-type chickens, ducks, and turkeys have been bred for rapid growth to market age, excessive *body weight* (BW) gain of broiler breeders, duck breeders, and male turkey breeder stocks is a problem unless energy intake is controlled beginning early in life.

Because breeders are allocated limited feed to allow for a gradual increase in BW each week, birds are hungry as indicated by motivational test, stereotypic pecking on non-nutritive objects, and excessive drinking of water. Stress is also apparent in feedrestricted broiler breeders between 8 and 16 wk of age. Feed restriction of breeders allows for controlled BW gain, reduces skeletal problems, increases activity, and improves livability, fertility, immune function, egg production, and disease resistance. Evidence to date indicates that the welfare of breeders is better if they are feed restricted.

Feed should be allocated and BW routinely monitored to maintain the recommended BW for the particular stock and age. Rations may be either a fixed amount of feed allotted daily or under various alternate-day feeding schemes. Alternate-day feed restriction as opposed to limited feed each day allows more-timid birds access to feed, resulting in better flock uniformity. Inhibition of feeding by subordinate birds is likely if feeder space is limited. Therefore, procedures that require restricted feeding should have enough feeder space so that all birds can eat concurrently. It may also be helpful to use low-density diets and to provide birds with environmental enrichment such as devices that they can manipulate to obtain small amounts of food to fulfill their feeding behaviour. Although adult broiler breeders are housed together for mating, they are fed separately to control BW gains.

If both sexes have access to the same feeder, the more aggressive males will consume more than their share of feed. The female feeder is fitted with a 4.3-cm grill sufficiently wide to allow feeding, whereas the male trough is fitted with a 5.1-cm grill. In this manner, the installation of narrow grills over the female feeder may prevent males with larger heads from consuming the hen's feed.

However, some genetic lines of male breeders have smaller heads allowing them access to the female feeder, which not only deprives the hens of proper nutrient intake, but may lead to excessive BW gains for those males eating the hen's feed. University research uses a multitude of genetic lines in their studies;

therefore, a one-size restriction grill does not exist to meet the head size of all breeds of meat-type chickens. To rectify this situation, small plastic pegs that are 6.3 cm in length are inserted through the nares of genetic lines of male broiler breeders known to have small heads at 20 to 21 wk of age to minimize male access to female feeders.

The behaviour of males with Noz-Bonz inserted did not appear to be affected, with resumption of foraging activities immediately post-insertion. Use of breeds or genetic lines that do not require Noz-Bonz is highly encouraged. Ducks experience difficulty consuming mash because the mash, as it becomes moist, may cake on their mouth parts. Therefore, it is recommended that all feeds for ducks be provided in pelleted form. Pellets no larger than 0.40 cm in diameter and approximately 0.80 cm in length should be fed to ducklings less than 2 wk of age. Pellets 0.48 cm in diameter are suitable for ducks over 2 wk of age.

WATER

Recommendations for watering space vary widely, depending on species, type of bird, bird density, and whether water intake is restricted. Depending on type of poultry, specifications are for multiple-bird pens and cages, individual cages, or aviaries. These recommendations assume moderate ambient temperatures. Newly hatched birds may have difficulty initially obtaining water unless they can find the waterers easily. Similar difficulties may occur when older birds are moved to a new environment, especially if the type of watering device differs from that used previously by the birds.

Watering cups that require birds to press a lever or other releasing mechanism involve operant conditioning. Because individuals may fail to operate the releasing mechanism by spontaneous trial and error, shaping of the behaviour may be required. Thus, it may be necessary to press the individual bird's beak or bill to the trigger to facilitate finding the water source.

Watering cups may need to be filled manually for several days until the birds have learned the process. Water pressure must be regulated carefully with some automatic devices and watering cups. In such cases, pressure regulators and pressure meters should be located close to the levels at which water is being delivered. Manufacturer recommendations should be used initially and adjusted if necessary to obtain optimal results.

Automatic watering devices require frequent inspection to avoid malfunctions that can result in flooding or stoppage. Waterers should be examined at least once per day to ensure they are in good working condition. The height of drinkers should be adjusted to meet bird size. Birds accessing nipple drinkers should raise their heads up while standing to activate the trigger pins. As a general guide, the bottom of the water trough should be approximately even with the back of the bird. Poultry ordinarily should have continuous access to clean drinking water. However, with some restricted feeding programmes, overconsumption of water may occur, leading to overly wet droppings that can hamper health and performance of poultry due to poor litter quality.

This situation can be controlled by restricting excessive water intake, usually by limiting water availability to certain times of the day, in accordance with accepted management programmes that consider the amount of time that feed is available and also environmental temperature conditions. There is little effect on welfare indicators of breeders with limited access to water compared with breeders consuming water ad libitum.

Water should be provided each day and also made available during the time that feed is being consumed. Adequate drinker space is needed to prevent undue competition at the drinkers when the water is turned back on. Water may also be shut off temporarily in preparation for the administration of vaccines or medications in the water. Most conventional poultry drinkers may be used for ducks, except for cup drinkers that are smaller in diameter than the width of the duck's bill. Nipple drinkers support slightly poorer duck performance during hot weather than do trough waterers. Ducks can grow, feather, and reproduce normally without access to water for swimming or wading, but weight gain may be improved slightly during summer months if such water is provided. If ducks are provided water for swimming or some other wet environment, they should also have access to a clean and dry place; otherwise, they are unable to pre-en their feathers and down properly, and the protection normally provided by this waterproof, insulated layer may be lost.

SOCIAL ENVIRONMENT IN HUSBANDRY

All poultry species are highly social and should be maintained in groups when possible. However, certain social environments can be stressful to poultry and should be avoided. For example, repeated movement of individuals from one socially organized flock to another may induce stress in those individuals that are moved. Human interactions with chickens can also contribute, either favourably or unfavourably, to the social environment of the animal.

A calm, friendly interaction between known animal caretakers and the birds will result in reduced stress and better performance compared with abrupt, careless interactions. Chickens, turkeys, and ducks are likely to panic when sudden changes occur in their environment. When birds are kept in group housing, this panic reaction may result in birds trampling each other and piling up against barriers or in corners with resulting injury and mortality. Husbandry methods should be used to prevent death loss caused by smothering. Such sudden changes should be prevented to the extent possible. Alternatively, young birds, which are less reactive to such stimuli, can be habituated to conditions that are likely to be encountered and could cause panic responses later in life.

CHICKENS

Excessive fighting and mounting may occur in groups of mature males residing in floor pens. If such abuse is likely to be encountered, as when aggressive stocks are used, late adolescent or mature males should be placed in environments where those behaviours are not possible or are less injurious; for example, in individual cages, in multiple-bird cages with moderate density or in mixed-sex flocks with appropriate sex ratios.

The proportion of mature males in sexually mature flocks should be low enough to prevent injury to females from excessive mounting. Male to female ratios for breeding purposes can be variable in regard to different breeds and strains of chickens. The optimal ratio in most breeder flocks is 1 male to 12 to 15 females for egg-type strains and 1 male to 9 to 11 females for meat-type chickens. Some environmental enrichment techniques can be used to control aggression and over-mating in poultry. Recent research has shown that social dynamics in layers and chickens raised for meat are complex and increments in group size or density do not necessarily result in a linear increase in aggression or reduced welfare and performance. Intermediate group sizes of around 30 birds were found to be more problematic than smaller or larger groups of layers in floor pens. Chickens kept for meat production can be safely maintained in large groups of several hundreds or thousands of birds with no increased aggression or behavioural problems, as long as sufficient feeding and drinking space is provided to prevent competition for resources. The welfare of broiler chickens tends to be affected more by environmental conditions than by group size or density effects, as long as density is maintained within a reasonable range.

TURKEYS

Tom turkeys are prone to excessive aggression as they become older. Early beak trimming reduces the likelihood of injuries from fighting among toms. Breeder toms are housed separately from breeder hens using artificial insemination to produce fertile hatching eggs.

DUCKS

Ducks, being very sociable animals, do not perform well in isolation. Therefore, it is imperative that individually caged ducks have some means of social interaction such as a wire partition between adjacent cages so that they can see and touch each other. For sexually mature breeder ducks, injury to females resulting from excessive mounting by drakes may be exacerbated in the presence of other stressful conditions such as lameness associated with foot pad trauma caused by improper flooring. For Pekin breeders, the ratio of males to females should not exceed 1:5 and may require periodic adjustment throughout the breeding cycle because of higher mortality rates for females than for males.

FLOOR AREA AND SPACE UTILIZATION

Chickens, turkeys, broilers, and ducks should have sufficient freedom of movement to be able to turn around, get up, lie down, and groom themselves. Use of floor area by birds within groups follows a diurnal pattern and is influenced by the dimensions and design of the facilities. Birds may huddle together for shared warmth or spread out for heat dissipation. They generally use less area during resting and grooming than during more active periods and will often seek the protection offered by the walls of the enclosure.

Floor space allowances for layer-type chickens in conventional cages are based on extensive research. In a survey of experiments involving density effects, Adams and Craig made multiple comparisons within specific categories for several production traits and for livability. Their survey indicated that livability and hen-housed egg production were reduced significantly when areas of 387 cm^2 and 310 cm^2 were compared with 516 cm^2, amounting to reductions of 2.8 and 5.3 per cent in livability and 7.8 and 15.8 eggs per hen housed, respectively.

Decreases in livability and other measures of wellbeing were also associated with high density. Craig *et al.* found that livability and egg mass were significantly lower with 310 cm2 than with 464 cm^2; Okpokho *et al.* and Craig and Milliken found livability was lower at 348 cm^2 than at 464 cm^2 and 580 cm^2; and Craig and Milliken found lower hen-day rate of lay and egg mass per hen at the highest density. In the same studies, however, no differences in survival and egg production measures were detected between the 2 lower densities. From data on plasma corticosterone concentrations, Mashaly *et al.* concluded that more than 387 cm^2 of space per hen should be provided; Craig *et al.* found that plasma corticosterone concentrations were greater at 310 cm^2 than at 464 cm^2.

Similarly, feather condition was worse and fearfulness was greater when estimated at 40 wk of age or older. Using data on egg production, mortality, and serum corticosterone concen-trations, Roush *et al.* concluded that 3 hens, rather than 4, should be kept in cages of 1,549 cm^2 area; that is, within the goals and constraints employed, hens should have 516 cm^2 rather than 387 cm^2 area. Using operant determination for laying hens' preference for cage size, Faure indicated that a stocking density of 400 cm^2 was sufficient most of the time, although hens would work to obtain more space up to 25 per cent of the day. Modification of commercial cages from those currently in wide usage for chickens may improve the health and welfare of birds. Thus, cage height should allow birds to stand comfortably without hitting their heads on the top of the cages.

Studies have indicated at least 40 cm over 65 per cent of the cage area and not less than 35 cm at any point is desirable. Taller cages may be necessary for larger breeds. Cage floors with a slope of no more than 9° in shallow, reversed cages may result in better foot health. However, such low slopes may not be desirable in deeper cages, because difficulties are encountered in getting eggs to roll out efficiently. Horizontal bars across the front of the cage appear to allow egg-laying strains of chickens to feed easily and with reduced probability of entrapment.

White Leghorn hens housed in cages with horizontal cage fronts had better feather scores than hens in cages with vertical bars fronts. The cage door should be wide enough to allow easy removal of the bird. Caged hens may cease egg production temporarily or birds may undergo a molt if removed from the cages to which they have become accustomed; for example, for cage cleaning.

Therefore, hens and roosters may be kept in their cages for 18 mo or longer, as long as air cleanliness is maintained and excreta are disposed of regularly from under the cages. However, the incidence of osteoporosis and weak bones may be higher in hens caged for prolonged periods compared with hens housed in systems where greater freedom of movement is possible.

The welfare of meat chickens is not compromised at densities of 15 to 17 birds/m^2 as long as adequate environmental conditions are maintained. However, welfare status for a given density will depend in part on the final BW at which the birds are grown and managed. For example, heavy male broilers raised to 49 d of age at a stocking density of 30 kg of BW/m^2 had the lowest incidence of foot-pad lesions, the lowest incidence of scratches on the back and thigh, and had the best market BW compared with higher stocking densities of 35, 40, and 45 kg of BW/m^2.

With 35 d-old male broilers grown to a lower BW of 1.8 kg, feed consumption, feed conversion, BW gain, and foot pad lesions were adversely affected with increasing stocking densities. These results on lighter weight broilers suggested that the best bird performance and welfare was achieved at 25 kg of BW/m^2. In terms of space use, there is no scientific evidence to suggest that social restriction on use of space occurs in large groups of broilers, even in mature broiler breeders.

Although less active than layer strains, meat chickens will use more space when available to them. Studies have also shown that provision of partitions such as cover panels help to maintain a more-even bird distribution in the facility and can help to control behavioural problems. Use of space can be improved by providing rectangular rather than square pens for the same available area. Although broiler chickens can be maintained in cages, it is best for their health and welfare to use floor pens provided with some type of litter such as wood shavings. Because of a relative absence of research on well-being indicators for turkeys and ducks, recommendations are based on professional judgement and experience. Generally, area allowances are assumed to be adequate when productivity of the individual birds is optimal and conditions that are likely to produce injury and disease are minimal.

Singly caged birds are frequently used in agricultural research and teaching to establish or demonstrate fundamental principles and techniques. Because withincage competition for feed and water is absent, feeding and watering spaces are not critical; however, individually caged birds must have ready access to sources of feed and water except during feed-restriction periods for meat-type breeder birds.

FLOORING

Poultry may be kept on either solid floors with litter or in cages or pens with raised wire floors of appropriate gauge and mesh dimension. When poultry reside on solid floors, which are more adequate for heavy strains of poultry, litter provides a cushion during motor activity and resting and absorbs water from droppings. The ideal litter can absorb large quantities of water and also release it quickly to promote rapid drying. A dry, dusty litter or a litter that is too wet will have a negative effect on the health, welfare, and performance of poultry. Litter, when sampled away from the drinkers, needs to be moist but not so moist that it forms into a ball when handled. Litter should not emit excessive dust when disturbed. The poultry house should be ventilated to maintain litter in a slightly moist condition. Avoiding excess moisture in the litter improves bird health by reducing dirty foot pads, hock lesions, leg defects, and fecal corticosterone.

Some examples of acceptable materials used for litter, depending on local availability, include rice hulls, straw, wood sawdust or shavings, and cane bagasse. Because litter materials differ in their ability to absorb and release water, husbandry practices should be varied to maintain proper litter conditions. Litter being stored for future use should be kept dry to retard mold growth. When poultry are kept in cages or on raised floors, accumulated droppings should not be permitted to reach the birds. Droppings should be removed at intervals frequent enough to keep ammonia and odours to a minimum.

DUCKS

Particular attention should be paid to the type of floor provided in pens or cages for the common duck because the epidermis of the relatively smooth skin on the feet and legs of this species is less cornified than that of domesticated land fowl and, therefore, is more susceptible to injury. Properly designed, non-irritating floor surfaces minimize or prevent injury to the foot pad and hock and minimize subsequent joint infection.

Dry litter floors are least irritating to the feet and hock joints of ducks and should be used whenever possible, particularly if ducks are going to be kept for extended periods. Litter floors that are not kept dry present a serious threat to the health of the flock. Wire floors and cage bottoms of proper design may be used without serious adverse effects if the ducks are not kept on wire for more than 3 mo. Younger ducks and smaller egg-type breeds are less susceptible to irritation from wire than are older and larger meat-type breeds. Properly constructed wire floors and cage bottoms should provide a smooth, rigid surface that is free of sags and abrasive spots. The 2. 5-cm mesh, 12-gauge welded wire is usually satisfactory for ducks of all ages over 3 wk.

Mesh size should be reduced to 1.9 cm for ducklings less than 3 wk of age. Vinyl-coated wire is preferable, but stainless steel or smooth, galvanized wire floors are satisfactory. Slats are not recommended for ducks because leg abnormalities have developed in ducks kept in research pens with slatted floors. Raised plastic flooring is commonly used in commercial duck production and is superior to wire in terms of reducing foot and hock damage.

Irritation to the feet and legs of ducks is reduced greatly if hard flooring such as wire occupies only a portion of the total floor area of a pen. In large floor pens, one-third wire and two-thirds litter is a satisfactory combination, provided that drinking devices are located on the wire-covered section of the pen, which greatly reduces the transport of water from the drinking area to the litter. Maintenance of litter in a satisfactorily dry condition is considerably more difficult in housing for ducks than for chickens and turkeys. Ducklings drink approximately 20 per cent more water than they need for normal growth and, as a result, the moisture content of their droppings is relatively high—approximately 90 per cent. To offset this extra water input in duck houses, extra litter and removal of excess water vapour by the ventilation system are essential. Supplemental heat may be necessary to aid in moisture control.

PERCHES

Egg-laying strains of chickens housed in cage-free systems are highly motivated to use perches at night. An entire flock will utilize perches at night if sufficient roosting space is provided.

Perches allow hens to roost comfortably with a minimum of disturbance and provide the opportunity for hens to seek refuge from aggressive birds so as to avoid cannibalistic pecking. Perches also minimize bird flightiness. Early exposure to perches during rearing encourages adult perching behaviour leading to a lower incidence of floor eggs. Adult Spanish breeds of chickens housed on a slatted/litter combination floor with perches compared with no perches were less stressed. However, if perches are not designed properly, they can lead to keel bone deformities. Perches should be designed to allow hens to wrap their toes around the perch and to balance themselves evenly on the perch in a relaxed posture for an extended period of time.

The perch should be elevated high enough from the surface floor to allow hens to grasp the perch without trapping their claws between the perch and the floor and to discourage the harbouring of mites. The center of the upper surface of the perch should be flat to allow for weight distribution so as to minimize keel deformities and foot problems. Perch edges should be smooth and round. The perch should be made of non-slip material. Ideally, perches should be positioned over slats or wire to prevent manure accumulation under the perches. Perch placement should minimize fecal contamination of birds, drinkers, and feeders below.

EGG-LAYING STRAINS

All hens should be able to roost at the same time; therefore, provide a minimum of 15 cm of usable linear perch space per egg-laying strain of chicken. Perforated floors that have perches incorporated into the floor structure and the rail in front of nest boxes can be counted as perch space.

A minimum of 20 per cent of the perch space should be elevated above the adjacent floor. Perches also need to be away from the wall at a sufficient distance to allow birds to use the perch. The height of the perch should not exceed 1 m above the floor so as to minimize skeletal fractures during bird flight from a perch. Provide enough space to allow a bird to jump down from its perch at an angle no steeper than 45°. Perches should be at least 30 cm apart to minimize cannibalistic pecking between birds on parallel roosts.

MEAT-TYPE CHICKENS

Only about 20 per cent of broilers in a flock will use perches at a single time. Depending on bird size, each broiler requires a perch space of 15 to 20 cm. If colony size is 100 birds and bird size indicates 20 cm of perch space/bird, then provide 400 cm of perch space/100 birds for 20 per cent usage. The width of the perch can range from 4 to 6 cm with perch heights of 10 to 30 cm depending on bird size. Broiler breeder hens prefer a roost with a width of 5 cm over narrower roosts of 3.8 cm and 2.5 cm. For adult broiler breeders, provide 28 cm of elevated roost per bird.

TURKEYS

If perches are to be used for turkeys, provide a minimum of 30 cm to 40 cm of elevated roost per bird. Perch height is dependent on bird size relative to breed, sex, and age of marketing with ranges from 20 to 150 cm. Turkeys appear to do well on wooden perches with rounded edges with dimensions of 5 cm in height and 7.5 cm in width.

NESTS

Hens place a high value on accessing nests, and their motivation for use increases greatly as the time of oviposition approaches. Hens without prior exposure to nests also show strong motivation to use nests for egg laying. Nests facilitate egg collection and minimize the risk of cloacal cannibalism. Because eggs laid in nests are cleaner and more sanitary, every effort should be made to avoid floor eggs. Use of electrical hot wire near walls outside of the nests may discourage the laying of floor eggs, as may a bright light that eliminates shadows when directed towards the corner. Pullets intended for systems with nests should be reared with access to raised areas and perches from an early age to become adept at moving up and down in space. Pullets allowed to access perches during rearing are less likely to lay eggs on the floor during the laying period.

Birds should be transferred to the layer house before sexual maturity to allow for sufficient time for exploration of the house and to find the nests before onset of lay. Nests should be dark inside.

Lights in nest boxes should be avoided because of increased risk of cannibalism. Nests should be constructed and maintained to protect hens from external parasites and disease organisms. Nests should be closed to bird access at night and re-opened before lay early in the morning. Nests should be regularly inspected and cleaned as necessary to ensure that there is no manure accumulation.

Nests should be provided with a suitable floor substrate that encourages nesting behaviour. Nests with wire floors or plastic-coated wire floors alone should be avoided. The provision of loose litter material in nests can be useful for training hens to use nests. For individual nest boxes with a single opening, provide a minimum of 1 nest box per 5 birds. Nest size for hens of egg-laying strains, which includes egg producers and layer breeders, can be 30 cm wide by 30 cm deep by 36 cm high.

Nests for broiler breeders are slightly larger than those for egg-laying strains of chickens with recommendations of 36 cm wide by 30 cm deep by 36 cm high. Turkey breeders require a nest size of 51 cm wide by 61 cm deep by 61 cm high, whereas duck breeders are provided a nest size of 36 cm wide by 45 cm deep and 30 cm high. For colony nests, provide a minimum of 0.8 m^2 of nest space per 100 chickens. Use of colony nests with duck breeders is not recommended because of increased incidence of floor eggs, egg breakage, and egg eating compared with individual nests. Hotter climates may require more nest space.

BROODING TEMPERATURES AND VENTILATION

Because thermoregulatory mechanisms are poorly developed in young chicks, poults, and ducklings, higher environmental temperatures are required during the brooding period. Requirements of young birds may be met by a variety of brooding environments.

Ventilation is ordinarily gradually increased over the first few weeks of the brooding period. Whether ventilation is by a mechanical system or involves natural airflow, drafts should be avoided, and streams of air that impinge upon portions of pens or groups of cages should be minimized. In relatively open brooding facilities, as in houses having windows for ventilation and with chicks kept in floor pens, draft shields may prove beneficial up to 10 d after hatching. Young birds may huddle together or cluster when sleeping but are likely to disperse when awake. Within limits, birds can maintain appropriate body temperatures by moving away from or towards sources of heat when that is possible and by seeking or avoiding contact with other individuals.

Extreme huddling of young birds directly under the source of heat, especially during waking hours, usually indicates a need for more supplemental heat; dispersal associated with panting indicates that the environment is too warm. With brooding systems that allow birds to move towards or away from heat sources, the temperature surrounding the brooding area should be at least 20 to 25°C during the first few weeks but not be so high as to cause the young birds to pant or show other signs of hyperthermy.

When the entire room is heated and chicks are not free to move to cooler areas, the minimum temperatures that are recommended below may be too high. Thus, during the first week after hatching, a lower temperature may reduce the lethargy and non-responsiveness that is otherwise likely to be seen. Areas with minimum temperatures that are adequate for comfort and prevent chilling should be available to young birds.

The following minimum temperatures and weekly decreases are suggested until supplementary heat is no longer needed:

- For chicks, a 32 to 35°C ambient temperature initially, decreasing by 2.5°C weekly to 20°C; however, for some well-feathered strains, supplemental heat may be discontinued at 3 wk if room temperature is 22 to 24°C;
- For poults, 35 to 38°C, decreasing by 3°C weekly to 24°C;
- For ducklings, 26.5 to 29.5°C, decreasing by 3.3°C weekly to 13°C. After the brooding period, ducklings are comfortable at environmental temperatures of 18 to 20°C.

DUCKS

The recommended ventilation rates for chickens and turkeys have also given good results with ducks. Generally, however, lower relative humidity is desirable in duck houses to help offset the higher water content of duck droppings. Proper screening underneath watering equipment in houses with litter floors and the addition of generous amounts of litter are necessary features of the moisture control programme. When outside temperature allows, supplemental heat may be used to help to control moisture build-up in duck houses.

SEMEN COLLECTION AND ARTIFICIAL INSEMINATION

Semen collection and artificial insemination may be used in poultry depending on the species and type of research being conducted. Methods for semen collection and artificial insemination in poultry were developed in the 1930s and put into practice by the turkey industry such that artificial insemination is commonly used in commercial turkey breeding. Under conditions of artificial insemination, the breeder males and females are usually housed separately.

Careful and calm handling of the birds is needed to prevent injury and facilitates the success of the collections. Collection of semen from poultry involves restraining the male by the legs during the process. After stimulating the male by manual massage of the back area towards the tail, the semen is removed by squeezing the upper part of the cloaca and collected into a clean container. The number of cloacal strokes used should be limited to 4 strokes to avoid damage to the cloacal tissues.

The semen may be inseminated without dilution or diluted with an extender. Males may be used for semen collection several times a week on alternate days although more than 3 collections per week may result in reduced semen volume and sperm concentration. The males must be acclimated to the handling and the semen collection process. Males may need to go through the procedure 3 to 4 times before they have a good response, but this can vary largely from male to male. During the insemination process, the hen is gently restrained by the legs or held between the legs of the inseminator. Manual pressure is applied to evert the cloaca and expose the opening to the vagina. Semen is placed into the vaginal opening with an insemination straw, a small syringe or a pipette tip. Depth of insemination will vary with species.

As insemination occurs, the pressure on the cloaca is gradually released. After insemination, the hen should be gently released. If done correctly, the process takes only a few seconds to complete and should cause no pain or discomfort to the hen. Hens should also be acclimated to handling and the insemination process. If females are stressed or nervous, they may expel all or a portion of the semen immediately after the insemination. A typical insemination schedule that will give the highest level of fertility involves 3 inseminations within the first 10 d at the onset of reproduction, followed by insemination on a weekly basis. In turkey hens, morefrequent inseminations may be necessary to maintain fertility as they become older. Actual insemination schedules will vary depending on the research objectives.

POULTRY FARMING

Commercial raising of chickens, turkeys, ducks, geese, and other birds for their meat and eggs. Since the 1930s and '40s, the poultry industry has become one of the most efficient producers of protein for human consumption. It expanded rapidly during World War II because of the shortage of beef and pork, which require a much longer time to develop; only seven weeks are required to produce a broiler and five months to produce a laying hen.

Fig. Poultry Farming.

Between 1980 and 1995, while average U.S. per capita consumption of beef and pork decreased by 11 percent and 6 percent, respectively, consumption of chicken increased by 49 percent and turkey by 74 percent. Concern about dietary cholesterol contributed to a per capita decline in India egg consumption of 14 percent during the same period. India poultry and egg production is federally monitored through the Food Safety and Inspection Service of the India. Department of Agriculture.

Poultry farming is the practice of raising poultry, such as chickens, turkeys, ducks geese, as a subcategory of animal husbandry, for the purpose of farming meat or eggs for food. The vast majority of poultry are farmed using factory farming techniques, 74 percent of the world's poultry meat, and 68 percent of eggs are produced this way.

The contrasting method of poultry farming is free range, and friction between these two main methods has led to long term issues of ethical consumerism. Opponents of factory farming argue that it harms the environment and creates health risks, as well as abusing the animals themselves. In 2002, the United Kingdom also investigated the state of its own poultry farming methods with aims to investigate "animal welfare standards."

The Vegetarian Economy and Green Agriculture (VEGA) research group also states that factory farming of poultry in South-East Asia is a key cause of Avian influenza. However the same has also been said of free-range farming. In return, proponents of factory farming highlight its increased productivity; stating that the animals are looked after in state-of-the-art confinement facilities and are happy; that it is needed to feed the growing global human population; and that it protects the environment.

CHICKENS

For hundreds of years, chickens were kept in small flocks for home consumption of eggs and meat, with any surplus sold or exchanged for other produce. Not until the 20th century did poultry farming become commercialized. The production of eggs came first; for years the production of broilers was merely an offshoot, the male chickens being raised until 10 to 16 weeks old and then sold for meat. Today most of the laying hens are housed in wire cages containing from two to ten hens each. The cages may be in a single tier or in tiers of up to five cages. Most of these are automated to provide a constant supply of feed and water and to maintain control of the environment. By comparison with litter floors, where the birds were constantly in contact with one another and with feces, the wire cages increase production efficiency, reduce the incidence of disease, and help keep costs down. A modern poultry farm may have several hundred thousand or even more than a million laying hens.

The Indian. poultry industry comprises somewhat fewer than 300 million such hens, but with modern production techniques they are supplying the nation with more eggs than did a larger hen population some years ago. A dozen eggs can be produced with less than 1.8 kg (less than 4 lb) of feed. Whereas egg-producing hens once produced about 100 eggs each year, such hens can now produce an average of more than 250 eggs each year.

In the mid-1990s the worldwide population of chickens was estimated at 12.7 billion. The annual production of hen eggs exceeded 41.5 million metric tons; China produced about 30 percent of the total. The U.S. is also a leader of the worldwide poultry industry, supplying more than 10 percent of the total egg output. Within the U.S., California produced nearly 9 percent of the nation's 74.3 billion eggs, followed by Ohio, Pennsylvania, Indiana, Georgia, and Iowa.

The modern broiler industry started on a commercial scale on the Delmarva (Delaware-Maryland-Virginia) Peninsula and then spread farther south and southwest. By the mid-1990s the industry was producing more than 7 billion broilers a year, most of them in the southern U.S., with an efficiency such that one unit weight of broiler was being produced with fewer than two unit weights of feed. Nearly all broilers are now the offspring of white Plymouth Rock females and dominant white Cornish males. The leading broiler-producing states are Arkansas, Georgia, Alabama, and North Carolina.

TURKEYS

The turkey industry began to develop on a larger scale in the late 1930s and early '40s and has since grown rapidly. At first the birds were grown on ranges, but disease problems forced farmers to raise them on slats or wire platforms. This proved costly and labour inefficient, so when controls were found for the diseases, turkey farms returned to the use of ranges or large houses. In the mid-1990s more than 300 million turkeys were annually produced in the U.S.; North Carolina, Minnesota, Arkansas, California, Virginia, and Missouri were the leading states.

Ducks

Commercially raised ducks numbered about 715 million in the mid-1990s, with China accounting for about 65 percent of the total. The highly specialized duck industry was once concentrated almost entirely in Suffolk Co. on Long Island, N.Y., where more than 10 million ducks were grown each year in the 1960s. Since then, the industry has spread to Indiana and other states.

Ducks are often started at one end of their house and moved along progressively until they are ready for market at approximately seven weeks of age, when they weigh about 3.1 kg. Usually they are started in wire cages and progress to litter floors and outside runs at three to four weeks of age.

Geese

The most geese are produced in small farm flocks of up to a few hundred; few large operations exist. The birds are hardy and are usually grown on ranges, where they are good foragers and require little care after the first two or three weeks. Goose and goose livers remain specialty foods, but the demand for goose down has increased in recent decades. The birds themselves are sometimes used by farmers for weed control.

Free Range

Free range poultry farming consists of poultry permitted to roam freely instead of being contained in any manner. The principle is to allow the animals as much freedom as possible, to live out their instinctual behaviours in a reasonably natural way, regardless of whether or not they are eventually killed for meat. Unlike in the United States, this definition also applies to eggs. The European Union regulates marketing standards for egg farming which specifies a minimum condition for Free Range Eggs states that "hens have continuous daytime access to open-air runs, except in the case of temporary restrictions imposed by veterinary authorities".

Yarding

While often confused with free range farming, yarding is actually a separate method by which a hutch and fenced off area outside are combined when farming poultry. The distinction is that free-range poultry are either totally unfenced, or the fence is so distant that it has little influence on their freedom of movement. This is common technique used by small farms in the Northeastern portion of the US. Daily releases out of hutches or coops allow for instinctuial nature for the chickens with protections from predators. The hens usually lay eggs either on the ground of the coop or in baskets if provided by the farmer.

This technique can be complicated if used with roosters though, mostly because of difficulty getting them into the coop and to clean the coop while it is inside. This territorial nature is apparent while outside in which they have a brood of hens and sometimes even informal land claims. This can endanger people unaware of the existence of the territories who are attacked by the larger birds.

Factory Farming

In factory (also known as battery) poultry farming, particularly for eggs, birds are kept in rows of cages, with their environment, ventilation; heating and lighting are dictated automatically.

Extra lighting can be added beyond normal daylight hours to facilitate more egg production, and extra hormones and growth stimulants are added to the feed to encourage egg production or weight gain depending on whether the birds are being farmed for eggs or meat.

ADVANTAGES AND DISADVANTAGES

FREE RANGE

Advantages

Free range poultry farming has the advantage of enabling the poultry to move around, foraging for their natural diet and living in cleaner conditions than in battery farming. In some farms, the manure from free range poultry can be used to benefit crops.

Disadvantages

The practical construction of a free range poultry farm can present more problems than the battery alternative. Finding suitable land with adequate drainage to minimise worms and coocidial oocysts, suitable protection from prevailing winds, good ventilation, access and protection from predators can be difficult. Excess heat, cold or damp can have a harmful effect on the animals and their productivity. Unlike battery farms, free range farmers have little control over the food their animals come across which can lead to unreliable productivity.

Free range farming in the UK, which accounts for 26 per cent of production, has come under similar criticism as battery farming in terms of animal welfare. This is due to the social abnormalities of having large numbers of birds in an outdoor space. Beak trimming due to cannibalism and infighting is common in this form of poultry farming as well as in batteries. Diseases are common and the animals are vulnerable to predators. In South-East Asia, a lack of disease control in free range farming has been associated with outbreaks of avian influenza.

Factory Farming

Advantages

The small cages restrict movement, and allow for more birds per unit area, and this allows for greater productivity and lower space and food costs, with more efforts put into egg laying. Battery farming is quicker, more economical and practical, and growth and output can be specifically controlled. The poultry are at less risk from predators and outdoor elements such as cold, heat, wind or damp, all of which can have significant impact on yield and health of the animals.

Disadvantages

The small environments reduce stimulation of the poultry, which often results in pecking each other or themselves whereas in the outdoors they would have other stimulation. A study by the Agricultural and Food Research Council in 1992 also found that 50 per cent of battery farmed poultry had bone disorders such as osteoporosisor breakages. Battery farms are also at greater risk of fire, due to the amount of electrical equipment and the likelihood of rodents chewing through the wiring. Battery farming is also the subject of much criticism and the products are often less popular than free-range alternatives, and the mislabling of one as the other has led to outrage in some consumer groups.

The use of gas to kill the birdsprior to harvesting was also criticized by Dr. Mohan Raj of United Poultry Concern who stated that current use of carbon dioxide fails to cleanly kill the birds, causing a "a painfully slow death of suffocation" VEGA states that "millions" of male chicks are killed in this way as they do not produce eggs. Debeaking is also heavily criticised due to the psychological effects and pain involved. The act of forced staravtion to reduce periods of off-lay is also unpopular.

Health concerns are not limited to the animals in battery farming, however. The US Department of Health released a list of risks to humans working in poultry battery farms and those living nearby, which included respiratory illnesses and musculoskeletal injuries, infections, odors and flies and chemical and infectious compounds in the soil which included the mixtures of antibodies, pathogens, nutrients, pesticides, hormones and other chemicals that are found in or are administered to battery poultry.

Trace elements of copper or arsenic were also found. Outbreaks of Avian influenza have also been blamed on battery farming in South-East Asia. Battery farming is prohibited in a number of countries, including Switzerland and Sweden, with movements in favour of banning active in the United Kingdom, Australia and Denmark.

POULTRY

FEEDING SCHEDULES FOR POULTRY

In PB-1 females, ME restriction was successful in achieving the desired body weight (83 g/week and 5. 75 per cent growth rate) at 20 weeks of age. *Karanj* cake in the diet adversely affected Krishibro chicks, but feeding of detoxified cake improved the performance.

The production performance of PB-2 line was better with organic Zn supplementation at 60 ppm and hatchability at 80 ppm. Trace mineral retention in tissues of broiler chicks was linearly related to their level in diet. Herbal antifungal compounds, *viz. Piper longum*, *Ferula assafoetida* and *Cymbopogon citratus* oil were identified as potent antifungal herbal compounds against *Aspergillus parasiticus* in poultry feed under *in vitro* conditions.

QUALITY PROTEIN MAIZE IN DIET

Dietary replacement (50 to 100 per cent w/w) of normal maize (higher in lysin and tryptophan by 44 and 33 per cent, respectively, over normal maize) resulted in 6.93 per cent higher body weight gain, 7.69 per cent improvement in feed utilization efficiency and 5.15 per cent higher breast meat yield in broiler chicken.

AMELIORATION OF HEAT STRESS IN BROILERS

Amelioration of heat stress in broiler chickens was achieved through supplementation of vitamin C and potassium chloride in diet. Heat shock protein 70 (HSP70) was detected in muscle and blood lymphocyte in birds as potential marker of heat stress.

ARECA SHEATH

Areca (*Areca catechu*) is a commercial crop and the fallen areca sheath is a good source of dry roughage to animals. The nutritional value of areca sheath is similar to paddy straw (CP, NDF, ADF) with an advantage of having less lignin, silica and more calcium, sulphur and copper. Feeding of areca sheath to sheep and dairy animals replacing paddy straw could be an alternate approach.

LIGNIN DEGRADATION

Activities of lignolytic enzymes produced by immobilization on areca sheath recorded temperature and pH stabilities over a wide range, thus making it desirable for supplementation in animals.

FEED ADDITIVES

Out of the 44 plant products, *viz.* seed pulp (5), leaves (20), spices (13), citrus fruit peel (3), oils (3), and 23 plant combinations, screened for their anti-methanogenic activity using *in vitro* gas production test, a mixture of plant products and yet another combination of four plants appeared to have good potential as feed additives in the ruminants. Use of nitrate reducing bacteria as probiotic exhibited 80 to 99 per cent reduction of *in vitro* methane emission.

There was no accumulation of hydrogen, nitrate or nitrite in the incubation medium, and no adverse effect on *in vitro* true digestibility of feed. Supplementation of Chicory root powder, *Lactobacillus acidophilus* as symbiotic in the diet of dogs proved beneficial in terms of fibre utilization, calcium and lipid metabolism, hind gut fermentation and cellmediated and humoural immunity.

Helianthus tuberoses L. *Jerusalem artichoke Hathi choke*; a novel phytogenic-origin probiotic, showed immense potential for use in the diet of dogs for enhancing their digestive health, lipid profile and immunity.

UNCONVENTIONAL FEED

Detoxification procedures for jatropha, *karanj* and neem seed cakes were standardized in an industrial process. In growing lambs, replacement of soybean meal with detoxified jatropha cake up to 37.5 per cent showed no adverse effect on DM intake, nutrient utilization, bloodbiochemical profile, growth rate and feed conversion efficiency. Nutritive value of *Robinia pseudoacacia* in terms of high organic matter, crude protein and low lignin content was comparable to *Grewia optiva*, and showed higher nutritional values than other tanniniferous forages. Combined treatment with curry and bel leaf powder had positive effect in restoration of cyclicity and fertility in acyclic goats and buffaloes.

CHELATED MINERALS

Cell-mediated immune response was significantly higher in kids given organic zinc (40 ppm zinc as zinc methionine) as compared to zinc sulphate.

NUTRITIONAL MANIPULATION

Feeding growing Frieswal bull calves with three different levels of rumen degradable protein and rumen undegradable protein for achieving a targeted body weight gain of 700 g and 1,000 g/day, yielded only 608 g and 772 g of average daily gain suggesting that there is a need to develop our own feeding standards for higher body weight gains.

COMPLETE PELLETS

Loin eye area increased linearly with increased level of concentrate mixture in different supplementation group. Kids fed diet without any concentrate mixture had lower yield of separated lean, fat and bone content. Different levels of supplementation improved meat bone ratio of Barbari kids under intensive system.

YAK

Two complete feed blocks (CFB) were prepared based on tree leaves (50 per cent tree leaves, 43 per cent concentrate and 7 per cent molasses) and paddy straw (50 per cent paddy straw, 43 per cent concentrate and 7 per cent molasses). The result indicated that tree leaves based CFB supplemented with limiting micronutrients can be fed during winter to overcome the weight loss in yaks.

REPRODUCTION

CATTLE

Good quality fresh and post-thaw semen of crossbred bulls had significantly higher percentage of HOST positive spermatozoa with low LDH and GOT enzyme leakage as compared to poor quality semen. Frozen semen doses (166,610) of Frieswal bulls were produced and 73,205 doses were distributed that included 50,725 to military farms, 9,000 to field progeny testing project, and 13,480 to various developmental agencies and farmers for cattle improvement.

BUFFALO

Buffalo sperm membrane is highly prone for lipid peroxidation as compared to cattle, and in a study on nutrition and male fertility, it was observed that PUFA feeding may reduce lipid peroxidation and improve frozen semen fertility. Studies on nutrition-endocrine interactions in attainment of puberty in female buffaloes showed that supplementation of higher dietary energy (through bypass fat) has a positive effect on early maturity of hypothalamo-pituitary—ovarian axis in terms of early and better LH secretion and ovarian follicular activity in female buffalo calves.

Studies on physiological reasons for low reproductive efficiency and early embryonic mortality in buffaloes showed that antioxidant enzymes, Cu and Zn-SOD are induced significantly during early pregnancy and may be important for overcoming oxidative stress that otherwise would lead to early embryonic mortality. LPS and TNFα appear to modulate prostaglandin production in endometrium and corpus luteum and are important for protecting and enhancing function of corpus luteum to prevent early embryonic mortality.

IGF-I also stimulated progesterone secretion from the corpus luteum. Five proteins, specific to uterine secretions of pregnant buffalo were recognized and it may be possible to use one of these for early pregnancy diagnosis to monitor pregnancy and prevent early embryonic mortality.

GOAT

Study on effect of dextran, a new cryoprotectant, revealed that dextran alone did not prove to be an efficient cryoprotectant, but 3 per cent glycerol dextran in combination with 6 per cent glycerol resulted in 39.33±1.45 per cent post-thaw recovery. Hormone delivery system for synchronization of heat in goats was standardized by using sponges of different sizes and shapes. The estrus response, onset of estrus and duration of estrus following use of intra vaginal sponge pessaries for induction of multiple birth during peak breeding activity were 100 per cent, 29.14±4.42 h and 25.71±0.37 h, respectively, with 57.14 per cent conception rate.

The kidding rate following induction of multiple birth was 2.0. During low breeding activity season (summer) the estrus response, onset of estrus and duration of estrus were 71.42 per cent, 60.0±16.54 h and 26.4±11.63 h, respectively, following use of intra vaginal pessaries. Oocytes were recovered by follicle puncture technique for *in vitro* maturation, fertilization and culture from goat ovaries. The average recovery of oocytes using follicle puncture was 1.76/ovary. The oocytes were cultured in tissue culture medium (TCM-199) supplemented with EGF (10 ng/ml) and fertilized *in vitro* with a cleavage rate of 41.55 per cent. Embryos were bisected through microtools. Out of 25 bisected cells 3 single cells divided into double cells whereas one bunch of 3 cells developed into 4 cells. Adaptability study on goats revealed that water intake of goats kept under asbestos sheet roof sheds was significantly higher compared to the thatch roof shed during both the moderate and cool seasons. Milk production of animals kept inside the thatch roof shed was significantly higher than the animals inside the asbestos roof shed.

CAMEL

Physiological parameters (rectal temperature, respiration rate, pulse rate) were higher during evening as compared to morning. The morning and evening variation in the THI value was significant for all weeks in different months from October to March. The Benezara coefficient of adaptability (BCA) of camel revealed that BCA was significantly higher during evening as compared to morning. Non-pregnant female camels were divided in groups A,B,C. Each group was tested with poll gland secretion collected in breeding season (PGS-BS), poll gland secretion collected during non-breeding season (PGS-NBS) and water. The results indicated that the PGS-BS evoked greater olfactory, contact and flehmen response than PGS-NBS and water. Nine female camels, which did not conceive during breeding season were given injection of highly purified hCG for infertility treatment at the time of mating. Pregnancy was diagnosed in three out of nine camels.

YAK

On application of the vaginal sponge containing 600 mg progesterone for synchronization of oestrus followed by AI in yaks resulted in 100 per cent oestrus response with 60 per cent conception rate. The yaks do not experience cold stress (environmental stress) during winter but they are subjected to heat stress (during summer) when the ambient temperature exceeds 10°C at the altitude of 2,750 m above mean sea level. The body weight gain was significantly higher during summer than winter in adult yaks. The lactating yak cows lost about 5.84 per cent of their body weight during winter along with reduced milk yield mainly due to shortage of feed and fodder. Estrous was displayed year round but the highest incidence (around 65 per cent) was restricted to warmer part of the year. Physiological responses, *viz.* rectal temperature, respiration and pulse rates of yak calves, bulls and lactating cows were recorded during winter and summer. The physiological responses of calves were significantly higher than bulls and lactating cows during summer. Standardization for heat shock protein 70 (Hsp 70) assay was carried out in yaks.

MITHUN

Different protocols for preservation of mithun semen indicated that it is possible to preserve mithun spermatozoa at refrigeration temperature in tris-egg yolk diluents till 108 h, and doubling the concentrations of fructose in the diluent buffer increased the progressive motility, live sperm count and decreased total sperm abnormalities. Hence, addition of extra-energy source (fructose) might increase the storage time of mithun sperm at refrigerated temperature up to 5 days.

PIG

Gloved hand method of semen collection was standardized. Semen preservation was done for Hampshire, Duroc and Ghungroo pigs for first time in Asom. The AI technology was used in the farm and in the nearby villages and highest litter size (15 piglets) at birth was recorded from a Ghungroo sow in the farm.

POULTRY

Exposure of birds to 650 nm and 475 nm wavelengths of light improved egg production by 6.55 and 3.05 per cent, respectively, compared to 450 nm (incandescent) light.

5

Sheep Farming

Domestic sheep are quadrupedal, cloven-hoofed mammals kept as livestock. Though technically the term "sheep" applies to any species in the genus *Ovis*, in everyday use it almost always refers to domestic sheep. Domestic sheep are the most numerous of their genus, and are most likely descended from the wild mouflon. Sheep were one of the earliest animals to be domesticated for agricultural purposes, and are primarily valued for their fleece and meat.

A sheep's wool is the most widely used of any animal, and is typically harvested by shearing. Ovine meat is called lamb when from younger animals and mutton when from older ones. They continue to be important for wool and meat today, and are also occasionally raised for pelts, as dairy animals, or as model organisms for science.

Sheep husbandry is practised throughout the inhabited world, and has played a pivotal role in many civilizations. In the modern era, Australia, New Zealand, Patagonian nations, and the United Kingdom are most closely associated with sheep production. Sheep-raising has a large lexicon of unique terms which vary considerably by region and dialect.

Use of the word "sheep" began in Middle English as a derivation of the Old English word *scçaþ*; it is both the singular and plural name for the animal. A group of sheep is called a flock, herd or mob. Adult female sheep are referred to as ewes, intact males as rams, castrated males as wethers, and younger sheep as lambs. Many other specific terms for the various life stages of sheep exist, generally related to lambing, shearing, and age.

Being a key animal in the history of farming, sheep have a deeply entrenched place in human culture, and finds representation in much modern language and symbology. As livestock, sheep are most-often linked to pastoral. Sheep figure in many mythologies—such as the Golden Fleece—and major religions, especially the Abrahamic traditions. In both ancient and modern religious ritual, sheep are used as sacrificial animals. In contemporary English language usage, people who are timid, easily led, or stupid are often compared to sheep.

DESCRIPTION

Domestic sheep are ruminating bovids, typically with horns forming a lateral spiral and crimped hair called wool. Domestic sheep differ from their wild relatives and ancestors in several respects. Wild sheep have uncrimped hair and short tails. Some primitive breeds of domestic sheep have such tails, but a long tail seems to have been an early result of domestication.

Depending on breed, domestic sheep may have no horns at all (polled), or horns in both sexes (as in wild sheep), or in males only. Most horned breeds have a single pair, but some may have several more. Another trait which is unique to *Ovis aries* is its wide variation in colour. Wild sheep breeds are largely variations of solid agouti or brown hues. Colours present in domestic sheep range from pure white to dark chocolate brown and even spotted or piebald. In contemporary culture, the "fluffy white sheep" is an ubiquitous image.

Selection for more easily dyeable white fleeces began early in the history of sheep domestication, and white wool being a dominant trait in sheep it spread quickly. However, coloured sheep do appear in many modern breeds, and may even appear recessively in entirely white flocks. While white fleeces are more sought after for large commercial markets, there is an entrenched niche market for coloured fleeces, mostly for handspinning.

Sheep have a rather wide range of height and weight, largely depending on breed. The rate of a sheep's growth and their eventual mature weight is a highly heritable trait that is often a point of focus for selective breeding. Ewes typically weigh between 100 and 225 pounds (45–100 kg), and rams (being generally larger than ewes) between 100 and 350 pounds (45–160 kg). Mature sheep have 32 teeth.

There are six molars and six premolars, all paired in the upper and lower jaws, together forming broad grinding surfaces at the back of the mouth. The eight incisors (as in other ruminants), are all in the lower jaw, biting against a hard, toothless pad in the upper jaw; together these are used to pick off vegetation. There are no canines, but there is a large gap where they would be, between the incisors and the premolars.

Until the age of four (when all the adult teeth have erupted), it is possible to distinguish the age of sheep from their front teeth: a pair of incisors erupts each year. The front teeth are gradually lost as sheep begin to age, making it progressively harder for them to feed and thus hindering the health and productivity of the animal.

For this reason, domestic sheep on normal pasture begin to decline slowly from four years on, and the average life expectancy of a sheep is 10 to 12 years—though some sheep may live as long as 20 years. Sheep have good hearing, and are sensitive to noise when being handled. Sheep have horizontal slit-shaped pupils, and possess a wide peripheral field of vision with some ability to distinguish colour.

With what is estimated to be approximately 270° to 320° of wide-angle vision, sheep can see behind themselves without turning their heads (though this ability may be impaired slightly by any wool present around the eyes). However, sheep have poor depth perception; shadows and dips in the ground may cause sheep to balk. In general, sheep have a tendency to want to move out of darkness and in to well-lit areas.

Sheep have an excellent sense of smell, and—like all species of their genus—have scent glands located on their face just in front of the eyes, and interdigitally on the feet. The purpose of these glands is not known for certain, but those on the face are likely to be related to breeding behaviour. The interdigital glands may also be related to sexual selection, but alternative reasons, such as secretion of a waste product or a scent marker to help lost sheep find their flock, have also been posited.

Sheep and goats are closely related (both are in the subfamily Caprinae), and it can sometimes be difficult to differentiate them by appearance alone. However, the genes of sheep and goats differ so greatly that cross-species hybrids rarely occur, and are always infertile. A hybrid of a ewe and a buck (a male goat) is called a sheep-goat hybrid, and is not to be confused with the genetic chimera called a geep.

Visual clues to differentiating sheep and goats include the beard and divided upper lip of goats, which sheep do not have. Sheep tails also hang down, even when short or docked, while the tails of goats are held upwards. Sheep breeds are also often naturally polled (either in both sexes or just in the female), while naturally polled goats are rare (though many are polled artificially). Males of the two species differ especially in that buck goats acquire a unique and very strong odor during the rut, whereas rams do not.

Behaviour

Sheep are prey animals with a strong gregarious instinct, and a majority of sheep behaviours can be defined in these terms. When sheep gather in larger social groups, it provides a mode of defence for a species without another means of doing so. All sheep exhibit some degree of natural flocking behaviour, varying with breed. The dominance hierarchy of wild *Ovis aries* and its natural inclination to follow a leader to new pastures were the pivotal factors in it being one of the first domesticated livestock species. Farmers exploit these behaviours to keep sheep together on open pastures and to move them more easily. Shepherds may also use sheepdogs in this effort, whose highly bred herding ability can assist in moving flocks. In regions such as Iceland, where sheep have no natural predators, none of the native breeds of sheep exhibit a strong flocking behaviour.

Sheep without a flocking instinct tend to scatter rather than stay together when threatened, in a similar manner to goats. It is virtually impossible to use herding dogs with such breeds. Sheep can also become hefted to one particular local pasture (heft) so they do not roam freely in unfenced landscapes. Hefting is an instinct most common in British upland breeds. Ewes teach the heft to their lambs, and if whole flocks are culled it must be retaught to the replacement animals. Flock dynamics in sheep are as a rule only exhibited in a group of four or more sheep.

Fewer sheep may not react as normally expected when alone or with very few other sheep. This is one reason why three is the minimum number of sheep in a sheepdog trial; it places the greatest pressure on a dog's herding abilities. For sheep, the primary defence mechanism is simply to flee from danger when their flight zone is crossed. Secondarily, cornered sheep may attempt to leap at and ram threats in order to escape.

This is particularly true for ewes with newborn lambs and rams (hence the verb form of the word ram). In displaying flocking, sheep have a strong lead-follow tendency, and a leader often as not is simply the first sheep to move. However, sheep do establish a pecking order through physical displays of dominance. Dominant animals are inclined to be more aggressive with other sheep, and usually feed first at troughs. Relationships in flocks tend to be closest among related sheep: in mixed breed flocks same-breed subgroups tend to form, and a ewe and her direct descendants often move as a unit within very large flocks. Sheep are frequently thought of as extremely stupid animals. A sheep's herd mentality and quickness to flee and panic in the face of stress often make shepherding a difficult endeavor for the uninitiated. Despite these perceptions, a University of Illinois monograph on sheep found them to be just below pigs and on par with cattle in IQ, and some sheep have shown problem-solving abilities; a flock in Yorkshire, England found a way to get over cattle grids by rolling on their backs.

If worked with patiently sheep may learn their names, and many sheep are trained to be led by halter for showing and other purposes. Very rarely, sheep are used as pack animals. Tibetan nomads distribute baggage equally throughout a flock as it is herded between living sites.

Reproduction

Sheep follow a similar reproductive strategy to other herd animals. A flock of ewes is generally mated by a single ram, who has either been chosen by a farmer or has established its dominance through physical contest with other rams (in feral populations). Most sheep are seasonal breeders, though some are able to breed (also called tupping) year-round. Ewes generally reach sexual maturity at six to eight months of age, and rams generally at four to six (ram lambs have occasionally been known to impregnate their mothers at two months).

Ewes enter in to estrus cycles about every 17 days, lasting for approximately 30 hours. In addition to emitting a scent, they indicate their readiness through physical displays towards rams. Sheep may alternatively display a preference for homosexuality, and the behaviour occurs in about eight percent of rams on average. Its occurrence does not seem to be related to flock hierarchy (as some homosexual behaviour is in mammals), rather the ram's typical motor pattern for intercourse is merely directed at rams instead of ewes.

The phenomenon of the freemartin, a female bovine that is behaviourally masculine and lacks functioning ovaries, is most commonly associated with cattle, but does occur to a less extent in sheep. The instance of freemartins in sheep may be increasing in concert with the rise in twinning (freemartins are the result of male-female twin combinations).

Without human intervention, rams fight during the rut to determine which individuals may mate with ewes. Rams, especially unfamiliar ones, will also fight outside the breeding period to establish dominance; rams can kill one another if allowed to mix freely. During the rut, even normally friendly rams may become dangerous to humans due to increased hormone levels.

Historically, especially aggressive rams were sometimes blindfolded or hobbled when interaction with people was necessary. Today, those who keep rams typically prefer softer preventative measures, such as moving within a clear line to an exit, never turning their back on a ram, and possibly using dousing with water or a diluted solution of bleach or vinegar to dissuade charges.

Without ultrasound or other special tools, ascertaining whether sheep are in fact pregnant is initially difficult. Ewes only begin to visibly show a pregnancy about six weeks before the birth, so shepherds sometimes merely rely on the assumption that a ram has been given enough time to impregnate all the ewes. However, a very commonly used piece of equipment for discerning which ewes have been mated with is a marking harness.

By fitting a ram with a chest harness that holds a special crayon (or *raddle*), ewes that have been mounted are marked with its colour. Dye may also be directly applied to the ram's brisket. After mating, sheep have a gestation period of around five months. Within a few days of the impending birth, ewes begin to behave differently.

They may lay down and stand erractically, paw the ground, or otherwise act out of sync with normal flock patterns. A ewe's udder will quickly fill out, and her vulva will swell. Vaginal, uterine or anal prolapse may also occur in problem cases, in which case either stitching or a physical retainer is used to hold the orifice in.

When birth is imminent, contractions begin to take place, and the fitful behaviour of the ewe may increase. A normal labour may take one to several hours: depending on how many lambs are present, the age of the ewe, and her physical and nutritional condition prior to the birth. Though some breeds may regularly throw larger litters of lambs, most produce either single or twin lambs. At some point, usually either at the beginning of labour or soon after the births have occurred, ewes and lambs may be confined to small lambing jugs. These pens, which are generally two to eight feet in length and width, are designed to facilitate both careful observation of ewes and to cement the bond between them and their lambs.

Ovine obstetrics is sometimes a problematic phenomenon. However, it is a myth that sheep cannot lamb without human assistance: many ewes give birth directly in pasture without aid. While the majority of births are relatively normal and do not require intervention, there are still a number of possible complications that may arise. A lamb may present in the normal fashion (with both legs and head forward), but may simply be too large to slide out of the birth canal.

This often happens when large rams are crossed with diminutive ewes (this is related to breed, rams are naturally larger than ewes by comparison). Lambs may also present themselves with one shoulder to the side, completely backward, or with only some of their limbs protruding. Lambs may also be aborted or stillborn. Some types of abortion in sheep are preventable through a vaccination administered before conception.

In the case of any such problems, those present at lambing (who may or may not include a veterinarian, most shepherds become accomplished at lambing to some degree) may assist the ewe in extracting or repositioning lambs. After the birth, ewes ideally break the amniotic sac (if it is not broken during labour), and begin licking clean the lamb. The licking both clears the nose and mouth, dries the lamb, and stimulates it. Lambs that are breathing and healthy at this point begin trying to stand, and ideally do so between an half and full hour, with help from the mother.

Generally after lambs stand, the umbilical cord is trimmed to about an inch (2.54 centimeters). Once trimmed, a small container (such as a film canister) of iodine is held against the lamb's belly over the remainder of the cord to prevent infection. In normal situations, lambs nurse after standing, receiving vital colostrum milk. Lambs that either fail to nurse or are prevented from doing so by the ewe require aid in order to live. If coaxing the pair to accept nursing does not work, one of several steps may then be taken. Ewes may be held or tied to force them to accept a nursing lamb. If a lamb is not eating, a stomach tube may also be used to force feed the lamb in order to save its life. In the case of a permanently rejected lamb, a shepherd may then attempt to foster an orphaned lamb onto another ewe.

Lambs are also sometimes fostered after the death of their mother, either from the birth or other event. Scent plays a large factor in ewes recognizing their lambs, so disrupting the scent of a newborn lamb with washing or over-handling may cause a ewe to reject theirs. Conversely, various methods of imparting the scent of a ewe's own lamb to an orphaned one may be useful in fostering. If an orphaned lamb cannot be fostered, then it usually becomes what is known as a bottle lamb—a lamb raised by people and fed via bottle. After lambs are stabilized, lamb marking—the process of ear tagging, docking, and castrating—is carried out. Vaccinations are usually carried out at this point as well. Ear tags with numbers are the primary mode of identification when sheep are not named; it is also the legal manner of animal identification in the European Union.

When performed at an early age, ear tagging seems to cause little or no discomfort to lambs. However, using tags improperly or using tags not designed for sheep may cause discomfort, largely due to excess weight of tags for other animals.

Castration is performed on ram lambs not intended for breeding, though some shepherds choose to avoid the procedure for ethical, economic or practical reasons. Ram lambs that will either be slaughtered or separated from ewes before sexual maturity are not usually castrated. Docking, which is a shortening—not a complete removal—of sheep's tails is practiced for health reasons. The tail is removed just below the lamb's caudal tail flaps.

Though docking is often considered cruel and unnatural by animal rights activists, it is considered by sheep producers large and small alike to be a critical step in maintaining the health of sheep. Long, wooly tails make shearing more difficult, interfere with mating, and make sheep extremely susceptible to parasites, especially those that cause flystrike. In Chuck Wooster's *Living with Sheep*, the author anecdotally relates his shearer's response when told of his opposition to docking: "There's nothing natural about a sheep walking around with a dozen pounds of wool on it, and nothing natural about a sheep with a wooly tail.

Natural's got nothing to do with it. If lambs are seen walking around with tails covered by maggots and flies, cut the tails off. Both castrating and docking can be performed with several different instruments. An elastrator places a tight band of rubber around an area, causing it to atrophy and fall off in a number of weeks. This process is bloodless and does not seem to cause extended suffering to lambs, who tend to ignore it after several hours. In addition to the elastrator, a Burdizzo, emasculator, heated chisel or knife are sometimes used. After one to three days in the lambing jugs, ewes and lambs are usually sufficiently stabilized to allow reintroduction to the rest of the flock.

Health

Sheep may fall victim to chemical, pathogenic and physical means of sickness and injury. As a prey species, a sheep's system is designed to hide the obvious signs of illness as much as possible, in order to prevent being targeted by predators. However, there are some signs of ill health that can be ascertained without any special testing. Sick sheep may be unwilling to eat, vocalize excessively, and be generally listless. Throughout history, much of the money and labour of sheep husbandry has gone in to dealing with sheep ailments in order to avoid the heavy losses sustained from them.

Shepherds in history often created remedies through experimentation on the farm. In some countries, sheep do not have the economic impact to necessitate often expensive clinical studies designed to approve drugs specifically for usage with sheep. In such instance, many shepherds resort to extra–label usage of drugs approved for other animals.

In the 20th and 21st centuries, a minority of sheep owners have turned to alternative treatments such as homeopathy, herbalism and even traditional Chinese medicine to treat sheep veterinary problems. Despite some favourable anecdotal evidence from sheep producers, the effectiveness of alternative veterinary medicine has been met with skepticism in scientific journals. The need for traditional anti–parasite drugs and antibiotics is widespread, and is a primary impediment to certified organic farming with sheep.

Many breeders take a variety of preventative measures to ward off problems before they arrive. The first is to ensure that all sheep purchased for the initial flock and replacement animals arc hcalthy to begin with. Many buyers avoid outlets which arc known to be clearing houses for animals culled from healthy flocks as either sick or simply inferior. This can also mean maintaining a closed flock, and quarantining new sheep for a month. Two fundamental preventative programmes are maintaining good nutrition and stressing sheep as little as possible.

Handling sheep in loud, erratic ways causes them to emit cortisol, a stress hormone. Elevated amounts of cortisol can lead to a weakened immune system, thus making sheep far more vulnerable to disease. Signs of stress in sheep include: excessive panting, teeth grinding, restless movement, wool eating, and wood chewing. Preventing the presence of harmful chemical agents is also an important part of a sheep owner's work. Common culprits of poisoning include pesticide sprays, inorganic fertilizer, motor oil, as well as radiator coolant (the ethylene glycol antifreeze in this is sweet-tasting).

A common form of preventative medication for sheep are vaccinations and treatments for parasites. Both external and internal parasites are the most prevalent malady in sheep, and impede the productivity of flocks even if they are not fatal. Various worms comprise most of the internal parasites, which are ingested during grazing, incubate within the sheep, and are expelled through the digestive system (beginning the cycle again).

Oral parasite medicines known as drenches are given to a flock to treat worms, sometimes after a count of eggs in the sheep's feces has been taken to assess infestation levels. Afterwards, sheep may be moved to a new pasture to prevent ingesting the same parasites. External sheep parasites include: lice, sheep keds, nose bots, sheep itch mite, and maggots. Keds are blood-sucking parasites that can cause general malnutrition and decreased productivity, but are not fatal. Maggots are those of the bot fly and the blowfly. Fly maggots cause the extremely destructive condition of flystrike. Flies lay their eggs in wounds or wet, manure-soiled wool, and when the maggots hatch they burrow in to a sheep's living flesh, eventually resulting in death if unchecked. In addition to other treatments, crutching (shearing wool from a sheep's rump) is a common preventative method. Nose bots are flies that inhabit a sheep's sinuses, causing breathing difficulties and discomfort. Common signs are a discharge from the nasal passage, sneezing, and frantic movement such as head shaking. External parasites may be controlled through the use of backliners, sprays or immersive treatments (sheep dips). A wide array of diseases caused by bacteria affect sheep, but a few are especially prevalent and some may even be transmittable to humans. Diseases of the hoof, such as foot rot and foot scald may occur, and are treated with footbaths and other remedies. These painful conditions cause lameness and retard a sheep's ability to feed. Ovine Johne's Disease is a wasting disease that affects young sheep. Blue tongue disease is an insect-borne illness causing fever and inflammation of the mucous membranes.

Soremouth (also known as scabby mouth, contagious ecthyma or orf) is a skin disease leaving lesions that can be transmitted to persons handling infected flocks. More seriously, the organisms that can cause spontaneous enzootic abortion in sheep are easily transmitted to pregnant women. Also of concern are scrapie and foot and mouth disease, as both can decimate entire flocks. The latter poses a significant risk to humans. During the 2001 outbreak of foot and mouth disease, hundreds of sheep in the U.K. were culled, and some rare British breeds were at risk for extinction from the pandemic.

Predation

Other than parasites and disease, predation is a threat to sheep health and thereby the profitability of sheep raising. Sheep have very little ability to defend themselves, even when compared with other prey species kept as livestock. Even if sheep are not directly bitten or survive an attack, they may die from panic or from injuries sustained. However, the impact of predation varies dramatically with region.

In Africa, Australia, the Americas, and parts of Europe and Asia predators can be a serious problem. In contrast, some nations are virtually devoid of sheep predators. Many islands that are known for extensive sheep husbandry are suitable largely because of their predator-free status. Worldwide, canids—including the domestic dog—are responsible for the majority of sheep deaths.

224,200 sheep were killed by predators in 2004, comprising approximately 37 per cent of all ovine deaths for that year. The sheep lost in that year represented a sum total of 18.3 million dollars for sheep producers. Coyotes were responsible for 60.5 per cent of all deaths, with the next largest being domestic dogs at 13.3 per cent. Other North American predators of sheep included cougars (5.7 per cent), bobcats (4.9 per cent), eagles (2.8 per cent), bears (3.8 per cent), and foxes (1.9 per cent). Wolves, ravens, vultures, and other animals together made up the remaining 7.1 per cent of deaths.

As all NASS statistics on sheep only take into account sheep after docking, the American Sheep Industry Association estimates that an additional 50-60,000 lambs were killed (before docking) that were not a part of the count. The number of sheep lost to predators may also be higher when considering that reports are generally only made when there is a reasonable expectation that a producer will be financially reimbursed for the loss.

The only widespread potential predators of sheep are cougars and jaguars, both of which are known to prey on livestock regularly. South American canids such as the Maned wolf and foxes of the genus *Pseudalopex* are also blamed for sheep deaths, but no evidence for a statistically significant amount of predation by these species has ever been presented. Though large, the South African sheep industry is significantly hindered by the innumerable predators present in the country. Other African nations that rely on sheep face a similar problem.

The main Australian predator of sheep is the dingo, which is a large-enough danger to sheep to precipitate the construction of the world's largest fence: the Dingo Fence. In contrast, New Zealand has no large carnivores whatsoever. The sole animal known to attack sheep in New Zealand is the rare, unusual kea parrot endemic to the country's South Island. The U.K. and Ireland once both had wolves and bears, but today only the Red fox and birds of prey remain a tangible threat to sheep in the British Isles.

Sheep producers have used a wide variety of measures to try and combat predation throughout history. Pre-modern shepherds had only the most basic of tools: their own presence and livestock guardian dogs. Whereas sheepdogs herd sheep, guardian dogs are trained to integrate in to flocks and protect them from predators. The ability of these dogs to do so is a transference of the canine pack social structure on to a flock. Following their invention, the focus in dealing with predators shifted to the nearly exclusive use of guns, traps, and poisons to kill predators both defensively and preemptively.

The population of predator species plummeted worldwide, pushing some to extinction or significantly reducing their original ranges. With the appearance of the environmental and conservation movements, and subsequent state, provincial, national and international legislation, simply exterminating predator species failed to be a viable option for protecting flocks.

While killing predators caught directly in the act of attacking sheep may be legal, hunting predators is usually not. Poisons, such as compound 1080 and sodium cyanide, that were previously either placed in bait or in special sheep collars are now banned or restricted in most countries.The 1970s saw an ensuing resurgence in the use of livestock guardian dogs and the development of new methods of predator control, many of them non–lethal. Donkeys and llamas have been used since the 1980's in sheep operations, using the same basic principle as livestock guardian dogs.

Donkeys and llamas—which have a natural dislike of all canines—need no special training to deter predators. However, intact males may kill ewes by attempting to breed them, and must be kept without others of their kind in order to bond with the sheep. Dogs, even those from specialized breeds, need some training to be effective and can be used in concert with other dogs. Dogs are the only guardian animal known to be effective with bears and cougars; donkeys and llamas are usually ineffective as they are themselves vulnerable to predation by bears and cougars.

However, dogs have been killed by wolves and may even attract wolf packs. Whatever species is used for guarding, individual animals may be temperamentally ill–suited to working with sheep. In addition to animal guardians, contemporary sheep operations may use non–lethal predator deterrents such as motion–activated lights and noisy alarms. While these devices have been shown to be successful, predators can become habituated to them.

Sheep were among the first animals to be domesticated by mankind, with sources providing a domestication date between nine and eleven thousand years ago in Mesopotamia. The species has several characteristics—such as a relative lack of aggression, a manageable size, early sexual maturity, a social nature, and high reproduction rates—that made it particularly amenable to taming.

Today, *Ovis aries* is an entirely domesticated animal that is largely dependent on man for its health and survival. Small feral populations of sheep exist, but exclusively in areas that are devoid of predators (usually islands). No feral sheep population has ever reached the scale of feral horses, goats, pigs, or dogs.

Determining the exact genetic relation of domestic sheep to its wild ancestors is a complex problem, and no particular lineage has received complete acceptance in the scientific community. The most common theory currently supported by DNA analysis indicates that *Ovis aries* is jointly descended from both the European (*O. musimon*) and Asiatic (*O. orientalis*) species of mouflon.

It has also been proposed that the European mouflon is an ancient breed of domestic sheep turned feral rather than an ancestor, the lack of a fossil record for the species may support this hypothesis. The urial (*O. vignei*) was once often thought to have been a forebear of domestic sheep; they occasionally interbreed with mouflon in the Iranian part of their range.

However, both the urial, argali, and snow sheep possess a different number of chromosomes than other Ovis species, making any direct influence highly implausible. Contemporary evaluations of sheep DNA show no evidence of urial ancestry. DNA analysis comparing European and Asian breeds of sheep shows a significant percentage of genetic variance between the two. Two explanations for this phenomenon have been posited. The first is that there is a currently unknown species or subspecies of wild sheep that contributed to the formation of domestic sheep. A second theory suggests that this variation is the result of multiple waves of capture from wild mouflon, similar to the known development of other livestock.

Initially, sheep were kept solely for meat, milk and skins. Archaeological evidence from statuary found at sites in Iran suggests that selection for wooly sheep may have began circa 6000 BC, but the earliest woven wool garments have only been dated to two to three thousand years later. By that span of the Bronze Age, sheep with all the major features of modern breeds were widespread throughout Western Asia. However, one chief difference between ancient sheep and most modern breeds is the technique by which wool could be collected. Primitive sheep cannot be shorn, and must have their wool plucked out by hand in a process called "rooing".

This is due to the fact that their kemps are still longer than the soft fleece. The fleece may also be collected from the field after it falls out. This trait survives today in unrefined breeds such as the Soay. The Soay, along with other Northern European breeds with short tails, unshearable fleece, diminutive size, and horns in both sexes, are more closely related to ancient sheep than most breeds found today. Originally, weaving and spinning wool was a handicraft practiced at home, rather than an industry.

Egyptians, Babylonians, Sumerians, and Persians all depended on sheep; and though linen was the first fabric to be fashioned in to clothing, wool was a prized product. The raising of flocks for their fleece was one of the earliest industries, and flocks were a medium of exchange in barter economies.

From Southwest Asia, sheep husbandry spread quickly in to Europe. Practically from its inception, ancient Greek civilization relied on sheep as primary livestock, and were even said to name individual animals. Scandanavian sheep of a type seen today—with short tails and multi-coloured fleece—were also present early on. Later, the Roman Empire kept sheep on a wide scale, and the Romans were an important agent in the spread of sheep raising throughout the continent.

Pliny the Elder, in his Natural History (*Naturalis Historia*), speaks at length about sheep and wool. Declaring "Many thanks, too, do we owe to the sheep, both for appeasing the gods, and for giving us the use of its fleece.", he goes on to detail the breeds of ancient sheep and the many colours, lengths and qualities of wool. Romans also pioneered the practice of blanketing sheep, in which a fitted coat (today usually of nylon) is placed over the sheep to improve the cleanliness and luster of its wool.

During the Roman occupation of the British Isles, a large wool processing factory was established in Winchester, England circa 50 AD. By 1000 AD, England and Spain were recognized as the epicenters of sheep production in the Western world. As the original breeders of the fine-wooled merino sheep that have historically dominated the wool trade, the Spanish gained great wealth. Wool money largely financed Spanish rulers and thus the voyages to the New World by conquistadors.

The powerful *Mesta* (its full title was *Honrado Concejo de la Mesta*, the Honourable Council of the Mesta) was a corporation of sheep owners mostly drawn from Spain's wealthy merchants, Catholic clergy and nobility that controlled the merino flocks. By the Seventeenth century, the *Mesta* held in upwards of two million head of merino sheep.

These flocks followed a seasonal pattern of transhumance across Spain. In the spring, they left the winter pastures (*invernaderos*) in Extremadura and Andalusia to graze on their summer pastures (*agostaderos*) in Castile, returning back again in the autumn. Spanish rulers eager to increase wool profits gave extensive legal rights to the *Mesta*, often to the detriment of local peasantry. The huge merino flocks had a lawful right of way for their migratory routes (*cañadas*). Towns and villages were obliged by law to let the flocks graze on their common land, and the Mesta had its own sheriffs which could summon offending individuals to its own tribunals.

Exportation of merinos was also a punishable offense that required royal permission, thus ensuring a near-absolute monopoly on the breed until Napoleon's invasion of that country in the mid-18th century. After the breaking of the export ban, fine wool sheep began to be distributed worldwide. The export to Rambouillet by Louis XVI in 1786 formed the basis for the modern Rambouillet breed. After the Napoleonic Wars and the global distribution of the once-exclusive Spanish stocks of Merinos, sheep raising in Spain reverted to hardy coarse-wooled breeds such as the Churra, and was no longer of international economic significance.

The sheep industry in Spain was an instance of migratory flock management, with large homogenous flocks ranging over the entire nation. Comparatively, the ovine model used in England was quite different but had a similar importance to economy of the British Empire. Up until the early 20th century, owling (the smuggling of sheep or wool out of the country) was a punishable offense, and to this day the Lord Speaker of the House of Lords sits on a bench known as the Woolsack.

The high concentration and more sedentary nature of shepherding in the U.K. allowed sheep especially adapted to their particular purpose and region to be raised, thereby giving rise to an exceptional variety of breeds in relation to the land mass of the country. This greater variety of breeds also produced a valuable variety of products to compete with the superfine wool of Spanish sheep. The growth of the sheep industry in Britain quickened in the 15th and 16th centuries due to booming wool prices and the greater pasture available due to a diminished population (the result of the Black Death). More landholders were willing and able to turn their lands over to pasture for sheep, and an influx of Flemish immigrants skilled in cloth making and dyeing during the reign of Edward III helped to cement the wool trade.

By this time, Britain began moving from an open field system to one of enclosure. The unilateral closing of commons by landed gentry had enormously detrimental effects on local communities—entire villages were sometimes depopulated to make way for pasture—and was widely protested. By the time of Elizabeth I's rule, sheep and wool trade was the primary source of tax revenue to the Crown of England.

An important event not only in the history of domestic sheep, but of all livestock, was the work of Robert Bakewell in the 1700s. Before his time, breeding for desirable traits was often based on chance, with no scientific process for selection of breeding stock. Bakewell established the principles of selective breeding—especially line breeding—in his work with sheep, horses and cattle, and his theories were an influence on the subsequent work of Gregor Mendel and Charles Darwin. His most important contribution to sheep was the development of the Leicester Longwool, a quick-maturing breed of blocky conformation that formed the basis for many vital modern breeds.

Today, the sheep industry in the U.K. has diminished significantly. Even the British taste for mutton (lamb is still popular) has declined, which prompted the creation of the Mutton Renaissance Campaign by the Prince of Wales.

No ovine species native to the Americas has ever been domesticated, despite being closer genetically to domestic sheep than many Asian and European species, having the same number of chromosomes. The first domestic sheep—most likely of the Churra breed—arrived with Christopher Columbus' second voyage in 1493. The next Transatlantic shipment to arrive was with Hernán Cortés in 1519, landing in Mexico. No export of wool or animals is known to have occurred from these populations.

However, Churras were introduced to the Navajo tribe of Native Americans, and became a key part of their livelihood and culture. The modern presence of the Navajo-Churro breed is a result of this heritage. The next transport of sheep to North America was not until 1607 with the voyage of the *HMS Susan Conant* to Virginia. However, the sheep that arrived in that year were all slaughtered because of a famine, and a permanent flock was not to reach the colony until two years later in 1609. Two decades later, the colonists had grown their flock to a total of 400. Sheep were also eventually purchased from the Dutch colonizers of Manhattan.

In 1662, a woolen mill was built in Watertown, Massachusetts, and by the 1640s there were approximately 100,000 head of sheep in the 13 colonies. Early on, the British government banned further export of sheep to the Americas, or wool from it, in an attempt to stifle any threat to the wool trade in the British Isles. One of many restrictive trade measures that precipitated the American

Revolution, the sheep industry in the Northeast grew despite the bans. Gradually throughout the 1900s, sheep production moved westward, and today the vast majority of flocks reside on Western range lands. During this westward migration of the industry, competition between sheep and cattle operations grew more heated, eventually erupting in to range wars. Other than simple competition for grazing and water rights, cattlemen believed that the secretions of the foot glands of sheep made cattle unwilling to graze on places where sheep had stepped.

Another effect of the westward movement of sheep flocks in North America was the decline of wild species such as Bighorn sheep. Most diseases of domestic sheep are transmittable to wild ovines, and such diseases, along with overgrazing and habitat loss, are named as primary factors in the plummeting numbers of wild sheep. Sheep production peaked in the United States during 1940s and 50s at more than 55 million head. Henceforth and continuing today, the number of sheep in North America has steadily declined with wool prices and the lessening American demand for sheep meat.

In South America, especially in Argentina, Chile, Brazil, Uruguay, and Peru, there is an active modern sheep industry. Sheep keeping was largely introduced through immigration to the continent by Spanish and British peoples, for whom sheep were a major industry during the period. South America today has a large number of sheep, but the highest-producing nation kept only just over 15 million head in 2004, far less than most centers of sheep husbandry.

Brazil is, however, not an export nation of any consequence in the sheep industry and sheep operations often find themselves in stiff competition from the immense cattle industry in the country. The most influential region internationally is that of Patagonia, and Patagonian shepherds are often employed by U.S. sheep producers. With few predators and almost no grazing competition, the region is prime land for sheep raising. The most exceptional area of production is surrounding the La Plata river in the Pampas region.

Sheep production in Patagonia peaked in 1952 at more than 21 million head, but has steadily fallen to less then ten today. Most operations focus on wool production for export from Merino and Corriedale sheep; the economic sustainability of wool flocks has fallen with the drop in prices, while the cattle industry continues to grow.

Australia and New Zealand are crucial players in the contemporary sheep industry, and sheep are an iconic part of both countries' culture and economy. New Zealand has the highest density of sheep per capita (sheep outnumber the human population 12–1), and Australia is the world's indisputably largest exporter of sheep and cattle. In 2007, New Zealand even declared February 15 their National Lamb Day, an official holiday celebrating the country's history of sheep production. The First Fleet brought the initial population of 70 sheep from the Cape of Good Hope to Australia in 1788.

The next shipment was of 30 sheep from Calcutta and Ireland in 1793. All of the early sheep brought to Australia were exclusively used for the dietary needs of the penal colonies. The beginnings of the Australian wool industry were due to the vision and efforts of Captain John Macarthur. At Macarthur's urging 16 Spanish merinos were imported in 1797, effectively beginning the Australian sheep industry. By 1801 Macarthur had 1,000 head of sheep, and in 1803 he exported 245 lbs (111 kg) of wool to England. Today, Macarthur is generally thought of as the father of the Australian sheep industry.

The growth of the sheep industry in Australia was explosive. In 1820, the continent held 100,000 sheep, a decade later it had one million. By 1840, New South Wales alone kept 4 million sheep; flock numbers grew to 13 million in a decade. While much of the growth in both nations was due to the active support of Britain in its desire for wool, both worked independently to develop many new high-production breeds: the Corriedale, Coolalee, Coopworth, Perendale, Polwarth, Booroola Merino, Peppin Merino, and Poll Merino were all created in New Zealand or Australia.

Wool production was a fitting industry for colonies far from their home nations. Before the advent of fast air and maritime shipping, wool was one of the few viable products that was not subject to spoiling on the long passage back to British ports. The abundant new land and milder winter weather of the region also aided the growth of the Australian and New Zealand sheep industries.

Flocks in Australia have always been largely range bands on fenced land, and are aimed at production of superfine wool for clothing and other products as well as meat. New Zealand flocks are kept in a fashion similar to English ones, in fenced holdings without herders. Though wool was once the primary income source for New Zealand sheep owners, today it has shifted to meat production.

The Australian sheep industry is the only sector of the industry to receive strident international criticism for its practices. The practice of mulesing, in which skin is cut away from an animal's perineal area without anesthesia to prevent cases of flystrike, has been condemned widely as painful and unnecessary. In response, a programme of phasing out mulesing is currently being implemented.

Most of the sheep meat exported from Australia are either frozen carcasses to the U.K. or live animals to the Middle East. Shipped on converted oil tankers in what has been called crowded, unsafe conditions by critics, live sheep are desired by Middle Eastern nations to meet the requirements ritual halal slaughter. Opponents of the export—such as PETA—say that sheep exported to countries outside the jurisdiction of Australia's animal cruelty laws are treated with horrendous brutality and that halal facilities exist within Australia to make export of live animals redundant. A few celebrities and companies have pledged to boycott all Australian sheep products in protest.

ECONOMIC IMPORTANCE OF SHIP FARMING

Domestic sheep have been important to many economies, given that sheep provide a wide array of raw materials. Wool continues to be a vital textile, although in the late 20th century wool prices began to fall dramatically as the result of the popularity and cheap prices for synthetic fabrics. For many shepherds, the cost of shearing is greater than the possible profit from the fleece, making subsisting on wool production alone practically impossible without farm subsidies. Fleeces are used as material in making alternative products such as Wool insulation. In the 21st century, the sale of meat is the most profitable enterprise in the sheep industry.

Sheepskin is likewise used for making clothes, footwear, rugs, and other products. Byproducts from the slaughter of sheep are also of value: sheep tallow was once used in candlemaking, sheep bone and cartilage has been used to furnish carved items such as dice and buttons as well as rendered glue and gelatin. Sheep intestine can be formed into sausage casings, and lamb intestine has especially been formed into surgical sutures, as well as strings for musical instruments and tennis rackets. Sheep droppings have even been sterilized and mixed with traditional pulp materials to make paper. Of all sheep byproducts, perhaps the most valuable is lanolin: the water-proof, fatty substance found naturally in sheep's wool and used as a base for innumerable cosmetics and other products.

In the modern era, sheep retain considerable importance in the economies of several countries. Countries such as China, India, Iran and Sudan have extremely large flocks that are generally of national and regional importance. Australia, New Zealand and other countries focus on global exports. Sheep play a major role in the economies of several smaller countries, where the number of sheep may not be great and impact on the global economy is negligible, but is still important locally. Such flocks may be a part of subsistence agriculture rather than a system of trade.

As Food

Sheep meat and milk were one of the earliest staple proteins consumed by human civilization after the transition from hunting and gathering to agriculture. Sheep meat prepared for food is known as either mutton or lamb. "Mutton" is derived from the Old French "*moton*", which was the word for sheep used by the Anglo-Norman rulers of much of the British Isles in the Middle Ages.

This became the name for sheep meat in English, while the Old English word *sceap* was kept for the live animal (eventually forming the word English speakers use today). More recently, "mutton" has been limited to the meat of older sheep; "lamb" is used for that of younger ones, usually those under a year old.

In the 21st century, the nations with the highest consumption of sheep meat are the Persian Gulf states, New Zealand, Australia, Greece, Uruguay, the United Kingdom and Ireland. These countries eat between 14 and 39 lbs. (approximately 3 to 18 kg) of sheep meat per capita, per annum. By comparison, countries such as the U.S. consume only a pound or less (under 0.5 kg). In addition, such countries rarely eat mutton, and may even restrict their habits to the more expensive cuts of lamb: mostly the chop and the leg.

In those that do favour sheep meat, it is often the product of a past history of sheep production, even if they produce little or no sheep today. In such countries in particular, dishes comprising alternative cuts and offal may be popular or traditional. Sheep testicles—called animelles or lamb fries in culinary terminology—are considered a delicacy in many parts of the world. Perhaps the most infamous dish of sheep meat is the Scottish haggis, composed of a sheep's heart, liver and lungs cooked within its stomach.

Though it may have been in antiquity, sheep's milk is no longer drunk directly on any appreciable level. Sheep's milk is, however, valuable in the creation of some popular varieties of cheese and yogurts. Containing 6 per cent fat and double the solids of cow's milk, it can be crafted in to such dairy products with relative ease. Well-known sheep milk cheeses include the Roquefort of France, the Brocciu of Corsica, Manchego from Spain, the Pecorino Romano of Italy and Feta of Greece. Sheep milk does contain 4.8 per cent lactose, which may affect those who are intolerant.

Though not common model organisms, sheep have had played a small, yet influential role in scientific experimentation. Sheep are generally too large and dissimilar to humans to make ideal research subjects, but the few contemporary instances of sheep experimentation has caused a significant impact on science and culture at large. In particular, the Roslin Institute of Edinburgh, Scotland used sheep for genetics research that produced groundbreaking results. In 1995, two ewes named Megan and Morag were the first mammals cloned from differentiated cells.

A year later, a Finnish Dorset sheep named Dolly was the first mammal ever to be cloned from an adult somatic cell. Following this, Polly and Molly were the first mammals to be simultaneously cloned and transgenic.

In the 21st century, controversy raged over a scientific study at the Oregon Health and Science University, which investigated the biological mechanisms producing homosexuality in rams. Organizations such as PETA actively campaigned against the study, accusing scientists of trying to cure homosexuality in the sheep. OHSU and the involved scientists vehemently denied such accusations.

Sheep have had a strong presence within many cultures, especially in areas where they form the most common type of livestock. A wide symbology relates to sheep in art, religion and language. The raising of sheep for wool and meat became a major industry in the U.K. Australia and New Zealand and remains significant. As a result, sheep and sheep shearing have become an important part of the cultural traditions of these countries in particular.

Sheep have become an integral part of many linguistic traditions. In the English language, to call someone a sheep or ovine may allude that they are timid and easily led, if not outright stupid. Male sheep are often used as symbols of virility and power, such as for the St. Louis Rams and the Dodge Ram.

Sheep are key symbols in fables and nursery rhymes like The Wolf in Sheep's Clothing, Little Bo Peep, Baa Baa Black Sheep, and Mary Had a Little Lamb. Novels such as George Orwell's Animal Farm, Philip K. Dick's Do Androids Dream of Electric Sheep? and Thomas Hardy's Far From the Madding Crowd utilize sheep, as either characters or symbols.

The light-hearted detective novel *Three Bags Full* focuses on a flock of anthropomorphic sheep out to solve the murder of their shepherd. Poems like William Blake's "The Lamb", and songs such as Pink Floyd's "Sheep" use sheep for metaphorical purposes. In more modern popular culture, Exploding sheep are a video gaming phenomenon. The 2007 film *Black Sheep* exploits sheep for horror and comedic effect, ironically turning sheep into blood-thirsty killers. The Academy Award-winning film *Babe*, adapted from Dick King-Smith's *The Sheep Pig*, revolves around sheep husbandry and utilizes sheep as central characters. Counting sheep is popularly said to be an aid to sleep, and some ancient systems of counting sheep persist today. Sheep also enter in colloquial sayings and idiom frequently with such phrases as "black sheep".

To call an individual a black sheep implies that they are an odd or disreputable member of a group. This usage derives from the recessive trait which causes an occasional black lamb to be born in to an entirely white flock.

These black sheep were considered undesirable by shepherds, as black wool is not as commercially viable as white wool. Stereotypically, sheep are proverbial for their willingness to follow others of their kind. Citizens who accept overbearing governments have been referred to by the portmanteau neologism of sheeple. Somewhat differently, the adjective "sheepish" is also used to describe embarrassment.

6

Fish Farming

Fish farming is the principal form of aquaculture, while other methods may fall under mariculture. It involves raising fish commercially in tanks or enclosures, usually for food. A facility that releases juvenile fish into the wild for recreational fishing or to supplement a species' natural numbers is generally referred to as a fish hatchery. Fish species raised by fish farms include salmon, catfish, tilapia, cod, carp, trout and others.

Fig. Fish Farming

Increasing demands on wild fisheries by commercial fishing operations have caused widespread overfishing. Fish farming offers an alternative solution to the increasing market demand for fish and fish protein.

FISH

CAPTURE FISHERIES

The marine fish landings of India during the year 2009 has, provisionally, been estimated as 3. 16 million tonnes. The pelagic finfishes constituted 52 per cent, demersal fishes 28 per cent, crustaceans 16 per cent and molluscs 4 per cent of the total landings. The sectorwise contributions during the year

2009 were mechanized 74 per cent, motorized 22 per cent and artisanal 4 per cent. The west coast accounted for 56 per cent of the total landings. The northeast region, comprising West Bengal and Odisha contributed 20 per cent to the total production; southeast region consisting of Andhra Pradesh, Tamil Nadu and Puducherry contributed 24 per cent. On the west coast, the northwest region comprising Maharashtra and Gujarat recorded 26 per cent of the total, and the southwest region comprising Kerala, Karnataka and Goa contributed 30 per cent.

CRUZING FOR HILSA FISHERIES IN HOOGHLY

The investigations on distribution of hilsa in Hooghly from Nabadweep to Kakdweep estuary were done through cruzing. It established that depth and width of the river stretch affect the fishing effort and hilsa catch. The zones with <3 m depth had rare fishing activity, while fishing intensity was higher in deeper water (10 m depth). Hilsa was evenly distributed over the observed >100 km freshwater stretch. Movement of hilsa in shoals was missing during non-spawning season. Hilsa was available throughout the year with peak in monsoon. The observations may lead to possible occurrence of a resident stock of hilsa in Hooghly in addition to regular seasonal influx of migrant groups from sea to freshwater zone for breeding.

IMPACT ASSESSMENT OF FISHING BAN

To assess the impact of fishing ban on reservoir ecosystem, massbalanced models were used on different ecological groups before and after introduction of fishing ban in Wyra reservoir of Andhra Pradesh using Ecopath software. It is the first attempt in India on modeling a tropical productive reservoir, which contributes to fishery, particularly the commercially important prawn fishery.

The results of the assessment indicated:

- Slight increase in system turnover,
- Better organization/increased efficiency,
- Decreased tendency for eutrophication (low sum of all flows into detritus),
- Superior stability of total biomass/TST,
- Higher total primary production/total respiration,
- Increased resilience (higher overhead per cent), and
- Higher maturity and stability (higher primary production/biomass and lower ascendancy per cent).

FORMULATION OF SHRIMP FEED

Yard trials revealed that 3 per cent Ca with 2 per cent P supplementation resulted in a significantly higher digestibility of Ca and P as well as weight gain compared to other groups receiving lower levels of Ca and P. Feeds for high salinity, indicated that high growth can be achieved at 40 and 33.5 per cent dietary crude protein levels compared to lower dietary protein levels. Based on the better protein efficiency ratio and apparent protein utilization at high salinity (40 per cent), protein utilization efficiency was better at 33.5 per cent dietary protein level.

REDUCING FISH MEAL IN SHRIMP FEED

One of the bottlenecks to the approach of using plant proteins for reducing the level of costly fish meal in shrimp diet, is the high cellulose content of these feeds. *Bacillus macerans,* a potential cellulolytic bacteria, isolated from gut of *Lates calcarifer,* was identified for testing as feed supplement in plant protein based shrimp feed.

After 48 h of incubation, crude protein content, free glucose concentration and microbial count were maximum in the feed and dry matter loss was less. Growth, survival, protein efficiency ratio were significantly higher and feed conversion ratio was significantly lower in shrimps fed with feed supplemented with live cellulolytic bacteria as compared to control feed.

FISH FEED FOR MARINE ORNAMENTAL FISHES

The CMFRI has launched a dry formulated feeds named CadalminTM *Varna* for marine ornamental fish like clown and damsel fish. These feeds contain 40 per cent protein, 9 per cent fat, 39 per cent carbohydrates, 7 per cent ash and less than 2 per cent fibre, colour imparting nutrients like carotenoids from natural sources, nutraceuticals and microbial products. They are slow sinking crumbles available in three particle sizes, 0.25 mm, 0.75 mm and 1 mm produced through twin-screw extrusion technology, which is the state-of-the-art in aquatic feed production. The feed costs only ₹400/kg in comparison to imported feed having a price in the range of ₹4,000/kg.

ADOPTION OF BIOSECURITY AND WSSV INCIDENCE

Disease surveillance in the two types of culture systems in West Bengal, shrimp monoculture and traditional bheries over a period of three years (2007–10) indicated that white spot disease (WSD), shell associated problems and gill associated problems were statistically significant among the two culture systems. Newly emerging diseases such as monodon slow growth syndrome was observed in both culture systems, whereas, white faecal disease was observed only in shrimp monoculture system.

WSD outbreak was significantly high where stocking was done by the seed, which has not been tested by PCR. Analysis of data regarding adoption rate of biosecurity protocols either single or in combination in shrimp farms in Chidambaram and Nagapattinam Districts of Tamil Nadu indicated that white spot disease incidence is at a higher rate when protocols of disinfection of implements and reservoir ponds, were not followed.

DIAGNOSTICS FOR MONODON SLOW GROWTH SYNDROME

One of the emerging shrimp diseases is the monodon slow growth syndrome (MSGS) with abnormal slow growth and marked size differences among individuals. An improved diagnostic nested RT-PCR with custom designed primer targeting RdRp gene of Laem-Singh virus (LSNV), which has been implicated in monodon slow growth syndrome (MSGS) was developed. Screening of farmed and wild broodstock samples with this improved diagnostic PCR showed a high prevalence of LSNV. High prevalence in broodstock could lead to the spread of this virus by vertical transmission.

MAJOR CATEGORIES OF FISH FARMS

There are two kinds of aquaculture: extensive aquaculture based on local photosynthetical production and intensive aquaculture, in which the fish are fed with external food supply. The management of these two kinds of aquaculture systems are completely different.

Extensive (pond) Aquaculture

Limiting for fish growth here is the available food supply by natural sources, commonly zooplankton feeding on pelagic algae or benthic animals, such as certain crustaceans and mollusks. Tilapia species filter feed directly on phytoplankton, which makes higher production possible.

The photosynthetical production can be increased by fertilizing the pond water with artificial fertilizer mixtures, such as potash, phosphorus, nitrogen and microelements. Because most fish are carnivorous, they occupy a higher place in the trophic chain and therefore only a tiny fraction of primary photosynthetic production will be converted into harvestable fish.

As a result, without additional feeding the fish harvest will not exceed 200 kilograms of fish per hectare per year, equivalent to 1 per cent of the gross photosynthetic production.

A second point of concern is the risk of algal blooms. When temperatures, nutrient supply and available sunlight are optimal for algal growth, algae multiply their biomass at an exponential rate, eventually leading to an exhaustion of available nutrients and a subsequent die-off. The decaying algal biomass will deplete the oxygen in the pond water because it blocks out the sun and pollute it with organic and inorganic solvents (such as ammonium ions), which can (and frequently do) lead to massive loss of fish.

In order to tap all available food sources in the pond, the aquaculturist will choose fish species which occupy different places in the pond ecosystem, e.g., a filter algae feeder such as tilapia, a benthic feeder such as carp or catfish and a zooplankton feeder (various carps) or submerged weeds feeder such as grass carp.

Intensive (closed-circulation) Aquaculture

In these kinds of systems fish production per unit of surface can be increased at will, as long as sufficient oxygen, fresh water and food are provided. Because of the requirement of sufficient fresh water, a massive water purification system must be integrated in the fish farm. A clever way to achieve this is the combination of hydroponic horticulture and water treatment. The exception to this rule is cages which are placed in a river or sea, which supplements the fish crop with sufficient fresh water. Environmentalists object to this practice.

The cost of inputs per unit of fish weight is higher than in extensive farming, especially because of the high cost of fish food, which must contain a much higher level of protein (up to 60 per cent) than, e.g., cattle food and a balanced amino acid composition as well. This frequently is offset by the lower land costs and the higher productions which can be obtained due to the high level of input control.

Essential here is aeration of the water, as fish need a sufficient oxygen level for growth. This is achieved by bubbling, cascade flow or aqueous oxygen. Catfish, Clarias ssp. can breathe atmospheric air and can tolerate much higher levels of pollutants than, e.g., trout or salmon, which makes aeration and water purification less necessary and makes *Clarias* species especially suited for intensive fish production. In some *Clarias* farms about 10 per cent of the water volume can consist of fish biomass.

Especially when fish densities are high, the risk of infections by parasites like fish lice, fungi (Saprolegnia ssp.), intestinal worms (such as nematodes or trematodes), bacteria (e.g., Yersinia ssp, Pseudomonas ssp.), and protozoa (such as Dinoflagellates) is much higher than in animal husbandry because of the ease in which pathogens can invade the fish body (e.g. by the gills).

The same holds for water pollution or depletion of oxygen in the water, which can ruin a fish crop within minutes. This means, intensive aquaculture requires tight monitoring and a high level of expertise of the fish farmer.

Intensive aquaculture was developed as a source for food fish. Raising ornamental cold water fish (goldfish or koi), although theoretically much more profitable due to the higher income per weight of fish produced, has never been successfully carried out until very recently.

The increased incidences of dangerous viral diseases of koi Carp, together with the high value of the fish has led to initiatives in closed system koi breeding and growing in a number of countries. Today there are a few commercially successful intensive koi growing facilities in the UK, Germany and Israel.

Some producers have adapted their intensive systems and attempted to make them biosecure in an effort to provide consumers with fish that do not carry dormant forms of viruses and diseases.

SPECIFIC TYPES OF FISH FARMS

Within intensive and extensive aquaculture methods there are numerous specific types of fish farms, each has benefits and applications unique to its design.

Integrated Recycling Systems

One of the largest problems with freshwater aquaculture is that it can use a million gallons of water per acre (about 1 m^3 of water per m^2) each year. Extended water purification systems allow for the reuse (recycling) of local water.

The largest-scale pure fish farms use a system derived (admittedly much refined) from the New Alchemists in the 1970s. Basically, large plastic fish tanks are placed in a greenhouse. A hydroponic bed is placed near, above or between them. When tilapia are raised in the tanks, they are able to eat algae, which naturally grows in the tanks when the tanks are properly fertilized.

The tank water is slowly circulated to the hydroponic beds where the tilapia waste feeds a commercial plant crops. Carefully cultured microorganisms in the hydroponic bed convert ammonia to nitrates, and the plants are fertilized by the nitrates and phosphates. Other wastes are strained out by the hydroponic media, which doubles as an aerated pebble-bed filter.

This system, properly tuned, produces more edible protein per unit area than any other. A wide variety of plants can grow well in the hydroponic beds. Most growers concentrate on herbs (e.g. parsley and basil), which command premium prices in small quantities all year long. The most common customers are restaurant wholesalers.

Since the system lives in a greenhouse, it adapts to almost all temperate climates, and may also adapt to tropical climates.

The main environmental impact is discharge of water that must be salted to maintain the fishes' electrolyte balance. Current growers use a variety of proprietary tricks to keep fish healthy, reducing their expenses for salt and waste water discharge permits. Some veterinary authorities speculate that ultraviolet ozone disinfectant systems (widely used for ornamental fish) may play a prominent part in keeping the Tilapia healthy with recirculated water. A number of large, well-capitalized ventures in this area have failed. Managing both the biology and markets is complicated.

Irrigation Ditch or Pond Systems

These use irrigation ditches or farm ponds to raise fish. The basic requirement is to have a ditch or pond that retains water, possibly with an above-ground irrigation system (many irrigation systems use buried pipes with headers. Using this method, one can store one's water allotment in ponds or ditches, usually lined with bentonite clay.

In small systems the fish are often fed commercial fish food, and their waste products can help fertilize the fields. In larger ponds, the pond grows water plants and algae as fish food. Some of the most successful ponds grow introduced strains of plants, as well as introduced strains of fish.

Control of water quality is crucial. Fertilizing, clarifying and pH control of the water can increase yields substantially, as long as eutrophication is prevented and oxygen levels stay high.Yields can be low if the fish grow ill from electrolyte stress.

Cage System

Fish cages are placed in open water resources to contain and protect fish until they can be harvested. They can be constructed of a wide variety of components. Fishes are stocked in cages, artificially fed, and harvested when they reach market size.

A few advantages of fish farming with cages are that many types of waters can be used (rivers, lakes, filled quarries, etc.), many types of fishes can be raised, and fish farming can co-exist with sport fishing and other water uses. Cage farming of fishes in open seas are also gaining popularity. Concerns of disease, poaching, poor water quality, etc., lead some to believe that in general, pond systems are easier to manage and simpler to start. Also, past occurrances of cage-failures leading to escapes, have raised concern regarding the culture of non-native fish species in open-water cages. Even though the cage-industry has made numerous technological advances in cage construction in recent years, the concern for escapes remains valid.

Classic Fry Farming

Trout and other sport fish are often raised from eggs to fry or fingerlings and then trucked to streams and released. Normally, the fry are raised in long, shallow concrete tanks, fed with fresh stream water. The fry receive commercial fish food in pellets.

While not as efficient as the New Alchemists' method, it is also far simpler, and has been used for many years to stock streams with sport fish. European eel (Anguilla anguilla) aquaculturalists procure a limited supply of glass eels, juvenile stages of the European eel which swim north from the Sargasso Sea breeding grounds, for their farms. The European eel is threatened with extinction because of the excessive catch of glass eels by Spanish fishermen and overfishing of adult eels in, e.g., the Dutch IJsselmeer, Netherlands. As per 2005, no one has managed to breed the European eel in captivity.

ENVIRONMENTALLY FRIENDLY METHODS

An alternative to open ocean cage aquaculture, one in which the risk of environmental damage is substantially eliminated is through the use of a recirculating aquaculture system (RAS). A RAS is a series of culture tanks and filters where water is continuously recycled. To prevent the deterioration of water quality, the water is treated mechanically through the removal of particulate matter and biologically through the conversion of harmful accumulated chemicals into nontoxic ones.

Other treatments such as UV sterilization, ozonation, and oxygen injection are also utilized to maintain optimal water quality.

Through this system, many of the environmental drawbacks of aquaculture are minimized including escaped fish, water usage, and the introduction of harmful pollutants. The practices also increase efficiency of feed utilisation and growth by providing optimal water quality parameters.

One of the drawbacks to recirculating aquaculture systems is water exchange. However, the rate of water exchange can be reduced through aquaponics, such as the incorporation of hydroponically grown plants and denitrification. Both methods reduce the amount of nitrate in the water, and can potentially eliminate the need for water exchanges, closing the aquaculture system from the environment.

The amount of interaction between the aquaculture system and the environment can be measured through the cumulative feed burden (CFB kg/ M^3), which measures the amount of feed that goes into the RAS relative to the amount of water and waste discharged.

Because of its high capital and operating costs, RAS has generally been restricted to practices such as broodstock maturation, larval rearing, fingerling production, research animal production, SPF (specific pathogen free) animal production, and caviar and ornamental fish production. Although the use of RAS for other species is considered by many aquaculturalists to be impractical, there has been some limited successful implementation of this with high value product such as barramundi, sturgeon and live tilapia in the US.

AQUACULTURE FISH

Aquaculture is the cultivation of aquatic organisms. Unlike fishing, aquaculture, also known as aquafarming, implies the cultivation of aquatic populations under controlled conditions. Mariculture refers to aquaculture practiced in marine environments. Particular kinds of aquaculture include algaculture (the production of kelp/seaweed and other algae); fish farming; shrimp farming, shellfish farming, and the growing of cultured pearls.

Aquaculture has been used in China since circa 2500 BC. When the waters lowered after river floods, some fishes, namely carp, were held in artificial lakes. Their brood was later fed using nymphs and silkworm feces, while the fish themselves were eaten as a source of protein. By a fortunate genetic mutation, this early domestication of carp lead to the development of goldfish in the Tang Dynasty.

The Hawaiian people practiced aquaculture by constructing fish ponds. A remarkable example from ancient Hawaii is the construction of a fish pond, dating from at least 1,000 years ago, at Alekoko. According to legend, it was constructed by the mythical Menehune. The Japanese practiced cultivation of seaweed by providing bamboo poles and, later, nets and oyster shells to serve as anchoring surfaces for spores. The Romans often bred fish in ponds.

The practice of aquaculture gained prevalence in Europe during the middle Ages, since fish were scarce and thus expensive. However, improvements in transportation during the 19th century made fish easily available and inexpensive, even in inland areas, causing a decline in the practice.

The first North American fish hatchery was constructed on Dildo Island, Newfoundland Canada in 1889, it was the largest and most advanced in the world. Americans were rarely involved in aquaculture until the late 20th century, but California residents harvested wild kelp and made legal efforts to manage the supply starting circa 1900, later even producing it as a wartime resource.

In contrast to agriculture, the rise of aquaculture is a contemporary phenomenon. About 430 (97 per cent) of the aquatic species presently in culture have been domesticated since the start of the 20th century, and an estimated 106 aquatic species have been domesticated over the past decade. The domestication of an aquatic species typically involves about a decade of scientific research. Current success in the domestication of aquatic species results from the 20th century rise of knowledge on the basic biology of aquatic species and the lessons learned from past success and failure.

The stagnation in the world's fisheries and overexploitation of 20 to 30 per cent of marine fish species have provided additional impetus to domesticate marine species, just as overexploitation of land animals provided the impetus for the early domestication of land species

In the 1960s, the price of fish began to climb, as wild fish capture rates peaked and the human population continued to rise. Today, commercial aquaculture exists on an unprece-dented, huge scale.

In the 1980s, open-netcage salmon farming also expanded; this particular type of aquaculture technology remains a minor part of the production of farmed finfish worldwide, but possible negative impacts on wild stocks, which have come into question since the late 1990s, have caused it to become a major cause of controversy.

TYPES OF AQUACULTURE

ALGACULTURE

Algaculture is a form of aquaculture involving the farming of species of algae. The majority of algae that are intentionally cultivated fall into the category of microalgae, also referred to as phytoplankton, microphytes, or planktonic algae. Macroalgae, commonly know as seaweed, also have many commercial and industrial uses, but due to their size and the specific requirements of the environment in which they need to grow, they do not lend themselves as readily to cultivation on a large scale as microalgae and are most often harvested wild from the ocean.

Freshwater Prawn Farming

A freshwater prawn farm is an aquaculture business designed to raise and produce freshwater prawn or shrimp for human consumption. Freshwater prawn farming shares many characteristics with and many of the same problems as, marine shrimp farming.

Unique problems are introduced by the developmental life cycle of the main species. The global annual production of freshwater prawns (excluding crayfish and crabs) in 2003 was about 280,000 tons, of which China produced some 180,000 tons, followed by India and Thailand with some 35,000 tons each. Additionally, China produced about 370,000 tons of Chinese river crab.

Integrated Multi-trophic Aquaculture

Integrated Multi-Trophic Aquaculture (IMTA) is a practice in which the by-products (wastes) from one species are recycled to become inputs (fertilizers, food) for another. Fed aquaculture (e.g. fish, shrimp) is combined with inorganic extractive (e.g. seaweed) and organic extractive (e.g. shellfish) aquaculture to create balanced systems for environmental sustainability (biomitigation), economic stability (product diversification and risk reduction) and social acceptability (better management practices).

"Multi-Trophic" refers to the incorporation of species from different trophic or nutritional levels in the same system. This is one potential distinction from the age-old practice of aquatic polyculture, which could simply be the co-culture of different fish species from the same trophic level. In this case, these organisms may all share the same biological and chemical processes, with few synergistic benefits, which could potentially lead to significant shifts in the ecosystem.

Some traditional polyculture systems may, in fact, incorporate a greater diversity of species, occupying several niches, as extensive cultures (low intensity, low management) within the same pond. The "Integrated" in IMTA refers to the more intensive cultivation of the different species in proximity of each other, connected by nutrient and energy transfer through water, but not necessarily right at the same location.

Ideally, the biological and chemical processes in an IMTA system should balance. This is achieved through the appropriate selection and proportions of different species providing different ecosystem functions. The co-cultured species should be more than just biofilters; they should also be harvestable crops of commercial value.

A working IMTA system should result in greater production for the overall system, based on mutual benefits to the co-cultured species and improved ecosystem health, even if the individual production of some of the species is lower compared to what could be reached in monoculture practices over a short term period. Sometimes the more general term "Integrated Aquaculture" is used to describe the integration of monocultures through water transfer between organisms. For all intents and purposes however, the terms "IMTA" and "integrated aquaculture" differ primarily in their degree of descriptiveness.

These terms are sometimes interchanged. Aquaponics, fractionated aquaculture, IAAS (integrated agriculture-aquaculture systems), IPUAS (integrated peri-urban-aquaculture systems), and IFAS (integrated fisheries-aquaculture systems) may also be considered variations of the IMTA concept.

Mariculture

Mariculture is a specialized branch of aquaculture involving the cultivation of marine organisms for food and other products in the open ocean, an enclosed section of the ocean, or in tanks, ponds or raceways which are filled with seawater. An example of the latter is the farming of marine fish, prawns, or oysters in saltwater ponds. Non-food products produced by mariculture include: fish meal, nutrient agar, jewelries (e.g. cultured pearls), and cosmetics.

SHRIMP FARMING

A shrimp farm is an aquaculture business for the cultivation of marine shrimp for human consumption. Commercial shrimp farming began in the 1970s, and production grew steeply, particularly to match the market demands of the U.S., Japan and Western Europe. The total global production of farmed shrimp reached more than 1.6 million tonnes in 2003, representing a value of nearly 9,000 million U.S. dollars.

About 75 per cent of farmed shrimp is produced in Asia, in particular in China and Thailand. The other 25 per cent is produced mainly in Latin America, where Brazil is the largest producer. The largest exporting nation is Thailand.

Shrimp farming has changed from traditional, small-scale businesses in Southeast Asia into a global industry. Technological advances have led to growing shrimp at ever higher densities, and broodstock is shipped world-wide. Virtually all farmed shrimp are penaeids (i.e., shrimp of the family *Penaeidae*), and just two species of shrimp—the *Penaeus vannamei* (Pacific white shrimp) and the *Penaeus monodon* (giant tiger prawn)—account for roughly 80 per cent of all farmed shrimp.

These industrial monocultures are very susceptible to diseases, which have caused several regional wipe-outs of farm shrimp populations. Increasing ecological problems, repeated disease outbreaks, and pressure and criticism from both NGOs and consumer countries led to changes in the industry in the late 1990s and generally stronger regulation by governments. In 1999, a programme aimed at developing and promoting more sustainable farming practices was initiated, including governmental bodies, industry representatives, and environmental organizations.

CRAYFISH CULTURE

Among the best known and most highly esteemed crustaceans are the lobsters *(Homarus* spp.), but they are rivalled as a delicacy by their smaller freshwater counterparts, the crayfishes or crawfishes. (Not to be confused with the marine spiny lobsters, family Palinuridae, sometimes marketed as crayfish or crawfish).

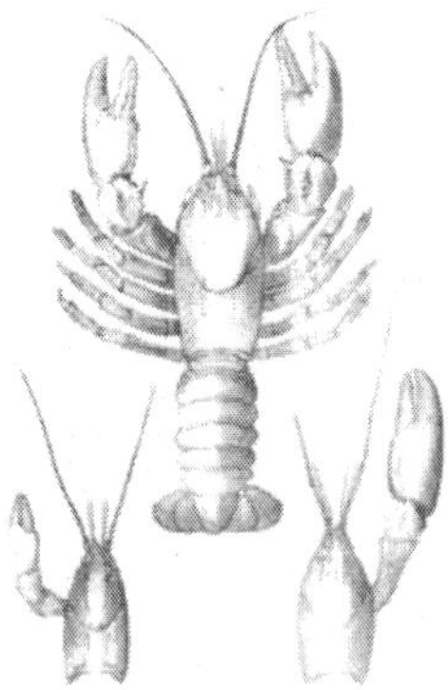

Fig. Cray fish

True crayfish comprise more than 300 species and are found on all the continents except Africa. Although crayfish are esteemed as a gourmet food in several European countries and are the primary source of protein for certain tribes in New Guinea, they are generally underutilized by man.

Crayfish have attained importance as a commercial food product in parts of Europe and the United States. These most enthusiastic consumers of crayfish are the French, and crayfish farms have been in operation in France since 1880. But the most important crayfish producing area is Louisiana, the only American state with a history of French culture. In a good year more than 800,000 kg of "wild" crayfish may be caught and marketed in Louisiana; during bad years the yield may be less than half that figure.

Fishery production is supplemented annually by 1.2 million kg produced on the 6000 to 7000 ha. of crayfish, farms in the state. Details are not available on techniques of crayfish culture in Europe; hence this report concentrates on practices in Louisiana.

NATURAL HISTORY OF THE PRINCIPAL CULTURED SPECIES

There are twenty-nine species of crayfish known to inhabit the waters of Louisiana, but only two are cultured. On most farms the dominant species is the red crayfish *(Procambarus clarkii),* but a few areas produce chiefly white crayfish *(Procambarus blandingi).* The natural history of the two species is similar. Mating occurs in open water in the late spring, when the water level in the Louisiana swamps is high. At that time the male crayfish deposits sperm in an external receptacle on the female. Shortly after the peak of the breeding season, female crayfish come out on shore and dig burrows near the water's edge. Since they are exposed to terrestrial predators at this time, areas well protected by emergent plants are preferred. Burrows are essentially vertical and usually 0.7 to 1.0 m deep, unless the water table is unusually high, in which case burrows half that depth may be found.

By the end of July, all adult females have constructed burrows, which are then usually occupied by a single male, along with the female. Each burrow is capped with a plug of dirt. Young crayfish and unpaired males also seek shelter in the mud or in naturally occurring holes during the summer, but they do not burrow. Some additional mating may occur in the burrows, but the crayfish are thought to be essentially inactive until September, when egg laying occurs.

Eggs and sperm are simultaneously released by the female and the fertilized eggs attached to the underside of her tail until hatching, which occurs in 14 to 21 days with red crayfish and 17 to 29 days with white crayfish. The adults die soon after hatching. An average hatch is about 400 red crayfish (maximum 700) and somewhat less for white crayfish. Fertilization and egg laying often occur in the burrow, but growth and survival of the 25-mm, free-swimming young is greatly enhanced if open water is available. In years when the water is low in the fall, hatching may occur in the burrow, in which case the young suffer from overcrowding and lack of food. Or the female may crawl off overland in search of water and die of dehydration or be captured by a predator.

Young crayfish usually seek shelter in dense plant growth near shore, and then move out to deeper water as mature. Both red and white crayfish are annual animals, and maturity may be reached in less than 6 months after hatching. Crabnasdyfish of all ages are omnivorous but prefer animal matter. Nevertheless, most crayfish, both in nature and in culture, must subsist on a predominantly vegetable diet.

Pond Culture in Louisiana

Crayfish culture is practiced in two types of water rice fields and artificial impoundments. The latter, though its large-scale use dates back little more than 20 years, now accounts for more than 80 per cent of the area devoted to crayfish farming.

Site Selection and Construction

As in most other forms of aquaculture, site selection is crucial in pond culture of crayfish. Due to the shallowness of crayfish ponds, it is essential that the land chosen be relatively flat. When the highest point in the pond bottom is covered by 0.3 m of water, depths of over 1 m should not occur in more than 25 per cent of the remaining area. It is not, however, of any particular value if the pond bottom is level. In fact, small elevations in the bottom have the favourable effect of increasing the area available for burrowing of early breeders.

Densely wooded areas are not favoured, since dense shrubbery and trees along pond banks hinder harvesting, shade out desirable aquatic plants, add leaves and debris which decay and reduce oxygen levels, and may keep temperature in the shallows below optimum.

It was in such environments, however, that pond culture of crayfish had its start in Louisiana and 2400 ha. of densely wooded swampland are still in use in an area just west of the Atchafalaya Basin, where conventional pond construction would be extremely difficult. These leveed swamplands are also heavily used by waterfowl, which are extensively hunted in Louisiana.

Soil quality surely has some effect on crayfish production but, in Louisiana at least, the primary consideration is that the soil hold water.

Water supply is of course important, but crayfish are very tolerant of naturally occurring physical and chemical conditions. Aspects of water quality which are of particular importance to the crayfish farmer are temperature and hardness. Optimum temperatures for red crayfish are thought to be 21 to 29°C, but growth is not drastically reduced until the temperature falls below 13°C.

Above 32°C, red crayfish burrow into the mud and become inactive. White crayfish do better at slightly lower temperatures and begin to burrow at temperature as low as 27°C. Growth is the primary reason for the culturist to be concerned with water temperature, since neither species is likely to suffer significant thermally induced mortality at temperatures common in Louisiana. For the same reason, dissolved oxygen concentration, though unlikely to be a cause of mortality, should be kept as high as possible.

Cultured crayfish have been observed to do well at pH levels as low as 5.8 and as high as 8.2. Crayfish in acid waters tend to have thinner shells. Soft water also results in thin, soft shells as well as poor growth and survival. It appears that the water in crayfish ponds should have a total hardness of at least 50 ppm, and that up to 200 ppm is desirable.

In recent years interest in crayfish farming in the coastal swamps of southern Louisiana has increased, and some attention has been paid to the effect of salinity on crayfish. Preliminary experiments indicate that crayfish will reproduce and grow fairly well at salinities of 6 to 10 per cent.

Tolerant as crayfish are to natural conditions, they are extremely sensitive to synthetic chemicals and other substances which may find their way into agricultural water supplies. Such commonly used compounds as Pyrethrum, creosote, orthodichlorobenzene, sodium cyanide, turpentine, orthocresole, cresylic acid, pine oil, nicotine, carbon bisulfide, phenothiazine, calcium cyanamid, and chlorinated hydrocarbon pesticides have all been shown to be toxic to crayfish. The last-named group presents particular problems because of the possible danger to human health from dosages sublethal for crayfish and because the crayfish farmer may find it difficult to prevent contamination of his water supply by pesticides used on nearby agricultural land. It is recommended that crayfish culturists located in the vicinity of intensive terrestrial farming operations alert neighbouring farmers and crop-dusting pilots of the location of their crayfish ponds and attempt to secure their cooperation in averting contamination. When spraying is going on nearby, pumping of water from streams and ditches into crayfish ponds should be suspended.

The danger of contamination by pesticides and other pollutants is greatly reduced if the water source is a deep well, but many farmers, for reasons of necessity or economy, use surface water. Whichever type of water is used, it is best if each unit of the farm is provided with its own pump. Units need not be discrete but may be portions of larger ponds separated by inside levees.

The smaller the units, the greater the ease of water quality control and general management. On the other hand, the more small units are built, the greater the construction cost per hectare of water surface. A large farm might consist of one or two 100-to 300-ha ponds broken up into 10 or so units. One-family farms as small as 8 ha exist but do not provide a full family income.

Detailed advice on levee construction may be obtained from the Soil Conservation Service, county agricultural agents, or commercial contractors. In general, levees should be high enough to keep out flood waters and wide enough to permit access to vehicles.

Water circulation has been found to be a deciding factor in the success or failure of many crayfish farms. To facilitate circulation it is necessary to provide not only an adequate inflow of water but good drainage as well. Drain pipes, which must be screened with 12-mm or smaller wire mesh, should be located as far away from pipes as possible to permit thorough mixing of water. Drains should be large enough to permit complete draining in 30 days.

Stocking

Although some ponds contain natural populations of crayfish, stocking is nevertheless necessary the first year, after which the population should be self-sustaining. The usual procedure is to stock ponds with adults in May to July, at which time crayfish are "tough" and prices low.

The pond should be flooded for at least 2 weeks before stocking and arrangements made to purchase freshly caught crayfish. Stock held in captivity for even so short a period as overnight should not be accepted. Large specimens, about 30 to 45/kg. are preferable.

Suggested stocking rates for red crayfish vary according to the amount of cover in the pond. At least 80 per cent of most lots of crayfish purchased for stocking in most parts of Louisiana will be red crayfish. If as occasionally happens, a stock of predominantly white crayfish is obtained. Crayfish of either species should be placed in the pond as soon as possible and kept cool and damp until that time. Predation on newly stocked crayfish will be reduced if they are released in densely vegetated areas or, if the pond is sparsely vegetated, in deep water far from shore.

The most important management techniques in crayfish culture involve manipulation of water level and quality. Each year ponds are drained in late June or early July, when the females have started burrowing. Slow draining is preferable, for fast draining will strand some crayfish which are not ready to burrow and expose them to predators. Similarly, slow draining allows young crayfish to seek hiding places for the summer.

Ponds are reflooded in September to ensure that the newly hatched young will have ample water. If growth rates are normal, they should be ready for harvest starting in late November or early December. The culturist should keep on an eye rainfall, which is the principal controlling factor on wild crayfish harvests, and time reflooding so that he can begin harvesting before the wild crop comes in. Early crayfish bring the best prices and may spell the difference between economic success and failure.

Ordinarily, harvesting lasts until the pond is drained the following summer. Between reflooding and the onset of the harvest season in ponds having heavy cover, water levels 50 cm or so lower than those recommended above can be maintained to furnish more shallow water for young crayfish.

During the harvesting season the water level should be kept fairly stable. This requires frequent pumping to replace water lost by evaporation. Pumping must also be resorted to replace deoxygenated water. Any abrupt reduction in harvest is usually a sign of deoxygenation or some such condition and calls for partial replacement of water.

Good Supply

Any animal food that cultured crayfish get is the result of natural production and their own foraging, but successful culturists encourage suitable food plants and discourage large, tough plants which interfere with harvesting, shade out smaller, more desirable plants, and are inedible by crayfish. Crayfish from ponds that do not contain adequate amounts of edible plants often have brown or black "fat" or livers, and tails which are not filled out, and are considered to be of inferior quality to well nourished crayfish with full tails and yellow livers.

Plants used as food and/ or cover in crayfish ponds must be capable of survival both while the pond is flooded and when it is dry. The most desired food and cover plant in crayfish ponds is alligator grass *(Alternanthera phylloxeroides).* It can be seeded into a pond by raking it out of ditches and scattering it in the pond during drawdown in June and July. Alligator grass may, on occasion, grow too thick and hinder harvesting. If harvesting is done by boat, boat trails may be disked or raked on the bottom during the summer. Although cattle should generally be kept out of crayfish culture areas, grazing them on the dry pond bottom may help retard alligator grass.

Water primrose *(Jussiaea* spp.) is now considered as good as or better than alligator grass and may be planted together with it. Water primrose has the advantages of not growing as thick as alligator grass and being more tolerant of cold weather. A number of other plants which occur naturally in crayfish ponds, including pondweeds *(Potamogeton* spp.), *Elodea,* and duckweed *(Lemna),* are fair food and/or cover plants and should not ordinarily be discouraged. It is certain that fertilization of dry ponds would benefit all of these and other plants, but techniques to encourage desired species have yet to be worked out.

Crayfish will also eat almost any soft terrestrial plant. Some farmers have planted sorghum or millet as feed for crayfish and claim increased yields.

It is likely that supplemental feeding with animal matter would increase production, but it is doubtful whether it would be economically feasible. Experiments with feeding young red crayfish at Auburn University showed that growth was better on a diet of *Elodea* and ground *Tilapia* than on either component alone. Research is currently being carried out at Louisiana State University with the aim of developing an economical, high-protein artificial feed for crayfish.

Problems

The most common pest plants in crayfish ponds are cattails *(Typha* spp.) They can usually be controlled only if caught early, at which time individual plants can be pulled out manually. If dense growths occur, disking and raking the pond immediately on drying may be helpful. Otherwise, it is necessary to call on state or federal agencies for assistance.

Outside help is also usually necessary to control water hyacinth *(Eichornia crassipes)* which, unchecked, may completely cover the surface of a pond. A number of emergent plants other than cattails are usually present but seldom reach problematical densities in ponds which are drained and dried annually.

Filamentous green algae, which occur principally during the winter, also may hinder harvesting operations. They are prone to die off suddenly in warm weather, creating a pollution problem. Partial removal of floating algal mats may be effected by vigourous circulation of the water.

Man is by no means the only animal which is fond is crayfish, and predation by fish, birds, raccoons, bullfrogs, snakes, turtles, salamanders, large water beetles, and others, is a problem confronting every crayfish culturist. The best protection against most of these predators is abundant cover in the pond, but fish, the most serious predators, may require additional measures. Annual draining and drying of course substantially reduce fish populations, but some of the worst predators, notably green sunfish *(Lepomis cyanellus),* bowfin *(Amia calva),* and bullheads *(Ictalurus* spp.), may survive in potholes until reflooding time. Rotenone at 2 to 3 ppm will destroy such fish. Great care should be exercised in its use, however, and a biologist consulted if possible, since 5 ppm of rotenone will kill some young crayfish.

It is standard practice to screen pond outlets to prevent fish from entering, but opinions differ as to the feasibility of screening inflow pipes. Many successful operators reason that most large fish are killed by the pump impellers and do not screen inflow pipes. In any event, the large 0.3-to 0.7-m pumps employed on most ponds move great volumes of water and make screening difficult.

It is recommended that, on new ponds supplied by surface waters, a temporary screen be installed on the inflow pipe and checked frequently during the first few days. If considerable numbers of large predatory fish are captured, a permanent screen should be employed. A suitable design consists of a cylinder of 25-cm weldwire about 1 m in diameter and 4 m long. Such a screen will allow young of predatory fish to enter the pond, but finer screens clog up very rapidly and require constant cleaning.

Harvesting

The most frequently used harvesting device is an 0.8 to 1.0 m long funnel shaped trap made of 21-cm mesh chicken wire and so constructed that it folds flat for hauling. Shorter traps are fully as effective for capturing crayfish, but dissolved oxygen concentrations may become very low in crowded, submerged traps, causing mortality. Long traps, however, may be propped up at an angle so that a small portion is above the surface of the water, and trapped crayfish can obtain air. Traps are commonly baited with fish heads or chunks of gizzard shad *(Dorosoma cepedianum),* which, is considered more effective than any other fish. Fresh fish may be scarce during the winter, and soybean cake, perforated cans of dog food, or almost any high-protein substance may be substituted. About 25 traps/ha are set out.

Most of the harvesting is done by professional fishermen hired by the culturist. This compounds the culturist's harvesting problems, since he is seldom able to maintain a crew of fishermen throughout a harvest season. But to realise maximum yields and fulfil contracts to buyers it is essential that harvesting be intensive throughout the season.

The problem is particularly acute during cold weather, when fishermen may be reluctant to run their traps frequently, and during the height of the wild crayfish season, when they may be able to make more money harvesting wild stocks on their own. The demand for labour is particularly great at the start of the season, in late November and early December, when many culturists prefer to harvest as heavily as possible to beat the peak of the wild crop, to utilize adults which have spawned and will die shortly and reduce the danger of crowding of the young crayfish. Crayfish have long been harvested as an incidental crop from rice fields in Louisiana, but it is only in the last 15 years that intensive cultivation has become common. Today, nearly 1000 ha. of rice fields are operated in 2-year rotation as crayfish ponds and pastures.

The approximate schedule for this sort of rotation is given. There are usually enough crayfish present in a Louisiana rice field to serve as brood stock. If not, adults may be stocked at 6 to 12 kg/ha during May. Some crayfish leave the drained fields in July and August, but most of them burrow into the moist soil.

Rice stubble resprouts about a week after harvest, at which time the fields are reflooded to a depth of 15 to 45 cm. Crayfish in rice fields feed mainly on rice stubble and various aquatic plants which find their way into the fields. Rice-crayfish farmers should not use rice seed treated with aldrin, which may be toxic to crayfish.

Harvesting crayfish from rice fields, although tedious, is easier and cheaper than in ponds. Due to the more regular bottom, nets may be employed. The usual type, called a drop net or umbrella net, consists of a rectangular piece of netting, each corner of which is connected by a wire or rod to a ring located over the centre of the net.

A pole or other lifting device is inserted in this ring and the net laid flat on the bottom. Crayfish are baited into it with beef pancreas or other offal and captured by lifting the pole. Ordinary farm labourers rather than professional fishermen may be hired to do this chore.

Growth to the minimum marketable size of 10 to 15 9 is rapid in either type of culture, requiring little more than 6 months. Larger 40-to 45-g crayfish, which bring the best prices, are harvested principally in the early part of the season and may be 8 to 14 months old. Well-managed ponds and rice fields commonly produce 400 to 700 kg/ha of crayfish, and yields of better than 1100 kg/ha have been achieved.

Marketing is by no means organized. Crayfish may be sold live, boiled; or as peeled tails to wholesale dealers, restaurants, or individuals. Production or peeled tails may be increased if a mechanical crayfish peeler, currently being developed at Louisiana State University, is perfected.

INDIAN CARPS

It has become a cliche that the people of India need protein. Even the most socially unaware persons have some grasp of the magnitude of India's problem. One can cite statistics *ad infinitum* to demonstrate that India is in a steadily worsening state of nutritional crisis, but however one defines the problem, it is clear that the solution, if there is ever to be one, will not come as any single panacea, but from effective population control coupled with increased and more efficient protein production by many means, including fish culture.

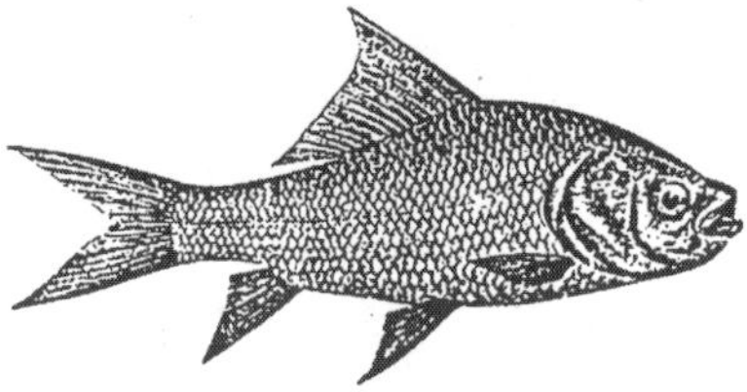

Fig. Indian crap

Fish culture shows promise in India, for there is a considerable demand for fish (total landings of fish in 2003 amounted to 7331 thousand metric tons), a number of native fish well suited to culture; a tradition of fish culture, and a great abundance of cultivable waters (an estimated 7,307,642 ha of fresh and brackish water).

As elsewhere in Asia, the most commonly cultured fish in India and Pakistan are members of the carp family (Cyprinidae). Indian cyprinids used in fish culture are popularly separated into two groups the most desirable species are referred to as major carps; the smallest less desirable species are called minor carps.

The minor carps persist in Indian fish culture largely because as fry they are extremely difficult to distinguish from the major carps, and are thus unintentionally stocked. Minor carps may also be deliberately stocked when stocks of major carps are scarce. In the south of India, in the states of Madras and Mysore, a third group, the Cauvery carps, largely supplant the major carps.

Indian carps have not yet become as popular for introduction and culture outside their native habitat as the common carp *(Cyprinus carpio)* or the Chinese carps, but the catla *(Catla catla)* has been introduced to Ceylon, experimentally cultured in Israel and the United States, and commercially raised in Malaysia. All the Indian major carps have been recommended for use in the Philippines.

The major carps comprise the catla, the rohu *(Labeo rohita),* and the mrigal *(Cirrhinus mrigala)*. Some authorities also include the calbasu *(Labeo calbasu).* These three or four species are often grown in polyculture, though their ecological niches are by no means as distinct and well defined as those of the Chinese carps, nor is Indian polyculture nearly as sophisticated as the ancient Chinese practice.

In general, the catla feeds on plankton and decayed macro vegetation on the surface and throughout the water column, the rohu is a column feeder on decayed vegetation with a taste for higher plants, the mrigal a bottom-feeding herbivore, and the calbasu a benthic omnivore. Certain of the Chinese carps have been experimentally included in Indian carp polyculture systems and in the future further combinations of Indian carps, Chinese carps, and other fish may be expected in both experimental and practical culture.

The niches of the Cauvery carps overlap; thus polyculture of these three species could probably be enhanced by introduction of other species. The species association for fish culture in southern India has already been diversified by the introduction of catla. The catla joins the fringe-lipped carp *(Labeo fimbriatus),* which consumes mostly filamentous algae and some zooplankton; the white carp *(Cirrhinus cirrhosa),* a plankton feeder with a preference for zooplankton, and the Cauvery carp *(Labeo kontius),* which competes with the fringe-lipped carp for filamentous algae, but also eats pieces of plants and detritus.

Minor carps present in fish ponds vary from region to region. Among those commonly found in one or another part of India are the reba *(Cirrhinus reba),* which feeds largely on phytoplankton and decayed plants; the nagendram fish *(Osteochilus thomassi),* a benthic and marginal feeder on filamentous algae and diatoms; and the sandkhol carp *(Thynnichthys sandkhol),* which consumes mostly phyto-plankton plus some zooplankton; also the bata *(Labeo bata),* the carnatic carp *(Barbus carnaticus), Barbus chola, Labeo boga, Labeo dyoechilus, Labeo gonius, Labeo nandina,* and *Barbus sarana.*

One more Indian cyprinid should be mentioned: the omnivorous copper mahseer *(Barbus hexagonolepis).* Not exactly a "minor" carp, the copper mahseer attains a maximum length of 90 cm. It is cultured in some parts of India but is not included in the traditional Indian polycultural scheme.

For many years a major obstacle to the development of culture of the Indian carps has been the inability of culturists to consistently breed them in captivity. All species spawn naturally in rivers and will not reproduce in standing water, although the major carps may be spawned in specially constructed reservoirs, or bunds, where there is enough current to approximate river conditions.

The copper mahseer will also spawn in running water ponds provided the temperature is about 20°C, or ripe fish may be hand stripped and the eggs fertilized artificially. Little or no effort has been expended on hand stripping or reservoir spawning of the Cauvery carps or minor carps.

The technique of bund spawning can be applied only where the drainage from an extensive catchment area can be accumulated in a natural depression. The low end of such a depression is blocked off by a strong embankment, so that during the rainy season intermittent streams will inundate the depression. Or a dam may be built in the uplands to concentrate runoff in one place, from which it is channelled into the bund.

During the dry season a minimum of 1.5 m of water is maintained in a pool perhaps 3000 m^2 in area, which is stocked with catla, rohu, mrigal, and calbasu in a 4: 2: 1: 1 ratio at a sex ratio of 2 males per female; males may be distinguished by the rough dorsal surface of the pectoral fin. The exact number of fish stocked depends on the extent of the spawning area; about 1 fish per 15 m^2 is appropriate. The pool is usually fished just before the start of the monsoon, and the numbers of each species and sex estimated, so that losses can be rectified by stocking. The spawning area is a flat piece of land on which grass is grown during the dry season so that the water will not be easily muddied when it is flooded. It may comprise the entire bund other than the permanent pool, or it may be just one corner of the bund. When the rains come and flood the spawning ground an outlet channel in the embankment is opened up so that water circulates through the bund. The breeders then move out of the pool and begin chasing in the shallow areas. Spawning usually occurs at night during the full moon or new moon. When spawning is completed, the eggs are collected by dragging seines, about 4 m long x 1.5 m deep, made of mosquito netting, over the spawning ground. Unfortunately a large percentage of the eggs is often destroyed by being trampled by the fishermen.

The eggs are placed in spawning pits dug as near as possible to the bund and supplied with bund water. These pits may be 1 to 1.3 m long, 0.6 m wide and 0.3 m deep and accommodate from 100,000 to 300,000 eggs. When the inlet to the pit is closed, the water temperature rises rapidly, accelerating hatching time to 12 to 18 hours, instead of the 36 hours that might be required at lower temperatures.

The rate of hatching in these pits is low due to bacterial decay and lack of aeration in the stagnant water. Some improvement may be effected by suspending a hatching net or "hapa" in a hatching pit. The hapa is rectangular in shape and constructed as a net within a net. The eggs are placed in the inner net, which is made of material just fine enough to hold them. The newly hatched larvae pass through into the outer net of fine cloth and are nursed there until they are 4 to 5 days old, by which time they have become fry.

The technique of bund spawning may be used only in areas where the topography permits. Induced spawning with the aid of pituitary injections is more generally applicable. The first attempts to apply this technique to the Indian carps were made in 1956, but it is only in the last five or six years that consistent success has been achieved. Today all four major carps and a number of other Indian cyprinids are bred in this manner. Success has been found to hinge on the quality of pituitary glands used and the correct dosage.

Pituitaries used in induction should be taken from fully mature, ripe, or freshly spawned fish of the same species as the fish being spawned, or a very closely related species. Dosage is variable, depending on the stage of maturity of the breeders. Since best results are obtained only with fully mature fish, only dosages for such fish will be described.

Breeders are usually selected from 1.5-to 5.0-kg, 2-to 4 year-old fish stocked in spawning ponds a few months prior to the breeding season, which usually coincides with the southwest monsoon; then segregated at maturity. Maturity of males is easily determined; fully mature specimens ooze milt when the abdomen is gently pressed. Selection of females is more difficult, but fish with soft, rounded, bulging abdomens and swollen, reddish vents are preferred. A catheter may also be used to assess ripeness. Intramuscular injections are made on the caudal peduncle or near the shoulder region Females are injected two or three times, while males receive only one injection. The first injection, of females only, consists of 2 to 3 mg of pituitary extract/kg of body weight, after which the sexes remain separated for 6 more hours. Then males are given a dose equal to the first injection administered to females, while the females receive a second dose of 5 to 8 mg/kg.

Following this injection, the fish are placed together in groups of three (two males and one female) in covered breeding hapas, 1.6 to 6.5 m2 in surface area and 0.9 m deep fixed on bamboo poles in the marginal waters of ponds. Spawning ordinarily takes place within 3 to 6 hours, but if after 10 to 12 hours no spawning occurs, females only are given a third, slightly higher dosage of pituitary extract.

If the water temperature is near optimum (about 27°C for most species), 60 to 100 per cent success may be expected by this method. Once the eggs have hardened, 8 to 10 hours after spawning, they are transferred, in batches of 75,000 to 1,000,000, to 1.6-m2, 0.9-m deep hatching hapas set in marginal waters of ponds, where they hatch in 15 to 18 hours at 27 to 31°C. Spawners may be sacrificed to obtain their pituitary glands.

As induced spawning of cyprinids becomes more prevalent in India, selective breeding and hybridization of Indian carps can assume greater importance. Experimental hybridization of Indian carps with each other and with Chinese carps has already been done, but the resulting offspring have for the most part been unpromising if not incapable of survival.

One exception is the hybrid catla x -rohu, which combines the wide body of the catla with the small head of the rohu. These traits would give it an advantage at the market wherever consumers do not eat the head of the fish, while at the same time effectively giving the consumer more nourishment for his money. The catla x rohu hybrid is fertile and has produced a healthy F2 generation.

Despite recent improvements in induced spawning methods, the main source of stock for culture in India and Pakistan continues to be collections of eggs, larvae, fry, and fingerlings taken from rivers.

COLLECTION AND HATCHING OF EGGS

Eggs are collected only from the Halda River in the Chittagong area of Bangladesh. Drifting fertilized eggs are captured 12 to 14 hours after spawning, which generally takes place within 3 weeks before or after each full moon from April through July. The precise time of peak spawning is determined by test collections at the start of the season.

The number of eggs caught in these collections enables the collectors to forecast the peak of spawning so that they can fish about that time or shortly after, when the maximum number of eggs is available. The net used in egg collecting is simple, consisting of no more than a rectangular piece of mosquito netting 11 to 12 m long and 2.7 m wide with a bamboo pole attached at each end. Such nets are usually operated by two men in a boat stationed at right angles to the current, but they may also be moored in the river or used by men wading.

Captured eggs may be placed in a compartment of the collecting boat for some time, even up to hatching, but these results in high mortality due to congestion and inadequate aeration. Better methods of holding and hatching eggs involve nets suspended from a bamboo framework, bamboo baskets lined with fine cloth and suspended in the river, or hatching pits dug in the river bank and supplied with water through a system of pipes.

Suspended nets are more often used as a temporary device to accommodate the eggs immediately upon collection, after which the eggs are transferred to pits or baskets, which are believed to produce healthier larvae. A typical hatching pit was 448 cm long x 244 cm wide x 46 cm deep and could accommodate 120 to 300 kg of eggs (900,000 to 2,200,000 eggs).

The hatching rate in all these traditional devices is usually 25 to 50 per cent. Somewhat better results may be obtained using hatching pits in conjunction with the double net device already described in connection with bund spawning.

NURSERY PONDS

Perennial ponds have the serious disadvantage of harbouring a host of predators, parasites, and competitors of carp fry, which seldom can be entirely eliminated. For this reason, in stocking perennial ponds it is best to scatter the fish, rather than dumping them all in at once. Stocking at night is also advisable, since most fry predators feed visually and the fry are especially vulnerable to predation during the period of acclimatization to a new environment. Captured and conditioned fingerlings may be placed directly into growing ponds to fatten them for consumption or stocked in rearing ponds for further intensive care.

Fry, however, must be nursed for 12 to 15 days to insure their health and satisfactory growth. Both seasonal and perennial ponds, ranging from 7 m2 to 0.5 ha in area and 0.9 to 3.6 m deep, are used as nurseries. Sample stocking rates for nursing ponds. Temporary ponds are somewhat better, but the ideal is a pond which can be drained at the discretion of the culturist.

In the relatively few modernized nursery units in India, the ponds are drained and sun-dried before use, the bottom is plowed, and a short-season crop of legu-minous plants is grown as a source of nitrogen for the soil. After harvesting the legume crop, the plants are plowed under and the bottom is leveled.

Fertilization may be carried out using various kinds of manure or refuse, mixed with oil cake, applied at 200 to 325 kg/ha before the pond is filled. Vegetable manures are usually applied in heaps, weighted down with stones rather than scattered about the pond, so as to reduce the danger of widespread deoxygenation. Animal manure is placed in bags or baskets for the same reason. Dilute sewage is used as a fertilizer in some areas, notably in the Bidyadhuri Spill in Calcutta. In using sewage, care should be taken that the dissolved oxygen concentration in the pond does not drop below 3 ppm. Inorganic fertilizers, though too expensive for use by many culturists, have been used on an experimental basis.

They would appear to be especially promising for use in arid areas, where organic manures are relatively scarce and more appropriately used as a source of humus for the soil. Unfortunately, most of the research on inorganic pond fertilization in India has involved use of the N-P-K mixtures, with little testing of the individual elements.

Use of N-P-K mixtures may be wasteful, as limnological data from many parts of India show an abundance of potassium. The nitrogen-fixing blue-green algae so prevalent in Indian fish ponds, along with the occasional practice of growing legumes in dry ponds, may make the addition of nitrogen superfluous as well.

There is some experimental evidence that phosphates alone are as effective as or better than N-P-K mixtures in many situations, but the lack of adequate controls in this research leaves room for doubt. In parts of Bangladesh a 3: 1 mixture of cow dung and superphosphate is applied to ponds at 555 kg/(ha)(year). Clearly, pond fertilization practices in India and Pakistan can be improved, through research and by greater appreciation of the fact that the waters of the region vary widely in chemical characteristics, so that each body should be treated individually rather than according to some custom or general formula. If the soil is acid, it may also be necessary before filling to treat the pond with lime until a *pH* of 8 or 9 is reached. Large perennial ponds are generally acid, but 300 to 500 kg/ha of lime are usually sufficient to correct this situation.

In ponds which cannot be drained and dried, pest control is a major preoccupation of the culturist. Predators include fish such as the snakeheads *(Ophicephalus* spp.), frogs, and several va-rieties of insect and aquatic birds. Competitors of the carp fry include many species of small, commercially undesirable fish as well as tadpoles. Fishing is usually inadequate to control predator and competitor fish, and poisoning must often be relied on. A number of pesticides of plant origin are used, but the most common is derris root powder. Applied at 4 to 6 ppm, it eliminates virtually all fish as well as killing some aquatic insects and tadpoles.

Larger doses may be more effective, but the prescribed dosage is preferred since it initially only stuns the fish, and edible predators can be salvaged by quickly transferring them to clear water. Control of insects can be achieved by applying an emulsion of 56 kg of mustard or coconut oil and 18 kg of washing soap per hectare. These poisons lose their effectiveness within 2 to 12 days of application, at which time manuring or stocking may be initiated. Ducks and geese must be fenced out of nursery ponds.

Weeds may also cause a problem by competing for nutrients with plankton and by providing hiding places for predators. Manual or mechanical removal is preferable to chemical control, but the possibility of ecological control using shading or herbivorous fishes, should be considered. A lengthier treatment of weed control methods follows in the discussion of growing ponds.

PROBLEMS OF INDIAN CARP CULTURE

Depending on the plant species to be dealt with, a diversity of methods, including poisoning, is used in weed control. Many of the most effective poisons are chlorinated hydrocarbons such as 2,4–, which are scarcely to be recommended for introduction into ecosystems or human food supplies. Other poisons, such as copper sulfate, sodium arsenite, and anhydrous ammonia gas, lack the long-term cumulative effects and the dangers to man presented by the chlorinated hydrocarbons, but they may be toxic to fish. In all but the most desperate cases therefore, it is safer (and usually cheaper) to use mechanical or ecological methods of plant control.

For floating weeds, outright manual removal is the best treatment. Emergent plants may be controlled by cutting of leaves at weekly intervals for about 6 to 8 weeks before fruiting. Marginal weeds may be controlled by many methods, including plowing under, grazing by livestock, burning during the dry season, or deepening the margin of the pond. Rooted sub-merged weeds may be removed, albeit with considerable labour, by netting with strong nets, by dragging chains, or with the aid of various types of fork and rake.

The commonest agents of ecological weed control are herbivorous fish. The most commonly used species for this purpose is the grass carp, which in addition to destroying or reducing most submerged and emergent plants and adding to pond fish production, contributes to the nour-ishment of other fish by dropping partially digested plant remains in its feces. All four of the major carps and a number of the minor carps, although unable to utilize fresh macrophytes as food, consume the partially decayed plant material in grass carp feces. However, grass carp are not a panacea for weed problems, since there are some plants, such as Eichornia and Salvinia, which they eat only with reluctance if at all.

There is also the possibility that if grass carp find their way into natural waterways, they may be destructive of the native flora. Another herbivore, already intro-duced to Ceylon for use in weed control, is the tawes (Barbus gonionotus). A number of species of tilapia, including Tilapia melanopleura, Tilapia mossambica, Tilapia nilotica, and Tilapia zillii, may effectively control certain types of weed. However, T. mossambica has on at least one occasion been found to depress total yield in polyculture involving catla.

Another ecological method of weed control is shading. Trees around the border of a pond may provide sufficient shade to discourage marginal weeds, but to control offshore weeds in most ponds the culturist must rely either on floating plants such as duckweed (Lemna), individuals of which are small enough not to interfere with netting and routine management operations, or on creation of an algal bloom. A heavy bloom can be induced by repeated application of N-P-K fertilizers, but this may be prohibitively expensive.

An unusual combination of effects results from application of superphosphate or urea at 50 ppm or more. These substances at that concentration are toxic to most submerged plants, but they act as fertilizers and produce an algal bloom as well. The desirability of an algal bloom must of course be evaluated in terms of the characteristics of the individual pond and the feeding habits of the fish present.

Other pond management practices used in India and Pakistan include the treatment of foul water by doses of 1.5 ppm or less of potassium permanganate, raking the pond bottom to release accumulated gases, and addition of minute amounts of alum to settle suspended or colloidal matter and reduce turbidity.

7

Dairy Cattle and Livestock

DAIRY CATTLE

Dairy cattle include replacement heifer calves and yearlings, dry cows, lactating cows, and breeding bulls used for research and teaching purposes related to milk production. The basic requirements for safeguarding the welfare of dairy cattle are an appropriate husbandry system that meets all essential needs of the animals, and high standards of handling.

FACILITIES AND ENVIRONMENT

Physical accommodations for dairy cattle should provide a relatively dry area for the animals to lie down in and be comfortable and should be conducive to cows lying for as many hours of the day as they desire. Recent work indicates that blood flow to the udder, which is related to the level of milk production, is substantially higher when a cow is lying than when a cow is standing. Criteria for a satisfactory environment for dairy cattle include thermal comfort, physical comfort, disease control, and freedom from fear.

Cattle can thrive in almost any region of the world if they are given ample shelter from excessive wind, solar radiation, and precipitation. Milk production declines as air temperature exceeds 24°C or falls below–12°C for Holstein and Brown Swiss cows or below–1°C for Jerseys. Heat stress affects the comfort of cattle more than does cold stress. Milk production can be increased during hot weather by the use of sunshades, sprinklers, misters, and other methods of cooling as well as by dietary alterations.

Temperatures that are consistently higher than body temperature can cause heat prostration of lactating cows, but additional energy intake and greater heat production by the cow can compensate for lower temperatures, even extremely low ones. Consideration also needs to be given to humidity levels and wind chill factors in determining effective environmental temperatures.

Adaptation to cold results in a thicker haircoat and more subcutaneous fat, which also reduces cold stress. Because dairy animals adapt well to cold climates, maintaining indoor air temperature equal to or slightly above outdoor air temperature is quite tolerable to housed animals. Coincidentally, providing the ventilation rate necessary to maintain this minimum temperature difference leads to good air quality.

Protecting the animal from extreme drafts, providing dry lying places that contribute to a dry, fluffy, erect haircoat, meeting the nutritional needs of the animal, and allowing the animal sufficient freedom of movement are essential. The newborn dairy calf has a lower critical temperature of 8 to 10°C. The intake of high-energy colostrum permits rapid adaptation to environmental temperatures as low as–23°C and as high as 35°C in dry, individual shelters with pens or in hutches.

Calves may be housed individually in outdoor hutches or inside buildings in bedded pens or elevated stalls. If calves are exposed to low temperatures, they should be provided with dry bedding and protected from drafts. Proper ventilation is critical in closed buildings with multiple animals. Hutches should be sanitized by cleaning, followed by moving the hutch to a different location or leaving the hutch vacant between calves.

In hot climates or during hot summer weather, calf hutches need to be environmentally modified or shaded to ensure that the calf does not cxperience severe heat stress. Housing and handling systems vary widcly, depending on the particular use of the cattle in research and teaching. Recommended facilities for dairy cattle range from fenced pastures, corrals, and exercise yards with shelters to insulated and ventilated barns with special equipment to restrain, isolate, and treat the cattle.

Generally, headlocks, corrals, and sunshades are used in warm semi-arid regions. Pastures and shelters are common in warm humid areas. Naturally ventilated barns with free stalls are used widely in both warm and cold regions. To a lesser extent, insulated and ventilated barns with tie stalls are used in colder climates. Early research showed an economic advantage in providing housing for dairy cows during the winter instead of leaving them outside.

During good weather, to enrich the environment and to improve overall health and well-being, cows should be moved if possible from indoor stalls into the barnyard, where they can groom themselves and one another stretch, sun themselves, exhibit estrous behaviour, and exercise. Exercise decreases the incidence of leg problems, mastitis, bloat, and calving-related disorders.

Keeping cows out of mud and manure increases their productivity and reduces endoparasitic and foot infections. Current trends and recommendations favour keeping dairy cows on unpaved dirt lots in the southwestern United States and on concrete in the northern United States throughout their productive lifetimes.

Concrete floors should have a surface texture that provides good footing but does not cause injury. The concrete surface should be rough but not abrasive, and the microsurface should be smooth enough to avoid abrading the feet of cattle. Scraping a new concrete surface tends to remove microprojections formed during finishing Data are limited on the long-term effects of intensive production systems; however, concern has been expressed about the comfort, well-being, behaviour, reproduction, and udder, foot, and leg health of cows kept continuously on concrete.

As a safeguard, cows should be moved from concrete to dirt lots or pasture, at least during the dry period. An additional advantage is that the rate of detection and duration of estrus are higher for cows on recommended dirt lots or pastures than for cows on concrete. Exercise during the dry period does not adversely affect milk production, but does result in cows that are fit. Forced exercise after parturition reduces energy intake and milk production; therefore, forced exercise is not recommended.

Area

Between and within breeds, ages, and body conditions, critical dimensions of dairy cattle vary less with weight than with age. Body length and hip width are relatively uniform across breeds at weights between 180 and 450 kg. More than 94% of the dairy cattle in the United States are Holsteins, and area recommendations for female calves and heifers are usually related to age groupings for Holsteins. Average normal growth curves relate heart girth and live weight to age.

The length of individual stalls should be a little longer than the length of the animal, defined as the distance between the pin bones and the front of the shoulders or between the pin bones and the brisket. For stanchions and tie stalls, stall width to length ratio should be at least 0.7. The width of free stalls should be twice the hip width. Dairy cows prefer larger, more comfortable stalls and use free stalls 9 to 14 h daily. Free-stall systems may be adapted for feeding trials utilizing electronic gates. Free stalls are recommended for dairy cattle used in teaching, extension, and research programmes throughout much of the United States.

Bedding

Resting dairy cattle should have a dry bed. Stalls ordinarily should have bedding to allow for cow comfort and to minimize exposure to dampness or fecal contamination. When handled properly, many fibrous and granular bedding materials may be used, including long or chopped straw, poor-quality hay, sand, sawdust, shavings, and rice hulls. Inorganic bedding materials provide an environment that is less conducive to the growth of mastitis pathogens. Sand bedding may also keep cows cooler than straw or sawdust.

Regional climate differences and diversity of bedding options should be considered when bedding materials are being selected. Bedding should be absorbent, free of toxic chemicals or residues that could injure animals or humans, and of a type not readily eaten by the animals. Bedding rate should be sufficient to keep the animals dry between additions or changes. Any permanent stall surfaces, including rubber mats, should be cushioned with dry bedding.

Bedding material added on top of the base absorbs moisture and collects manure tracked into the stall, adds resiliency, makes the stall more comfortable, and reduces the potential for injuries. Bedding mattresses over hard stall bases such as concrete or well-compacted earth can provide a satisfactory cushion. A bedding mattress consists of bedding material compacted to 8 to 10 cm and enclosed in a fabric. Shredded rubber may be used and is recommended as mattress filler. Small amounts of bedding on top of the mattress keep the surface dry and the cows clean.

Ventilation

Ventilation permeates all aspects of the animal environment. Most often, ventilation is associated with respiratory health of animals: the quality of the air that animals breathe directly influences animal health and disease. Nevertheless, ventilation —directly and indirectly—affects many other aspects of animal health as well. Good ventilation in the lying area of lactating animals helps to keep bedding dry, a factor in favour of good mammary health.

Good ventilation along alleys helps to keep walking surfaces dry, a condition that contributes to healthy feet and a reduction in falling accidents. Good ventilation may lead to greater productivity; for example, maintaining air movement in the area of the eating area makes animals more comfortable, which is especially important during hot weather as an aid to maintaining dry matter intake. A comfortable, well-ventilated lying area encourages animals to lie down, an important contribution to many aspects of animal health.

During ventilation, outside air is brought into a barn where it collects moisture, heat, and other contaminants, all produced by the animals. Air is then exhausted to the outside. Ventilation is an air exchange process—contaminated air inside the barn is exchanged for fresh outside air. To determine ventilation rates, we focus on the moisture content of the air, as measured by relative humidity. But moisture is only one aspect; ventilation removes other undesirable contaminants as well. Ventilation is truly a process of dilution.

Air moved through a barn actually serves to dilute the inside air and, very importantly, to dilute all of its components. Dilution reduces concentrations of moisture and heat. Dilution also reduces concentrations of airborne disease organisms, harmful gases and dust, and undesirable odours. The dilution rate of ventilation is often expressed in air changes per unit time. For example, a ventilation rate of 4 air changes per hour implies that the entire volume of the ventilated space is replaced every hour.

In fact, some of the air may bypass the occupied zone in the barn, depending upon geometry of the space, the design of diffusers controlling inlet air, and so on. Thus, the effectiveness of ventilation is not often 1.0, but something less, perhaps 0.65. When ventilation is reduced below recommended levels—usually in a misguided effort to warm the barn using animal heat—less moisture is removed.

Sometimes the consequences of the resulting moisture buildup and lack of proper ventilation—usually condensation—are masked by:

- Insulating the barn,
- Using a greenhouse effect,
- Providing supplemental heat, or
- Dehumidifying the inside air.

For example, adding heat to the air reduces relative humidity, without the need for air exchange. It is quite possible to have substantial quantities of moisture added to the air and, if accompanied by heating of the air, have the relative humidity remain in an acceptable range. Thus, if relative humidity is the only measure of air quality, air quality may be deemed satisfactory.

However, even though excess moisture may not be apparent, the reduced dilution does indeed result in increased concentrations of airborne disease organisms, harmful gases and dust, and undesirable odours. If these increases are ignored, animal health problems are inevitable. Underventilation in winter is one of the most serious threats to the environment of animals.

Improper design and improper management of the ventilation may be reasons that wintertime ventilation is lacking, compromising animal health. Problems are most likely during winter, spring, and fall, especially during rainy weather and warmer days coupled with cold nights. Specific recommendations for ventilation system design are available. In general, minimum ventilation is provided by a continuous rate in winter, amounting to at least 4 to 6 air changes per hour.

Summer ventilation rates may range up to and above 90 air changes per hour. Maintaining good air quality is a fundamental aspect of that healthy environment with ventilation providing the key. Through ventilation, the air inside the barn is continually diluted, ensuring that the air the animal breathes has low concentrations of all contaminants that threaten the animal's health.

TYPES OF HOUSING

In colder climates, stanchion and tie-stall barns have served well for herds ranging up to 50 or 60 milk cows. However, stall barns are labour intensive, both for milking and feeding. Comfort or tie stalls are preferred over stanchions. To avoid contamination of the teat and reproductive tract orifices, manure removal must be more regular and thorough when cows are housed in tie stalls. Cow trainers and gutter grates are recommended to ensure cleaner stalls and cows. Free-stall barns are a type of loose housing with one free stall recommended for each lactating cow. Depending upon provisions for feeding, different groups of cows can be fed differently according to their particular nutritional requirements. This has led to barn arrangements that permit division of milking herds into groups, usually by production.

One free stall is recommended for each lactating cow. The stall base and bedding provide a resilient bed for cow comfort and a clean, dry surface to reduce the incidence of mastitis. Because cows prefer to stand uphill, the stall base should be sloped forward 3 to 4% from rear to front. Commonly used materials for the base include concrete, clay, sand, and stone dust; hardwood planks tend to rot. Rubber tires, if not firmly imbedded, tend to become loose.

In an ideal free stall, the stall bed and partition should define the lying position of the cow and accommodate natural lying and rising behaviour. Proper free-stall care includes daily inspection and removal of wet bedding and manure, in addition to adding dry bedding periodically. Neglected free stalls with excessive moisture or accumulations of manure can lead to an increased incidence of mastitis.

For stalls with bases that must be replenished such as sand, an upward slope of the base towards the front should be maintained. This upward slope helps position cows more squarely in the stall when lying down, which contributes to cleaner stalls and cleaner cows. Free-stall hardware and other components should be kept in good repair.

Corrals should be scraped as needed and concrete alleys should be scraped or flushed regularly to clean them effectively. Feedbunk areas should be scraped regularly and any leftover feed removed. Shades and corrals should be designed to minimize areas of moisture and mud. Pastures must be managed to avoid disease transmission. Stocking rates should maximize production per head unless forage supplementation is provided or unless production per unit of pasture area is to be studied.

This strategy minimizes the stress that may result from overgrazing and minimizes ingestion of plants from areas immediately surrounding those areas contaminated with excreta, thereby reducing the challenge of potential pathogens and helminth parasites. Some pathogenic microbes may survive more than 6 mo in fecal deposits. Shade should be provided during hot weather.

Special Needs Areas

Cows with special needs are associated with greater risk and thus require special consideration with respect to facilities:

- *Preparturition*: Cows that are near the time of calving benefit from a clean, dry environment and access to an appropriate dirt lot for exercise. Feeding facili-ties should be provided to prepare cows for the high-energy ration they will receive upon entering the milking herd. Free-stall housing situated for frequent observation and proximity to the maternity area is a desirable option.

- *Maternity*: In preparation for calving, cows should be moved to individual pens that are separate from other animals, especially younger calves. The environment should be well ventilated, and the pens should be maintained to be clean, dry, and well bedded. Recommended pen size is 3.7 m × 3.7 m or 3 m × 4.3 m. The maternity pen should have a stanchion on one side for cow restraint. A concrete curb between each stall aids sanitation. Deep bedding should be used on concrete floors to prevent cows from slipping. Grooved concrete is also recommended. Provisions should exist for lifting downer cows. Devices to aid and promote standing include hip lifters, slings, inflatable bags, and warm water flotation systems. Pen location should permit access by a tractor or loader to allow removal of downed cows. All downed cows should be promptly examined by a veterinarian and handled in a humane and appropriate manner. Each pen should be provided with adequate feeding space and fresh, clean water. Depending on local conditions, a calving pen may not be necessary. Cows can calve in a pasture area with lighting situated for observation. A calving pasture should be well sodded and drained, should be large enough to allow cows to move away from others in the group before calving, and should contain an adequate sheltered area. Use of a pasture pen can eliminate footing and bedding problems associated with calving pens.
- *Removing Calf*: Dairy calves are normally removed from their dams as soon as possible following birth. The cow and calf are more difficult to separate after 3 d. Therefore, early removal is recommended. To prevent transmission of Johne's disease, follow the National Johne's Education Initiative control programme.
- *Postcalving*: A cow that has recently calved should be placed in a special area for frequent observation before rejoining the milking herd. Individual feed intake and milk production should be monitored to determine whether the cow is progressing normally. Milk must be withheld from shipment as required by regulations. Free stalls or large, well-bedded pens may be used in this special area. For a larger herd, a special hospital and maternity barn, possibly equipped with a pipeline or portable milker, could house cows in this management category as well as cows that are calving or that have other special needs.
- *Treatment*: A treatment area in the barn is recommended for confining cows for artificial insemination, pregnancy diagnosis, postpartum examination, sick cow examination, surgery, and for holding sick or injured animals until recovery.
- *Dry-off*: Cows recently dried off should be separated from the milking

herd for feeding purposes. Recommended medical treatments should be performed, and cows should be observed frequently to ensure normal progress.

Lighting

Lighting recommendations for dairy cattle housed in indoor environments are the same as those for beef cattle in intensive environments.

FEED AND WATER

Except as necessary for a particular research or teaching protocol, dairy cattle should be fed diets that have been formulated to meet their needs for maintenance, growth, production, and reproduction. Feed ingredients and finished feeds should be wholesome, carefully mixed, and stored and delivered to the cattle to minimize contamination or spoilage of feeds. To ensure freshness, feeds that are not consumed should be removed daily from feeders and mangers, especially highmoisture feeds such as silage.

Feed should be far enough from waterers to minimize wetting of feed. Space should be adequate for feed and water. Feeders or mangers should be designed with smooth surfaces for easy cleaning and increased feed consumption. The recommended linear space per cow at the feed bunk is 61 to 90 cm, which should allow every animal uninterrupted feeding. Feeder design should permit a natural head-down grazing posture to promote intake, improve digestive function, facilitate normal tooth wear, and decrease feed-wasting behaviour.

At least one water space or 61 cm of tank perimeter should be provided for every 15 to 20 cows in a group. At least 2 watering locations should be provided for each group of cows. Each cow in tie stalls and stanchions should have its own water bowl or drinking cup. All calves should consume colostrum in amounts of 8 to 10% of body weight within 4 to 5 h after birth always before milk is fed, and another 2 to 3 L within 24 h of birth for a 36-to 45-kg calf. Colostrum should be monitored with a colostrometer for quality.

Mixed highquality colostrum pooled from several cows can be better than low-quality colostrum from a particular dam. However, it is currently suggested that individual cows be tested for colostrum quality for the use of their colostrum alone with avoidance of colostrum from known disease-carrying cows. Proper handling and storage of the colostrum is essential.

Until calves can consume dry feed at an adequate rate, they should be fed liquid feed in amounts sufficient to provide needed nutrients at a rate up to 20% of body weight at birth per day until weaned. Water should be given at times other than when milk or milk replacer is fed to avoid possible interference with curd formation. However, this is not a problem with most milk replacers currently fed. Fresh water should be provided at all times. Replenishment of water should follow milk or milk feeding by at least 15 minutes. Calves being raised as replacement heifers or for beef should be fed enough dry feed with sufficient fibre preweaning to stimulate normal rumen development. Calf research guidelines have been reported that permit uniformity in measuring and reporting experimental data.

Water intake affects consumption of dry matter and is itself influenced by individual behaviour, breed, production rate, type and amount of feed consumed, water temperature, environmental temperature, atmospheric vapour pressure, water quality, and physical facility arrangement. Nonlactating cows consume 3 to 15 kg of water/kg of dry matter consumed, depending on environmental temperature.

Lactating cows consume 2 to 3 kg of water/kg of milk produced plus that required for maintenance. Water should be available at all times; it should be checked daily for cleanliness and monitored regularly to ensure that it is free of contaminants that could potentially put zoonotic agents into the human food chain. Water sources should be readily accessible to all stock. Underfoot surroundings in watering areas should be dry and firm. Cattle should not be able to wade in drinking water.

SOCIAL ENVIRONMENT OF HUSBANDRY

Dairy cattle are social animals that exist within a herd structure and follow a leader. Cows exhibit wide differences in temperament, and their behaviour is determined by inheritance, physiology, prior experience, and training. Cattle under duress may bellow, butt, or kick; however, cows are normally quiet and thrive on gentle treatment by handlers. Cows learn to discriminate among people and react positively to pleasant handling.

Aversive handling leads to more incidents during handling and transport for calves than positive handling. Similarly, heifers and cows exposed to aversive handling took longer to traverse and more force to move than those handled more gently. Although the presence of an aversive handler reduced kicking during udder preparation, residual milk was 70% greater than for the control milkings.

Cows have higher milk yields if handlers touch, talk to, and interact with them frequently. Cows should have visual contact with one another and with animal care personnel. Handling procedures are more stressful for isolated cattle; therefore, attempts should be made to have several cows together during medical treatment, artificial insemination, or when cows are being moved from one group to another.

This was verified by increased heart rate, hypothalamic-pituitary-adrenocortical axis activity, and vocalizations. Pain sensitivity is reduced during isolation, suggesting a stress-induced analgesia. Care should be taken to minimize the negative impact of moving cows to new groups by avoiding frequent regrouping and by always moving more than one animal at a time to a new group. The use of a trainer cow can have a positive impact on adjustment to feedlot environments where many heifers are raised.

However, dairy calves had few indicators that repeated regrouping and relocations stressed calves. Aggression was rare and the calves seemed to habituate to the repeated mixing. Calves from larger groups after weaning had fewer incidences of displacement of other calves from the feed barrier, were more active, and had more positive interactions with familiar calves. Calves, like cows, prefer familiar calves to unfamiliar calves during stressful situations, and a familiar companion calf improved cows' reaction to separation.

Social status can affect health issues such as lameness. Low-ranking cows spent more time standing and standing half in cubicles than did middleand high-ranking cows. Standing half in cubicles correlated positively with the number of soft tissue lesions related to lameness. Dairy cattle have traditionally been kept in groups of 40 to 100 cows, although specific research protocols may require smaller or larger group sizes.

Variation in group size—small, medium and large—does not cause a problem *per se*. Expansion to a larger herd size, however, can affect management decisions because overcrowding with an insufficient number of headlocks or inadequate manger space per cow, irregular or infrequent feeding, and excessive walking distance to and from the milking parlor have a greater impact on behaviour and well-being than does group size. Cattle of all ages are gregarious.

Socially isolated cattle show clear signs of stress: increased heart rate, vocalization, defecation/urination, and cortisol levels. In addition, there are benefits of housing cattle together. For example, pairs of calves are more likely to play than isolated calves, a behaviour thought to be associated with positive welfare. Young calves should be kept in groups from 2 to 7 animals in order for animals to benefit from social contact, but larger groupings are associated with health problems and morbidity.

Management of resources is an important part of reducing aggression and other problems, such as cross sucking, in groups of animals. Adult dairy cattle should have 1 freestall/cow to reduce competition. Similarly, dairy cattle with more space at the feedbunk engage in fewer aggressive interactions and the reduction of competitive behaviour associated with more feeder space is particularly marked in post-and-rail feeder design.

Cross sucking in calves is an undesirable behaviour performed in groups. Calves are typically fed 10% body weight during the milk-fed period and there is clear evidence that this feeding level is insufficient. A combination of slower milk flow, hay feeding, and access to a nonnutritive artificial teat are also recommended to reduce cross sucking. Providing additional objects for oral manipulation, such as tires, has also been shown to reduce other problems such as stereotypic tongue rolling in calves.

Restraint and Handling

Vaccination schedules that are appropriate for the location and dynamics of the individual herd should be established with the advice of the attending veterinarian. Certain dairy cattle behaviours put at risk the health and well-being of herdmates as well as the humans handling the cattle. These behaviours can be reduced or modified by implementing principles of low-stress handling and restraint that include appropriate movement of people, well-designed facilities, optimal lighting, nonslip flooring, and smooth, quiet restraint devices.

Stanchions, head gates, and squeeze chutes can be modified to function optimally, but acclimation and positive reinforcement by individuals trained in low-stress handling can minimize the need for additional restraint by halters, rope, tail hold, and nose tongs. Hobbles and casting ropes should be used selectively and only when necessary. Chemical sedation is always preferable to excessive use of force or application of electrical prods. Information about calving management is given by Albright and Grandin.

First-calf heifers should be bred to calving-ease bulls and be of appropriate stature and body condition to minimize the chances of dystocia or the need for calving assistance. Optimal calving conditions in a clean, quiet environment with employees appropriately trained to follow calving protocols will result in more live calves and fewer calving injuries and illness.

Calving injuries should be assessed immediately so that appropriate footing is provided and proper treatment is implemented. Cows that are unable to stand should be moved to a soft-bedded pack and examined by a veterinarian within 2 to 4 h of calving. Calves require special handling and care from the time they are born. Colostrum should be fed or ingested within the first 5 h after birth always before milk is fed. Between 1.89 L and 3.79 L of colostrum are necessary to impart adequate immunity to the calf. In the absence of colostrum, a colostrum replacement product that delivers at least 125 g of immunoglobulin should be given by bottle, bucket, or tube feeder. Colostrum is rich in nutrients and provides the calf with vital immunoglobulins and other important immune factors. Clean navels can be dipped in a dilute chlorhexidine solution as soon as possible after birth. Good nutrition as supplied by a combination of milk starter grain, and fresh water along with proper handling and close monitoring starts a calf on its way towards a healthy life.

STANDARD AGRICULTURAL PRACTICES

All animals should be individually identified. Heifer calves should have supernumerary teats removed at an early age. Removal may be performed in the first 3 mo of life with a scalpel or sharp scissors. Older calves and heifers close to calving that have supernumerary teats should be examined by a qualified person. The removal of extra teats at this advanced age is necessary if they will later disrupt the milking process or be at risk of becoming infected. If so, they can be removed with proper restraint and use of appropriate anesthesia by a qualified and trained person. Milking procedures should follow National Mastitis Council guidelines. Routine breeding programmes should include housing and handling facilities that allow for effective implementation of artificial insemination programmes. Castration may be performed on male calves.

Foot Care

Lameness in dairy cattle is a major source of economic loss to the farmer and a serious cause of pain and discomfort to the cow. It is perhaps the most important condition affecting the welfare of cows on dairy farms. Lame cows suffer lowered milk production and reduced fertility, and are culled at 2 to 4 times the rate of healthy control cows.

The pain associated with lameness results in changes in the animal's gait that include:

- Arching of the back;
- Shortening of the stride length on the affected limb;
- Sinking of the dew claws on the unaffected contralateral limb;
- Head bob in a vertical plane;
- Reduction in walking speed, and frequent stops; and
- Swinging the affected limb in or out depending on the location of the painful lesion.

These alterations can be used to provide a locomotion score for each animal, and the most commonly used system in North America utilizes a 5-point system of scoring where 1 is nonlame and 5 is severely lame. Herd workers should be taught how to score locomotion so that they can identify cows with scores >2 for treatment by an attending veterinarian or hoof-trimmer.

Around 85% of lameness in dairy cattle is associated with lesions in the rear feet, particularly the outer claw, because of the overgrowth of horn resulting from the redistribution of weight as the cow walks on hard concrete surfaces, with a large udder occupying the space between her rear legs. This overgrowth of the outer claw may be removed and the weight transferred equally between the inner and outer claw by regular hoof-trimming.

Trimming to restore a normal toe length along the dorsal hoof wall of around 75 mm for mature Holstein cattle, combined with balancing weight between the inner and outer claw, lasts around 4 mo on average. Therefore, it is recommended that cattle be trimmed at 6-mo intervals, typically at the time of dry off and in mid-lactation around 90 to 150 d in milk. Some cows with pre-existing hoof disease may require attention more frequently.

Hoof lesions causing lameness may be broadly classified into 2 groups:

- Infectious and
- Claw horn.

Infectious lesions include digital dermatitis, interdigital phlegmon, and heel horn erosion. These lesions are associated with poor feet and leg hygiene and are a particular problem in free-stall environments, where the cow is exposed to alleyways contaminated with wet manure when she is not occupying a stall. Putative agents such as several species of *Treponema* and *Fusobacterium necrophorum* are involved in the pathogenesis of these conditions, but hydropic maceration of the skin of the interdigital space appears to be a prerequisite for the development of disease.

Infectious causes of lameness are controlled by improving leg hygiene by removing manure from the walkways and by the use of a topical antibacterial administered either directly to the lesion by a hand-held spray or via a footbath. The frequency of foot bathing is dependent on the degree of manure contamination of the cows, and a variety of chemicals are available for use, such as copper sulfate, zinc sulfate, and formalin.

Use of any of these chemicals should be done under veterinary direction. Claw horn lesions include sole hemorrhage, sole ulcer, toe and heel ulcer, and white line disease. These are clinical signs on the surface of the claw that represent the result of several possible causative pathways. Sinking of the third phalanx within the claw horn capsule, due to a breakdown in the connective tissue of the suspensory apparatus, may be caused by hormonal changes at calving time and nutritional events such as subacute ruminal acidosis.

Sinking of the third phalanx compresses the corium below, interrupting the flow of blood and nutrients to the cells responsible for horn growth. As a result, a defect develops that becomes apparent several months later as the sole horn continues to grow. Excessive removal of sole horn, either through poor hoof trimming or due to excessive wear from walking long distances on rough concrete will also contribute to lesion development.

Flooring surfaces should be nonslip, avoid excessive trauma to the claw surface and be dry. Concrete should be grooved to improve traction; a pattern that utilizes parallel grooves 3/4 inch wide and deep, spaced 3 inches on center appears to provide a good compromise between sufficient traction to reduce injury while limiting the amount of wear.

For transfer lanes between milking centers and the living accommodation, a 1-m wide strip of rubber flooring has been used successfully to reduce trauma and wear, and rubber flooring has been used in parlor holding areas to provide cushion for cows that have to stand for long periods of time. The severity of the claw horn lesions that develop is influenced by the time spent standing each day, which results in increased loading of the claw and increased compression of the tissues below the third phalanx.

Time spent standing may be increased by:

- Poor stall designs that fail to provide surface cushion, room to lunge, and sufficient resting area;

- *Overstocking*: Providing fewer usable stalls than there are cows in a pen;
- Excessively prolonged milking times;
- Time spent locked up away from the stalls for management tasks; and
- *Heat Stress:* Cows may stand more in an attempt to cool off.

In addition, lame cows struggle to use stalls with hard surfaces because the act of rising and lying down becomes more challenging due to foot pain. These cows stand more in the stall and fail to gain adequate rest for lesion healing. For this reason, deep sand-bedded stalls provide the gold standard in cow comfort. If sand stalls are unavailable, lame cows should be treated and returned to a bedded pack area for rest and recuperation until normal ambulation returns.

Failure to identify a claw horn lesion early in its course may result in deep digital sepsis. This is a complication caused by infection of the deeper structures of the claw, including the distal interphalangeal joint and tendon sheaths. Such animals are usually severely lame and require euthanasia or extensive surgery. Seeking veterinary assistance is recommended for individual cows that show signs of lameness or if a significant lameness issue exists for the herd.

Disbudding and Dehorning

A review of horn anatomy and growth and dehorning and disbudding of cattle was provided by the AVMA. The AVMA also provides guidance on use of sedation, anesthesia, and analgesia and alternatives to horn removal. The AVMA policy on dehorning/disbudding should be followed. Calves should be observed closely for 1 to 2 h following dehorning. No food or water should be offered until the sedation is completely worn off or reversed. Persistence of a depressed attitude, head-pressing, or an abnormal head tilt for more than 2 h should result in a complete examination.

Tail Docking

The bovine tail has several physiological and behavioural functions including dissipation of heat, and facilitation of visual communication among cattle and with human caretakers; the tail often serves as a primary mechanism of fly control. Removal of the lower portion of a cow's tail is commonly referred to as "tail docking" and the use of tail docking as a routine dairy farm management tool apparently originated in New Zealand.

New Zealand farmers responding to a 1999 survey believed removal of tails resulted in faster milking, reduced risks to the operator, and reduced rates of mastitis. Similar unsubstantiated claims have been made for the US dairy industry. Several European countries, some Australian states, and California have prohibited tail docking. Both the Canadian and American veterinary medical associations have policy statements that oppose the practice of tail docking for routine management of dairy cattle.

The policy statement of the *American Association of Bovine Practitioners* (AABP) indicates that scientific evidence to support tail docking is lacking and recommends that "if it is deemed necessary for proper care and management of production animals in certain conditions, veterinarians should counsel clients on proper procedures, benefits, and risks.

Scientific studies have been performed to evaluate both the potentially negative and positive aspects of tail docking. Important welfare issues that have been evaluated have included pain caused by tail docking, changes in fly avoidance behaviour, immune responses, and changes in levels of circulating plasma cortisol. Experiments that have been performed on both calves and preparturient heifers have consistently concluded that the process of tail docking does not induce significant acute or chronic changes in plasma cortisol or other selected physiological measures. Modest changes in general behaviour of calves that have been docked using rubber rings or cautery irons have been reported but these changes have not been associated with significant differences in normal feeding, ruminating, or grooming behaviours. Likewise, few significant differences in general behaviour of docked preparturient heifers have been noted.

However, greater changes have been observed in tail surface temperatures of docked heifers compared with heifers with intact tails, indicating that heifers may experience chronic pain similar to the phantom pain reported by human amputees. Research has demonstrated that tail-docked heifers flick their tails more often and are forced to use alternative behaviours such as rear leg stomps, feed tossing, and head turning to try to rid themselves of flies.

More flies settle on tail-docked cows than on intact cows, and the proportion of flies settling on the rear of the cow increases as tail length decreases. In another study, there were no significant differences in the numbers of stable flies found on the front legs of cows but docked cows had nearly twice as many flies on their rear legs compared with those with intact tails.

Fly avoidance behaviours were increased in the docked animals, whereas tail swinging was increased in the control animals. Foot stamping was identified only in docked animals and, overall, fly numbers and fly avoidance behaviours were increased in docked animals. Researchers have been unable to identify improvements in udder health or udder cleanliness for animals in commercial herds that have docked tails. In one study, the effect of tail docking on cow cleanliness and *somatic cell counts* (SCC) was evaluated over an 8-wk period for lactating cows that were housed in a free-stall facility.

Standardized cleanliness scores obtained from the rump, midline of the back, or rear udder were not significantly different between docked and intact animals nor was there any significant difference in SCC or the number of teats containing obvious debris. In another study, SCC, occurrence of *intramammary infections* (IMI), and udder and leg hygiene scores were evaluated over an 8-mo period for lactating dairy cows that had been blocked by farm and randomly allocated to tail-docked or control groups.

No significant differences were found in SCC or udder and leg hygiene scores. The prevalence of contagious, environmental, and minor pathogens was not significantly different between cows with docked or intact tails. Although current studies do not indicate that the process of tail docking modifies physiological indicators of stress, several studies have documented changes in fly avoidance behaviour and recent research has suggested that docked tails have enhanced sensitivity to heat.

No benefits to cattle welfare have been associated with tail docking. The routine use of tail docking in research or teaching herds should be discouraged, and alternatives to tail docking are recommended when appropriate. Any use of tail docking, other than for medical reasons, should be reviewed and approved by the IACUC.

SPECIAL CONSIDERATIONS OF MILKING MACHINE AND UDDER SANITATION

The milking facility should have a programme for regular maintenance of milking machines and follow the recommended mastitis control programme of the National Mastitis Council. Appropriate equipment and competent personnel should be available for milking. Personnel responsible for milking should receive ongoing training about proper milking procedures as the frequency of training has been associated with adequacy of milking performance.

Animal care facilities should be designed and operated to standards meeting or exceeding those of grade A dairies as defined in the Pasteurized Milk Ordinance. Areas where milking takes place must be designed and constructed in accordance with the 3-A Sanitary Standards Inc. Accepted Practices. Cows should be maintained in housing areas that provide for adequate hygiene to ensure that udders are visibly clean. Cows should be milked on a regular schedule that is appropriate for the goals of the herd or specific research project.

Written operating procedures should be established to control potential contamination of milk with antibiotics or other pharmaceutical agents. Antimicrobial treatments should be administered based on approved defined protocols. All extra-label treatments must be administered under the supervision of a veterinarian that has an appropriate veterinary-client-parent relationship.

Milking machine and udder sanitation are vital to an effective preventive programme against mastitis and follow guidelines as established by the NMC. Care should be used to minimize the excessive use of water before and during udder preparation. Emphasis should be placed on ensuring that cows enter the milking parlor with clean, dry teats. Udders, especially teat ends, should be clean and dry when teat cups are applied for milking.

The removal of foremilk before teat disinfection is encouraged as a means to detect mild cases of clinical mastitis. Teat sanitation, predipping, and wiping immediately before machine attachment reduce udder infection caused by environmental pathogens. Postmilking disinfection of teats is an essential management practice that greatly reduces the incidence of mastitis. Milkers handling cows should pay meticulous attention to their own personal hygiene and wash their hands thoroughly before milking and frequently during milking.

The use of clean nitrile or latex gloves during milking is highly encouraged to prevent contamination of the udder. Cows with subclinical cases of contagious mastitis should be milked last to reduce the spread of mastitis throughout the herd. Udder hair removal is recommended as a means to improve milking hygiene and udder health. Cleaning of milk handling equipment is accomplished by a combination of chemical, thermal and physical processes and cleaning regimens should be designed to meet appropriate regulatory standards.

Recommended cleaning and sanitizing practices are a balance between the cleaning temperatures, cleaning chemical concentration, contact time and mechanical action. Effective cleaning programmes for milking machines include use of hot water; use of disinfectant solutions and other chemical agents effective for removing mineral, milk fat, and protein deposits from equipment between milkings; disinfection of teat cups between cows; and flushing of teat cups with warm water, cold water, boiling water, or chemical disinfectant solution.

The most common routine in the United States is a combination of prerinse, alkaline detergent, acid rinse and premilking sanitize. Very small herds may utilize manual cleaning and disinfecting that involves hand-cleaning of some or all of the milk harvesting and storage equipment. Small to medium herds commonly use automatic washing equipment.

This equipment will automatically mix the chemicals with the appropriate water volume and temperature and circulate these solutions through the milking machine. On large farms, an attendant may be present to mix chemical solutions and operate valves for circulation. The effectiveness of milking system cleaning can be evaluated by examination of standard plate counts and laboratory pasteurized counts performed on bulk tank milk samples.

Stray Voltage

The term stray voltage describes a special case of voltage that develops on grounded metal objects on farms. If this voltage reaches sufficient levels, animals coming into contact with grounded devices may receive a mild electrical shock that can cause a behavioural response. At voltage levels that are just perceptible to the animal, behaviours indicative of perception such as flinches may result, with little change in normal routines.

Studies by numerous independent research groups in several countries are in agreement that the most sensitive cows begin to react to 50 or 60 Hz electrical current of 2 mA applied from muzzle to hooves or from hoof to hoof. This corresponds to a contact voltage level of about 1 V. As the voltage and current is increased, a greater percentage of cows will react with behavioural responses becoming more pronounced. Numerous studies have documented avoidance behaviours at levels above the first reaction threshold.

The median avoidance threshold for 50 or 60 Hz current flowing through a cow is about 8 mA. Even when the threshold is exceeded not all cows would be expected to show a behavioural response but as the voltage increases, signs in a herd would be expected to be more widespread and uniform. The scientific evidence strongly suggests there is no relationship between behavioural responses to stray voltage and physiological or hormonal responses.

There is no apparent relationship among behavioural modifications, milk production, and animal health. The only studies that have documented adverse effects of voltage and current on cows had both sufficient current applied to cause aversion and forced exposures. It is typical for voltage levels to vary considerably at different locations on a farm. Decreased water and feed intake or undesired behaviours will result only if current levels are sufficient to produce aversion at locations that are critical to daily animal activity.

These locations include feeders, waterers, and milking areas. Controlled research has shown that if an aversive voltage was administered to a water bowl once per second, water intake was reduced. However, when the same voltage was applied once every 10 min and once per day, no reduction in water intake was observed. If an aversive current occurs only a few times per day, it is not likely to have an adverse effect on cow behaviour.

The more often an aversive voltage occurs in areas critical to cows' normal feeding, drinking, or resting, the more likely it is to affect cows. No one sign is pathognomonic; a variety of signs has been reported in cows exposed to different levels of voltage. Documented signs are behavioural changes and decreased drinks of water per day and length of time per drink. The amount of water consumed may not be affected even when behavioural modification occurs.

Intermittent periods of poor performance, poor milk letdown, and incomplete or uneven milk-out, abnormal behaviour during milking, increased milking time, refusal of feed or water, increased SCC in milk, and increased mastitis are signs often attributed by farmers to stray voltage; however, none of these signs were evident in numerous controlled studies. These signs are often caused by other factors such as abusive cow handling, faulty milking machine, poor milking techniques and hygiene, and nutritional deficiencies.

Therefore, animal behaviour or other symptoms cannot be used to diagnose stray voltage problems. The only way to determine if stray voltage is a potential cause of abnormal behaviours or poor performance is to perform electrical testing. A thorough investigation of the entire production unit should be conducted to determine other sources of problems. Electrical systems should comply with wiring codes and standards at all times to protect both animals and people.

Whenever suggestive signs cannot be attributed to other causes, measurements should be taken to determine if a voltage potential exists, and the results recorded for future comparisons. A diagnostic confirmation of stray voltage must include a competent electrical measurement indicating at least 2 to 4 V between 2 points that a cow might contact, with some cows should exhibiting avoidance behaviours at this location. Voltage levels may need to be monitored at different times of the day and on different days because the threshold level may be exceeded intermittently.

All voltage readings should be made with a 500 to 1,000 Ω resistor across the 2 measuring leads to the cow contact points in addition to open circuit measurements. Readings without the use of a shunt resistor are meaningless. Although the resistance of cow and human tissues is similar, the contact resistance is generally lower for cows than for humans, particularly if cows are in a wet environment. The resistance of a cow's body plus the contact resistance with the floor is commonly estimated as 500 Ω. This is a reasonable value for a cow standing on a wet floor.

Cows standing on a dry surface will typically produce 1,000 Ω resistance or higher. Cows standing or lying on dry bedding will have a resistance many times higher than this. The resistance of a human can be as low as 1,000 Ω for wet hand-foot contact to >10,000 Ω for dry handfoot contact. The contact voltage to produce sensation can therefore be higher for humans than for cows, depending on the conditions of the contact points. If more than 1 V is detected at the cow contact points, it is advisable to have a qualified electrician or the local power supplier evaluate the situation.

Bulls

The feeding and watering of growing and mature bulls should meet requirements of the National Research Council. Bulls should be housed in clean, well-lit, and ventilated buildings or outside in facilities that protect them from inclement conditions and allow them to remain clean and dry. Young bulls kept in small and uniform groups should be observed carefully as they mature to make certain that one or more individuals are not injured.

A panel can be installed in the center of group-housing pens to allow subordinate bulls to escape aggressive behaviour of dominant pen mates. Aggressive behaviour increases with age, and group housing should be discontinued by around 3 yr of age. Smaller or subordinate bulls should be removed from the group, and a bull removed from a group for over a few hours should never be returned to the group.

Visual and vocal social interactions with other bulls may be stressful. The safety of humans and animals is the chief concern underlying bull management practices. By virtue of their size and disposition, bulls may be considered as one of the most dangerous domestic animals. Management procedures should be designed to protect human safety and to provide for bull welfare. Electroejaculation of bulls is sometimes necessary and should be performed by a qualified person using equipment that functions properly and is in good repair.

A programme of annual self-regulation should be followed for:

- Semen identification and sire health auditing service and
- Minimum requirements for health of bulls producing semen for artificial insemination.

CATTLE AND BUFFALO DEVELOPMENT

The broad frame-work of the cattle and buffalo breeding policy being followed since the mid-sixties envisaged selective breeding of indigenous breeds in their breeding tracts and use of such improved breeds for upgrading of the nondescript stock. While the States accepted the framework, appropriate implementation through field level programmes could not be done. Lack of interest in promoting Breed Organisation/Societies and related farmers' bodies contributed to the gradual deterioration of indigenous breeds.

Government intervention for breed improvement is not available to majority of owners of indigenous breeds of cattle. Eventually, the availability of good quality bulls needed for natural mating in breeding tracts became scarce, leading to further deterioration of indigenous breeds in these tracts. Production of quality indigenous bulls has been a long-neglected area and would require a major thrust in order to harvest the best male germplasm available in the country. The present production capacity of frozen semen doses is about 30 million against the estimated requirement of 65 million doses annually.

Except for a few pockets in important breeding tracts and in sperm stations, indigenous bulls of unknown pedigree and with poor quality semen are generally used. Crossbreeding, which was to be taken up in a restricted manner and in areas of low producing cattle, has now spread indiscriminately all over the country. Continuous emphasis on cross breeding with exotic breeds even in the tracts of indigenous breeds led to the nearextinction of some of the known breeds. Further, the indiscriminate use of contaminated semen or infected bulls results in the spread of sexually transmitted diseases like Infectious Bovine Rhinotracheitis (IBR) at an alarming rate.

MILK PRODUCTION

Milk production in India remained more or less stagnant from 1950 to 1970. Thereafter, it increased rapidly, reaching 84.6 million tonnes (mt) in 2001-02 (anticipated). But the Ninth Plan target of milk production (96.49 mt) was not achieved. The per capita availability of milk increased from 112 gm per day in 1973-74 to about 226 gm per day in 2001-02. However, it is still below the world average of 285 gm per day. Investment in the dairy sector in the Ninth Plan decreased significantly compared to the Eighth Plan.

Out of 168 Milk Unions, 58 Milk Unions (34.5 per cent) were running in loss as of March 2000. So far, the Government policy in the dairy sector has been to give preference to the establishment of milk processing plants linking rural milk producers to urban consumers through a network of cooperatives. Restrictions on establishing new milk processing capacity under Milk and Milk Products Order (MMPO) has now been removed. No policy measures have been undertaken so far to give a fillip to the unorganised sector involved in the production of Indian dairy products (like ghee, paneer, chhena, khoa etc.), which have tremendous potential in the export market in Asian and African countries.

EGG PRODUCTION

The Indian poultry industry has come a long way–from a backyard activity to an organised, scientific and vibrant industry. It is estimated that the egg production in the country is about 33. 6 billion numbers against the Ninth Plan target of 35 billion numbers. The most notable growth among the livestock products has been recorded in eggs and poultry meat. Since 1970-71, their output has grown at 5.87 per cent per annum.

The significant achievement in poultry development has come from the initiatives taken up by the private sector for commercial pure-line breeding. However, despite the huge investment made, mostly by the private sector, the poultry-processing sector is incurring losses. The status of the poultry sector as to; whether it falls under agriculture or industry, is somewhat ambiguous and, therefore, it has remained deprived of various benefits available to these sectors. Poultry farming should be declared as an agricultural activity.

The poultry production model in vogue (high input-high output using commercially developed strain of birds) has been primarily responsible for the rapid growth in production of eggs and broiler meat in the country, but it is successful mainly in large-scale units (more than 1,000 units of birds). Due to high feed cost, non-availability of credit and marketing support, most of the small farmers have become contract farmers and are exploited by middlemen.

Government intervention, by way of various support mechanisms, is now needed for the promotion of poultry in rural areas. Indigenous poultry breeds, including the improved strains that can survive with low quality raw feed and better resistance against diseases, can be reared under free range conditions by rural unemployed youth and women for some additional income and employment.

MEAT PRODUCTION

In India, meat production is largely a byproduct system of livestock production utilising spent animals at the end of their productive life. Meat production was estimated at 4.6 mt in 1998. Projects sanctioned during the Seventh and Eighth Plans for improvement/modernisation of abattoirs and carcass utilisation centres for fallen animals are still to be completed.

GOAT DEVELOPMENT

Despite the least attention from the planners, goat population in India has increased at the fastest rate among all major livestock species during last two decades. However, instead of increasing the goat population, emphasis should be laid on productivity per animal, organised marketing and prevention of emergence of new diseases like Peste des petits ruminants (PPR) which has led to higher mortality and abortion in goats. The goat improvement programme is to be given a push through extending credit to the poor landless farmers.

SHEEP PRODUCTION

During the last four decades, there has not been much increase in the sheep population. Production of wool has increased from 43.3 million kg in 1996-97 to 49.0 million kg (anticipated) in 2001-02. The Ninth Plan target of wood production (54.0 million kg) was not achieved. The fine wool production in the country is around 4 million kg against the demand of around 35 to 40 million kg. Indian wool is primarily used for the production of carpet, drugget, wall hangings etc. To enhance the quality and quantity of carpet wool, shepherds need incentives like credit, health coverage, breed improvement programmes and timely disposal of wool and surplus animals at a reasonable price.

PIG DEVELOPMENT

Pig husbandry is the most important activity in the animal husbandry sector in the northeastern region inhabited by tribal people. The region also has a substantial pig population, which constitutes around 25 per cent of the country's pig population. The bulk of the population is, however, of the indigenous type whose growth and productivity is very low. The major difficulty in pig development is the acute shortage of breeding males.

ANIMAL HEALTH

Since the Second Plan, efforts have been made to control diseases namely, Rinderpest, Foot and Mouth Disease, Haemorrhagic Septicemia, Black Quarter and Anthrax. Although Rinderpest has been eradicated from the country, the prevalence of the other diseases continues to be one of the major problems in the animal production programme. Some of the emerging diseases like PPR, Bluetongue, Sheep Pox and Goat Pox, Classical Swine Fever, Contagious Bovine

Pleuropneumonia, New Castle Disease (Ranikhet Disease) are causing substantial economic losses. The programme for creation of diseasefree zones was sanctioned in the Ninth Plan but was not implemented. The Department of Animal Husbandry and Dairying is also not well equipped with the necessary infrastructure and qualified technical manpower to execute the programmes and perform its mandatory duties and responsibilities like disease diagnosis and accreditation as per the international standards, development of an effective surveillance and monitoring system for diseases, mass immunisation against the most prevalent diseases etc.

Dovetailing the Animal Research Institutes of the Indian Council of Agricultural Research (ICAR) with the Department would not only improve its efficiency but also provide it with an effective delivery machinery to carry out its regulatory and certification authority functions, including the conservation of endangered breeds of livestock. The suggestion for the establishment of an Indian Council for Veterinary and Fisheries (ICVFR) by carving out the animal science and fishery institutes from ICAR has not yet materialised.

ANIMAL STATISTICS

The Livestock Census Scheme suffers from quantitative as well as qualitative problems. The present arrangements for conducting the Livestock Census in the States and Union Territories are not satisfactory in relation to timely collection of data and reporting. The Integrated Sample Survey Scheme for estimation of production of major livestock products also needs improvement.

CONSERVATION

The last few decades have witnessed serious erosion, and even extinction, of some indigenous animal breeds in the country. Many existing breeds are facing varying degrees of threat, endangerment and are heading towards eventual decimation. In all States, crossbreeding of cattle is now occupying a dominant position in the production programme and, in this process, the native cattle breeds, which are well adapted, have suffered wilful neglect resulting in their progressive elimination from the production system. India is bestowed with rich domestic animal biodiversity, having 30 breeds of cattle, 12 breeds of buffalo, 20 breeds of goats, 40 breeds of sheep, eight breeds of camel, six breeds of horses, three breeds of pig and 18 of poultry. Besides, there are other species like equine, mithun, yak, turkey, ducks, etc. Indigenous breeds/types are rich in variability and are endowed with many positive traits like superior disease resistance, better tolerance to high heat and humidity and other characteristics suitable to particular agro-climatic environments. It has also been noted that indigenous breeds are more efficient in feed conversion particularly the crop residues and naturally available low quality roughages.

Indigenous breeds at risk are:

- Cattle:
  - Red Sindhi,
  - Sahiwal,
  - Tharparkar,
  - Punganur and
  - Vechur.
- Buffaloes:
  - Nili-Ravi,
  - Bhadawari and
  - Toda.
- Sheep:
  - Nilgiri,
  - Muzaffarnagri,
  - Malpura,
  - Chokla,
  - Jaisalmeri,
  - Munjal,
  - Changthangi,
  - Tibetan,
  - Bonpala from Sikkim and
  - Garrole sheep
- Goat:
  - Beetal,
  - Jamunapari,
  - Chegu,
  - Changthangi,
  - Surti and
  - Jakhrana.
- Camel:
  - Bacterian,
  - Jaisalmeri and Sindhi.
  - Yak Mithun
- Poultry:
  - All the 18 indigenous breeds of poultry are facing extinction.
  - The three important breeds are Aseel, Kadaknath and Naked Neck,

It has been globally recognised that conservation and improvement of native animal genetic resources are essential for sustainable development in agriculture and animal husbandry. The conservation and improvement programme should be decentralised and each State/adjoining States where a breed exists should take necessary steps with the active involvement of institutions, Gaushalas, Non-government Organisations (NGOs) and Breed Societies. The efforts should, however, be effectively coordinated centrally.

Given the severity of the resource constraint, all the Central sector and Centrally sponsored schemes were subjected to zero-based budgeting during the Ninth Plan. The objective was to retain only those schemes that are demonstrably efficient and essential. The schemes that are similar in nature would be converged to eliminate duplication and resource flow would be linked to performance. Out of 41 schemes, 23 schemes were weeded out, one scheme was transferred and six schemes were merged.

Table. Average Annual Growth Rate of Milk and Egg Production 1950-51 to 2000-01

Year	Milk (%)	Eggs (%)
1950-51 to 1960-61	1.64	4.63
1960-61 to 1973-74	1.15	7.91
1973-74 to 1980-81	4.51	3.79
1980-81 to 1990-91	5.68	7.80
1990-91 to 2000-01	4.21	4.46

LIVESTOCK SECTOR AND THEIR CAUSES

OVERGRAZING

Overgrazing of rangeland reduces the density of vegetation and the amount of organic matter generated. This, in turn, increases soil erosion from wind and water and decreases soil fertility through loss of nutrients. In arid and semi-arid areas these impacts may also contribute to desertification. Fortunately, ecosystems in these areas demonstrate considerable resilience and often recover when grazing pressure is reduced, either through traditional methods or through modern management practices.

POLICY AND LEGAL PROBLEMS

In areas that traditionally rely on grazing, the health of rangeland is generally best maintained by traditional pastoralist practices, which regulate grazing location and herd size in accordance with drought cycles and the supply of fodder. In fact, government policies or donor interventions that disrupt or discourage these practices may be a root cause of degradation. A variety of government policies may restrict the movement of livestock within a range area and prevent livestock managers from moving stock from areas that have been depleted of fodder to better supplied areas.

Two particular policy-based problems are:

1. *Land Tenure Insecurity:* Lack of confidence in secure title to rangeland (especially on communal lands) has been shown to reduce the

incentive to manage the land sustainably. Many national governments have either implicitly or explicitly claimed ownership of range and wildlands and ignored traditional or customary claims.

2. *Privatization of Communal Resources:* Where national governments have privatized, or are privatizing, formerly state-owned or communal lands, new owners may erect fencing or prevent herds from crossing or grazing on their property.

WELLS AND BOREHOLES

Traditionally, access to water on critical grazing lands has been controlled to limit livestock populations and prevent herds from outgrowing the forage supply in dry areas. Thus, new wells or boreholes, generally sponsored by donors or governments, may undermine traditional livestock management systems practices by allowing herds to grow beyond sustainable levels for surrounding areas.

Overgrazing and degradation are most noticeable in the immediate vicinity of the boreholes or wells, but their effects can extend (in gradually decreasing severity) over a considerable radius. Boreholes also reduce pressure on livestock owners to decrease herd size during drought and may discourage movement of herds to other rangelands, disrupting historic wet season/dry season grazing patterns. Larger herd sizes and reductions in pastoral movement may prove to be a recipe for severe degradation of soil and vegetation.

WET-SEASON GRAZING

Poor timing in the use of rangeland can also damage the soil. Wet-season grazing can compact the moist earth, reducing its ability to absorb moisture This increases erosion from water run-off.

POOR BALANCE OF LIVESTOCK SPECIES

Each species or breed of livestock has foraging preferences and will graze favoured areas and plants while neglecting others. Browsing animals, such as goats and camels, prefer the leafy tops of shrubs. By contrast, grazers tend to consume ground-level grasses and leafy plants. A poor balance between browsers and grazers can change the mix of plants in ways that degrade the area. For example, too many grazers can diminish the number and populations of herbaceous plant species and allow woody plants to become dominant, changing, possibly irreversibly, the character and utility of the ecosystem.

DAMAGED HABITAT AND REDUCED BIODIVERSITY

Livestock production can damage habitats and reduce biodiversity in wildlife and domestic stock, in vegetation, and in aquatic and wetland ecosystems.

HARM TO WILDLIFE AND DOMESTIC STOCK AND LOSS OF WILDLIFE HABITAT

The loss of habitat caused by livestock production in grazing and mixed farming systems may be one of the greatest threats to wildlife. Human population growth and density, and the accompanying increase in livestock, often leads producers to expand livestock grazing ranges into wild lands and the conversion of wild lands to mixed farming use.

SLAUGHTER OF WILDLIFE BY LIVESTOCK MANAGERS

Another danger to wildlife is intentional slaughter by livestock managers. Fear that the wildlife will prey on livestock and damage crops is a common motivation, as is the belief that the wildlife are competing with livestock for fodder, the desire to prevent spread of disease to livestock, and concern for human safety.

EXTINCTION OF LOCAL LIVESTOCK BREEDS

Systematic livestock production may result in loss of genetic diversity in livestock species. This is unfortunate because genetic diversity is a measure of a species' robustness. Local breeds may have traits conferring resistance to emergent or future pathogens, or have other favourable adaptations to local environments. The consistent replacement of local breeds with more productive imported ones can contribute to the extinction of that breed and of all the genetic diversity harboured within its population.

HARM TO VEGETATION

CLEARING OF FOREST AND WILD LANDS

Vegetation is typically altered or destroyed when forests/wild lands are cleared or are burned to promote new growth. This changes local ecosystems and may contribute to global warming. Fires to burn vegetation are dangerous and degrade air quality.

LOSS OF RANGELAND FERTILITY

Ironically, mixed farming systems may reduce the fertility of rangeland while helping to solve a farmland problem. Traditional farming practices cause a net loss of nutrients in farm soils; that is, when crops are harvested and sold (or even when human waste is deposited in a latrine) nutrients that make the soil fertile may be lost.

Mixed farming reduces the extent of this loss, by transferring nutrients from the range to the farm in the form of manure. The gain in fertility for the farm is, of course, a net loss for rangeland. Over time, the altered nutrient balance can reduce the productive capacity of the range and/or lead to changes in the composition and density of plant species.

LOSS OF FARM FERTILITY

Damage from mixed farming often occurs when poverty and population growth pressure change the crop/grazing land ratio, where other nutrient sources are not available. As the land area available for each grazing animal shrinks, overgrazing becomes prevalent leading to soil erosion and nutrient loss. Harvesting livestock further reduces nutrients available for crop production.

DAMAGE TO RIPARIAN SOIL AND VEGETATION

Livestock in grazing and mixed farming systems often graze very heavily in riparian areas along streams and lakes. Results include trampling, loss of vegetation, soil disturbance, and soil compaction, erosion and/or sedimentation which can severely damage riparian habitats, increase siltation and adversely affect watersheds. Pesticide contamination from treatments to protect livestock from insectborne infections (*e.g.*, livestock "dipping") may ultimately reach the aquatic environment. Here it can be toxic to aquatic organisms, as well as people or animals who depend on these sources for drinking water.

INTRODUCTION OF INVASIVE PLANTS

New breeds or fodder crops can introduce invasive non-native plants into a region. The manure, coats and hooves of newly introduced breeds can carry plant seeds. Most non-native plants are not invasive, but when they are, the results can be devastating.

CONTAMINATION FROM MANURE

Livestock manure contains relatively high concentrations of nutrients, solids, enteric bacteria and other microorganisms, and organic material. The manure from industrial livestock operations is often discharged or "leaked" into lakes or streams, because it cannot be economically transported to replenish crop fields. When this occurs, the nutrients can cause eutrophication (rapid plant growth in water bodies), solids can create sedimentation, and organic material leads to oxygen depletion (BOD) of the water. Manure from mixed farming, if applied in a concentrated fashion, can lead to similar problems.

DEGRADE WATER QUALITY AND REDUCE WATER SUPPLIES

Where water is scarce, either chronically or seasonally, the diversion of water to sustain livestock potentially limits its availability for other purposes. This is of particular concern in arid and semi-arid regions, where the construction of boreholes to supply livestock can lead to unsustainable withdrawal rates and the dangerous depletion of aquifer reserves. Stockpiled manure can contaminate bodies of water, causing myriad adverse effects. These include eutrophication, oxygen depletion, sedimentation, contamination with enteric bacteria and possibly other pathogenic organisms, toxic pollution from pesticides, and contamination of groundwater and aquifers with both nitrates and pesticides. Moreover, high concentrations of nitrate in potable water supplies represent a potential health hazard, especially for children.

HARM TO HUMAN HEALTH

Using water for farm animals may make it less available for the many uses that influence health, such as bathing, washing, cooking, and drinking. Moreover, excessive contamination by enteric microorganisms, toxic pesticides or nitrates in may render water unfit for human consumption and may be especially dangerous to children. Pesticides or other vector control treatments used on livestock represent threats to the health of livestock managers, their families, and others exposed directly or through water use. These substances may be toxic, cause birth defects, alter children's proper development, promote cancer, or slowly poison one or more organ systems.

CLIMATE CHANGE/GLOBAL WARMING

Approximately 17 per cent of global methane is produced by livestock digestion and manure. Methane is a greenhouse gas that has 56 times more global warming potential than carbon dioxide over a 20-year time horizon. Furthermore, clearing wild or forest land for new fields reduces the land's ability to act as a carbon sink. Setting fire to vegetation to clear land or promote new growth for livestock to graze on may also contribute to global warming.

ODOUR

Concentrated manure stored at industrial livestock facilities can generate strong and unpleasant odours, damaging the quality of life of nearby residents. This problem is most evident when facilities are located in densely populated areas.

SECTOR PROGRAMME DESIGN—SOME SPECIFIC GUIDANCE

The following questions and suggestions are intended to help project designers and managers identify factors and practices that may cause—or prevent—adverse environmental impacts.

Bear in mind that the first priority of most livestock managers and farmers is household food security and family welfare. Sustainable practices must always be balanced against these immediate demands.

CONSIDER CLIMATE, TERRAIN, AND ECOSYSTEM

Since environmental impacts from livestock production vary, depending on the specific climates, terrains and ecosystems involved, project designers need to address these characteristics during the initial design phase:

- What is the climate in the project area (arid, semi-arid, temperate, subhumid, humid)? What do historical records of rainfall patterns and flooding indicate? Is the proposed livestock management practice compatible with this climate?

- What terrains are found in the project area(s) (alluvial plain, highland, rocky desert, wetland, etc.)? Do they have any known vulnerabilities to livestock grazing? For example, are there many unprotected streams or rivers? Are there slopes with limited topsoil sensitive to erosion?
- Will the project encompass or border on protected or ecologically sensitive areas? Are there any threatened or endangered species in the area? Would the proposed project directly or indirectly threaten wildlife or native vegetation? For example, does the project require expansion of grazing into protected areas or make livestock more vulnerable to wildlife predators, triggering reprisals by farmers?

EVALUATE POLICY, LEGAL, CUSTOMARY AND CULTURAL CONTEXT

The policy, legal, and cultural contexts of a project merit attention, since, these factors may limit programme options or erect substantial barriers to success.

POLICY/LEGAL

- Do livestock owners and managers have legal and recognized ownership and responsibility for land and grazing resources? What are the current or proposed national land tenure policies, and how are they implemented at the local level? Are they effective in encouraging sustainable management of grazing land and resources?
- What is the tenure status of current or proposed rangeland—Is it owned by individuals or the community? Does the government have claims?
- What wildlife protection laws exist and how are they implemented in the region?

CULTURE/CUSTOM

- What role does livestock play in local culture and customs, and how might the proposed project affect these practices? Would the proposed project disrupt traditional grazing patterns?
- If there are customary land tenure arrangements, what are they and how would the proposed livestock management system work within these arrangements? For example, will livestock herders—who often come into conflict with farmers, particularly during droughts—have a means of working out disputes with farmers?
- If livestock management arrangements are communal, how would these be affected by and/or affect the proposed development activities?

ASSESS CURRENT AND PROPOSED SPECIES AND BREEDS

Introduction of a new breed into an area should be approached with caution. The new breed may bring with it diseases that can decimate local livestock herds and wildlife. In addition, the foraging habits of a new breed may lead to a sharp decline in available forage and biodiversity.

A new breed's reproductive habits can lead to a herd's uncontrolled growth.

Weeds can be accidentally introduced along with a new animal species, and they may displace desirable vegetation. The long term full costs and benefits of introducing a given new livestock species into a particular environment should be assessed.

For example, large animals roam over extensive areas in search of food, often require a greater financial investment, can be more difficult to control, and have lower reproductive potential than small animals. It is important not to underestimate the value of breeds that are well adapted to the environment. Livestock tend to overgraze favoured areas and plants while neglecting others.

Native plants may not be able to survive heavy grazing. While unforaged plants tend to lose vigour and nutritional value as they mature. The introduction of new plant species (whether accidentally or intentionally) may quickly result in replacement of native plants. Even when grazing pressure is reduced, exotic plant species sometimes retain their dominance. Ask the following questions when introduction of a new breed or species is proposed:

ABOUT CURRENT SPECIES AND BREEDS

- Which wild and domestic species are already present in the area, and in what numbers?
- How have they been used in local farming systems and traditions?
- What are the feeding preferences of local livestock and wildlife? What is the balance between browsers and grazers? Do domestic species compete for resources with one another and with wildlife?
- Have population sizes of wild or domestic species changed recently?
- Could local breeds satisfy the project's needs?

ABOUT PROPOSED SPECIES AND BREEDS

- If new species or breeds are being considered, how will their production complement or conflict with local species or breeds, wildlife, and other local resource users?
- How would they fit in local herding systems?
- Are they well suited to the local climate and environment?
- Are they resistant to local livestock diseases?
- Have alternative species or breeds been considered for possible introduction?

LIVESTOCK FISH FARMING

POULTRY-CUM-FISH FARMING

This system utilises poultry droppings of fully built- up poultry litter for fish culture. The fish production obtained is about 5000 kg/ha/yr. with 1250 kg chicken meat and 70000 no. of eggs. Approximately 500-600 no. of birds is reared in a 1 ha pond. The Rhode Island or Leghorn variety birds are more preferred over others. They require 0.3-0.4 square meterspace/bird. Hoppers are used to feed them and to minimise feed wastage. The poultry birds (layers) are fed with starter, grower, and brooder feed according to their age. In India it is practised in Andhra Pradesh, Bihar, Haryana, Kerala, West Bengal, Uttar Pradesh, Maharashtra, Orissa, and Tamilnadu.

DUCK-CUM-FISH CULTURE

The duck are commonly called as biologicals aerator. They are reared on the dyke of the pond in a low-cost house. This farming is practised in Tamilnadu, Assam, Bihar, Andhra Pradesh, Tripura, Orissa, Karnataka, Kerala and Uttar Pradcsh.

The 'Indian runner' and 'Khaki campbell' varieties are found more suitable in this culture. About 300 no. of ducklings (some spp. are reared 450-500 in no.) are reared to fertilize the 1 ha. pond. The duck not only act as live aerator by splashing water with their webbed feet but also control the aquatic weed (Lemna, Azolla, etc.), aquatic insects, molluscs, tadpoles, etc. Each duckling require about 0.3-0.5 square meter area as living space. The total production from such type of culture is about 3500-5000 kg fish, 18000-18500 eggs and 600 kg of duck meat. The duck droppings are used as manure for primary production.

PIG-CUM-FISH CULTURE

This system has certain advantages over others. The 30-35 pig's waste may produce 1 tonne of Ammonium Sulphate and 40-45 pigs are adequate to fertilize 1 ha water area under polyculture. Each pig requires about 3-4 sq.m floor space.

This system of integration is very common in China, Taiwan, Vietnam, Thailand, Malaysia, Hungary and some European countries. The White Yorkshire, Hampshire and Landrace are the popular breed of pig for integration with fish. Pigs need clean housing which should provide adequate protection from adverse climates. The pigs are fed on pig mash which is made up of rice bran, rice polish, wheat-bran, broken maize, ground-nut oil cake, fish- meal mineral mixture, salt, etc. The spoiled vegetables can also be mixed in it. This system provides about 3000-4000 kg/ha/yr fish, 4500 kg/yr pig meat and 800 no. of piglets every year.

CATTLE-CUM-FISH CULTURE

It is a common practice all-over the world. The cow excreta is most abundant in terms of availability and a healthy cow may excrete over 4000-5000kg dung and 3500-4000 litre urine on an annual basis. The BOD of cow manure is lower than other livestock manure. About 5-6 cows can provide adequate manure for 1 ha pond in addition to 9000 kg milk and about 3000-4000 kg of fish annually. Cow-shed should be built close to fish pond to simplify handling of cow-manure.

GOAT-CUM-FISH INTEGRATION

It is considered as poor man's cow and a goat's excreta is considered as a very good organic fertilizer. The goat excreta contains organic carbon-60 per cent, N-2.7 per cent, P-1.78 per cent, K-2.88 per cent and its urine is also equally rich in both N & P. At least 50-60 goats are essential to fertilize 1 ha pond. The goats should be provided with dry, safe, comfortable house protected from excessive heat. The goat breeds are Jamanapari, Beetal, Barbari for milk and Bengal, Sirihi, Deccani are used for meat purpose. Goats are selective feeders and consume Berseem, Napier grass, Cowpea Soybean, Mulberry, etc. This integration can provide 3500-4000 kg fish/ha/year without supplementary feeding and fertilizer.

RABBIT-FISH INTEGRATION

Rabbit meat is preferred by most of the health conscious consumers owing to its low fat in comparison to other meats. The important meat breeds are Soviet Chinchilla, Grey Giant, and White Giant, etc. Rabbits are reared in cage, hutch and floor system (floor should be cemented). Rabbit excreta contain organic carbon-50 per cent, N-2 per cent, P-1.33 per cent, and K-1.2 per cent. The rabbit excreta is high in nitrogen content and low in moisture, thus quality manure for sustained plankton production. It is estimated that excreta from 300 rabbits would be enough for 1 ha pond fertilization.

FISH PRODUCTION IN INTEGRATED FARMING SYSTEMS

Guided by major findings on livestock-fish farming published in a preliminary set of principles defining the various aspects of the culture technology and promotion of fish production in integrated farming systems is suggested and outlined, as a basis for further deliberation.

Principles pertaining to technology of livestock-fish farming are:

- Fish production in integrated systems is more complex than the conventional separate aquaculture system, requiring more knowledge and better management inputs
- Integrated farming systems may vary in the degree of intensification of the livestock and fish subsystems, varying from extensive, semiintensive to intensive subsystems.

- Extensive subsystems utilise natural feed produced without international fertilization; semiintensive subsystems require fertilization to produce natural feed and/or supplementary feed but with a significant component of diet supplied by natural feed; and in intensive subsystems all the nutritional requirements are provided by artificial feed given to fishes with natural feed contributing little or no nutrition.
- Fish cultured in an integrated farming system with livestock (especially duck; Edwards, 1986) benefit from a significant amount of the nutrition derived from natural food, which develops in the pond due to the fertilization by organic manures. This suggests the important role of pelagic algal-based food web which cultivates in fish biomass.
- The largest contribution of the manure to fish nutrition therefore, appears to be due to its fertilizing effect in the pond. Bacteria, breaking down the organic matter in the manure, release nutrients which lead to the production of phytoplankton and zooplankton.
- In a manure-fed fish pond, fish nutrition may also be derived from direct consumption of the manure. However, a detritus food web has a secondary role in fish biomass production biomass production, as compared with the algalbased food web.
- The direct nutritional value of manure for fish is by its content of spilled animal feed. A large proportion of the nutrients in livestock feed are not assimilated, but are voided in the excreta, particularly pig excreta (Edwards, 1985).
- Over-fertilization with manure may lead to poor water quality of fish pond, particularly leading to depletion of dissolved oxygen and fish kills.
- Management of water quality is needed to overcome fish kills due to oxygen depletion and extreme fluctuation of dissolved oxygen levels. The strategy is to promote a growing biomass of phytoplankton which will generate sufficient oxygen to maintain relatively high dissolved oxygen have on a 24-hr basis. It is essential to maintain a positive net photosynthesis
- The criteria for selection of fish species for stocking into manure-fed fish ponds should be based on the ability of fish species to:
  - Filter and feed on plankton (bacteria, phytoplankton and zooplankton)
  - Tolerate low levels of dissolved oxygen (< 2mg/l minimum as defined by "Criteria for the Protection of Aquatic Life")
- An optimal stocking density of fish species is critical in attaining high cumulative fish yields and in reaching the upper carrying capacity of a manure-fed fish pond.
- Determination of and recommendation of the optimal stocking density

of fish consider differences in local circumstances such as the fish species, manure-type and inputs to the pond, addition of other off-farm feed, and water quality of the pond.

- Ways to intensify fish production from integrated farming systems involve management inputs to:
  - Stock a higher initial fish biomass, followed by
  - Harvesting the fish intermediately when the growth curve of stocked fish starts to slow down
- There is a need for a more complex marketing system to handle the inputs and products from two subsystems as opposed to a single subsystem.
- Integrated farming on the one hand, enables the distribution of risk (both biological and economic), since two subsystems are involved as opposed to one in a single-commodity farming system; on the other hand, the failure of one subsystem can adversely affect the other.

FISH SPECIES SUITABILITY AND SELECTION

The main objective of manuring is to produce a good growth of plankton, mainly phytoplankton, but also zooplankton and bacteria which are protein-rich natural for fish.

On this basis, the most appropriate species of fish for a manure-fed pond are therefore those species which are able to filter and feed on plankton (bacteria, phytoplankton and zooplankton) from the water. The most common filter feeders are carps and mixed feeder on algae-detritus are certain species of tilapia. The most commonly raised carp species are the bighead carp Aristichthys nobilis Richardson and the silver carp Hypophthalmychithys molitrix Valenciennes.

The carps are usually raised in polyculture with the grass carp Ctenopharyngodon idella and Indonesian carp Puntius gonionotus Linnaeus controls micro- and macro-vegetation at the water's edge, the Hoeven's slender carp Leptobarbus hoevenii Blk and the common carp Cyprinus carpio Linnaeus which is a bottom feeder. An important observation is that the detritus-feeding silver carp benefited most from the pond conditions, increasing its biomass share from an initial 5.8 per cent to 20.7 per cent from a series of investigation on the integrated farming systems in the Transkei, South Africa. The Indian carps, catla Catla catla Hamilton, mrigal, Cirrhinus mrigala Hamilton Buchanam and rohu, Labeo rohita Hamilton have also been used in experimental duck/fish integrated farming systems in India.

Polycultures of carps have been widely assumed to be the most efficient way to obtain high fish yields but in Taiwan, the percentage of tilapia in the pond has increased markedly at the expense of the Chinese carps. The most commonly raised tilapia species are the blue tilapia Oreochromis aureus

Steindachner, the grey tilapia Oreochromis mossambicus Peters, the Nile tilapia Oreochromis niloticus Linnaeus or their hybrids.

Tilapia are more suitable than carp for manure-fed ponds since the former are more tolerant to low levels of dissolved oxygen than the latter. Air-breathing silver-striped catfish Clarias spp., are also integrated with pig farms. The detritusfeeding grey mullet Mugil cephalus L. may be excellent substitutes for the silver carp in integrated farming systems under saline conditions.

FISH YIELDS AND PRODUCTION RATES IN INTEGRATED FARMING SYSTEMS

There are few data on fish yields from ponds which received livestock manure (and spilled livestock feed). The fish yields in integrated livestock-fish farming systems reported in the literature vary greatly, with the actual and extrapolated equivalents ranging from 1.5 ton/ha/year to 18 ton/ha/yr. The variation in fish yield is due:

- The animal sub-system and its major manure input to the pond
- The fish species and stocking rates
- Intensification of culture management inputs *e.g.* addition of other off-farm feed

DIFFERENT TYPES AND AMOUNTS OF MANURE INPUTS

The rationale behind integrating fish with livestock is the large amount of nutrients (N-P-K) present in the feed that is recovered in the manure, with possible proportions of 72–79 per cent nitrogen, 61–87 per cent phosphorus, and 82–92 per cent potassium. These act as fertilizers in fish ponds to produce plankton which comprise high-protein natural food for certain species of fish.

Nitrogen and phosphorus are the nutrients most likely to be limiting for plankton growth in the pond but fish yield is probably more directly correlated to manure nitrogen content (Edwards, 1991) since nitrogen is more volatile than phosphorus. Based on the nitrogen content of the different manures, it was possible to estimate the stocking density of different livestock (laying duck, dairy cow, pigs and buffalo) in order to produce the same fish yield of 174.7 kg/ 200 m^2/year in 13 family-level farms.

Considering the livestock's manure characteristics (total live weight, age of livestock, feed characteristics of livestock, climate and management of livestock subsystem) and the varying N contents (per cent of the manure production/ animal/year, *i.e.*, 2.6 per cent for laying duck manure, 1.9 per cent for pig, 2.2 per cent for dairy cow and 1.1 per cent for buffalo, the estimated number of laying duck, pig, dairy cow and buffalo required to the same fish yields are 1335, 410, 40, and 85 individuals respectively. Prinsloo and Schoonbee noted the comparative effectiveness of the manure on the development of organisms in the food web and promotion of biological activity in fish ponds are: duck manure > pig manure > raw chicken manure > cattle manure > sheep manure.

Manure may contain up to 25 per cent crude protein, but more than half of this is usually non-protein nitrogen (*e.g.*, uric acid) which is not assimilated by fish; therefore manure itself is a poor feed for fish. Manures also contain less energy than conventional pelleted feeds. Though many species of fish consume manure directly, it is a low quality feedstuff compared to conventional pelleted feed and natural plankton. Another advantage of manure utilisation in fish ponds is that since carbon forms about 50 per cent of the biomass of plankton, the high organic content of the manure is an important source of carbon which is released by bacterial respiration. However, depending on the productivity and quality of water and soil, the effectiveness of livestock manure (enriched with cellulose-rich organic matter) in acting as nutrient sources for fish growth and in raising fish yield varies.

Two possibilities have been observed in Dor, Israel:

1. Manure would improve fish yields only under conditions of water and soil which are poor in essential nutrient minerals (*e.g.*, carbonates or organic matter), or of a low pH, fish ponds receiving very "soft" waters deficient in minerals, or ponds with a low primary production; and
2. Manure may be ineffective in raising fish yields above the rates achieved by daily applications of mineral (N-P-K) fertilizers, under conditions when the fish pond is receiving daily applications of mineral fertilizers, or where fish ponds are highly productive with very fertile soil naturally-enriched with nutrient minerals and carbonates, or fish ponds with nutrient-rich waters resulting in a high concentration of algal-based detrital organic matter.

LIVESTOCK PROTECTION

FOOT-AND-MOUTH DISEASE

During the period under the report, 799 outbreaks were recorded throughout the country. Maximum outbreaks were recorded in the eastern region where there was increase in the number of outbreaks compared to last year. Disease incidence in West Bengal directly affects the situation in NE region due to animal migration. Drastic reduction in outbreaks was noticed in southern region. FMD cases were not reported in Tamil Nadu, whereas Himachal and Punjab recorded a single case of FMD each. Maximum incidence of disease was reported in March and from August to November.

Out of 1,624 clinical specimens collected from 799 outbreaks, the virus could be detected in 1,067 samples (serotype O-991, A-38 and Asia 1-38) by using sandwich ELISA and multiplex PCR. Serotype O was responsible for maximum number of outbreaks (568) followed by A (33) and Asia 1 (9). Increase in serotype O incidence was noticed in all regions except northern and western region. Serotype A, which was absent in northern and central region last year appeared this year, making it the second most prevalent serotype affecting geographically all the regions of India.

Serotype Asia 1 was completely absent in southern, central and north eastern regions. Incidence of Asia1 serotype increased in western region. Though majority of the outbreaks involved cattle, disease was also reported in buffaloes, pigs, sheep and goats. A major shift was observed in genetic lineage of serotype O viruses circulating in India.

The 'Ind2001' lineage viruses gained upper hand after a gap of 8 years and outcompeted PanAsia II lineage. The cases due to PanAsia II, PanAsia 1 and Branch C-II viruses were not completely absent during the year. Branch-C II lineage viruses responsible for outbreaks during 2008 in West Bengal were recorded in Tripura, Asom, West Bengal and Odisha during 2009–10. In serotype A, it was observed that both VP359 deletion and non deletion mutants belonging to genotype VII are cocirculating in the field in recent times.

Precisely, the viruses from Punjab and Uttarakhand shared ancestry and clustered in the same VP359 deletion lineage, VIIf, whereas viruses from two different outbreaks from Odisha and those from Andhra Pradesh clustered together and revealed no deletion in the VP3 coding region. During the period, outbreak due to Asia1 serotype was recorded in Gujarat, Madhya Pradesh, Maharashtra, Uttar Pradesh and West Bengal. The Asia 1 field isolates were grouped with lineage C reiterating the supremacy of this lineage in the field since 2005. In vaccine matching exercise 33 virus isolates were subjected to one-way antigenic relationship study. All the isolates showed close antigenic match with respective in-use vaccine strains indicating appropriate antigenic coverage. Virus isolates (133) including 97 type O, 12 type Asia 1 and 24 type A field isolates were added.

In National FMD Virus Repository 1,687 (1,096-O, 317-Asia 1,259-A, 15-C) well characterized field isolates are available. Under National FMD sero-surveillance testing of 11,560 serum samples from 13 states revealed the presence of protective antibody titre in 42.3 per cent, 40 per cent and 30.5 per cent animal population against serotypes O, A and Asia1 respectively. This baseline response has improved due to application of vaccine. The current in-use vaccine strain IND R2/75 is found still the best and covers all the type O circulating outbreak strains in the country.

Classical Swine Fever

Classical swine fever (CSF), the topmost diseases in pigs, is one of the OIE listed diseases. The CSF or hog cholera is associated with high mortality, abortion, mummification of fetuses. It is of great concern in areas, where the pig population is in high density. CSF virus isolates maintained at the repository of PD-ADMAS, Bengaluru, were revived and processed for RT-PCR of NS5B genomic region. Phylogenetic tree based on the alignment of 408 nts of NS5B region was constructed. It showed three major branches, each representing three previously described genetic groups with further sub-branches.

The majority of the Indian CSFV isolates belong to subgroup 1.1. Clustering of the Indian field isolates away from the vaccine strain and other related strains revealed that the field isolates did not have any vaccine origin. Phylogenetic analysis based on 52 UTR classified subgroup 1.1 Indian isolates into two minor subbranches, however, no such classification could be seen from the analysis based on NS5b region; nevertheless in the latter analysis grouping of isolates was much more resolved. Lone isolate, IND294/08, is clustered within subgroup 2.2, ak into its grouping based on 52 UTR sequence.

Brucellosis

Serum samples from goat and sheep received from Gujarat, Rajasthan, Uttar Pradesh and Maharashtra revealed that among them 6.35 per cent were positive from brucellosis. Similarly, 1.8 per cent sheep and goat sera samples for Karnataka were positive for brucellosis.

Porcine brucellosis is one of the most important emerging zoonoses and currently, no control programme for porcine brucellosis exists and the epidemiological situation is still unknown.

Indirect-ELISA for screening brucellosis in swine was standardized using smooth lipopolysaccahride antigen from a standard strain *Brucella abortus* S99; 22.6 per cent were found positive, which is much higher than that found in sheep and goat. An indirect ELISA was developed to detect *Brucella* in any of the susceptible species. Expression of four recombinant *Brucella* proteins, *viz.* BP26, BP39, superoxide dismutase (SOD) and *Brucella lumazine* synthase (BLS) is being envisaged to overcome the problem of cross reactivity in diagnosis of this disease.

Leptospirosis

PCR based techniques were developed for differentiation of pathogenic and nonpathogenic *Leptospira* as well as for typing. Initially, *Leptospira* genus specific PCR was carried out to differentiate the *Leptospira* from other spirochete organisms, *viz. Borrelia*, *Treponema* and *Leptonema* etc., which amplified 331 bp products from the *Leptospira* organism only.

Sero Prevalence of IBR

Serum samples collected from Andhra Pradesh, Gujarat, Jammu and Kashmir, Karnataka, Madhya Pradesh Maharashtra, Meghalaya, Manipur,

Orissa, Punjab, Rajasthan, and Tamil Nadu, were subjected to Avidin Biotin ELISA. An overall apparent percentage positive was 34. The highest prevalence of 67 per cent was in Tamil Nadu, and the lowest in Meghalaya. Region-wise positive percentage were as follows: East zone 17 per cent, south zone 24 per cent, west zone 48 per cent, north zone 37 per cent, north-east 39 per cent, and central zone 25 per cent.

Herbal Anticoccidial and Antidiarrhoel Drugs

The extracts CIRG-5 and CIRG-6 from two different plants were evaluated *in vitro* against sporulation of coccidian oocysts and found 96 to 100 per cent effective to check the sporulation up to 96 h. Eight potential plant formulations of herbal extracts were subjected for clinical trials to control diarrhoea caused by *Escherichia coli* in kids of less than 1 month of age. Final product was developed from plant extract prototype CIRG-I and CIRG-2A and final products (formulation) @ of 10 mg/kg bwt for 1–2 days orally was highly efficacious to control EPEC diarrhoea in kid.

VACCINES AND THERAPEUTICS

- Live attenuated orf, buffalopox and camelpox vaccines met the thermo-stability and potency requirements of the biologicals.
- Bi-valent DNA vaccine containing HN and F genes of Newcastle disease virus was evaluated along with genetic adjuvant comprising IL-2, IL-4, IFNγ, and GMCSF gene constructs.
- An aroA gene knockout mutant of *Pasteurella multocida* serotype B:2 was constructed for the development of HS mutant vaccine.
- Chitosan coated cationic PLG nanoparticles were prepared and when tested for the transfection efficiency with commercial transfecting agent, indicated that these nanoparticles can be safely used for delivery of DNA vaccines.
- *Trypanosoma evansi* beta tubulin protein was expressed in both prokaryotic and eukaryotic expression systems.
- Two new target proteins of *Fasciola gigantica, viz.* leucine aminopeptidase and peroxiredoxin coding genes were cloned and successfully expressed in their native conformation in a prokaryotic expression system.
- Immunomodulators of microbial and herbal origin, *i.e. Mycobacterium phlei*, *Lactobacillus acidophilus* and *Tinospora cordifolia* were effective under field conditions to improve productive performance and mitigate the adverse effect of sub-clinical caecal coccidiosis and thereby improvement of body weight gain in broiler birds.
- Pharmacokinetics of gatifloxacin revealed that it has therapeutic potential for use in sheep.

- Medetomidine-butorphanol, thiopental and halothane combination provided better clinicophysiological and haemodynamic stability in general anaesthesia in buffaloes.
- Three TLR genes of yak (TLR 5, 7 and 9) and 2 TLR genes of mithun (TLR 4 and 7) were sequenced completely.

DIAGNOSTICS

- A recombinant lipoprotein 'P48' of *Mycoplasma agalactiae* was evaluated in an ELISA as a rapid immunodiagnostic tool in sheep and goats.
- A genus-specific PCR assay was standardized for rapid detection of *Arcobacter* sp. targeting 16S rRNA for confirmation of the isolates of this emerging pathogen.
- A 36 kDa protein was identified in different serovars of *Salmonella* during biofilm formation.
- Magnetic nanoparticles (MNPs) were synthesized by two different processes and were characterized for structural, morphological, size and magnetic properties. The MNPs were functionalized for immunoassay applications and evaluated for cellular toxicity in *in-vitro* experiments.
- A specific and sensitive diagnostic marker was identified for early detection of pancreatitis in dog.

MOLECULAR CHARACTERIZATION OF PATHOGENS/RECEPTORS

- Indian H5N1 isolates were characterized genetically and grouped with EMA *3/2.2.3 subclade,* which branched from two distinct nodes in the phylogenetic tree.
- The alignment of deduced amino acid sequences showed that all the Indian H5N1 viruses characterized contained the multiple basic amino acid motif at their HA cleavage site, which is characteristic of highly pathogenic avian influenza viruses.
- Two viruses from 2010 outbreaks in West Bengal contained amino acid asparagine at position 31 of the M2 protein, suggesting that both viruses are resistant to drug amantadine.
- An isolate of buffalo pox from man and animal recovered from an outbreak from Pune was characterized for B5R, H3L, D8L and A27L genes.
- An orf isolate from camel in a sporadic infection was characterized for B2L, GIF and VLTF-4 genes.
- Genotyping and partial sequencing of rotaviruses from man and pigs revealed human-pig reassortants as well as presence of novel combinations in India.
- Complete HN gene of velogenic Newcastle disease virus containing

kozak sequence was amplified and cloned into pcDNA3. 1 eukaryotic expression for enhanced expression.

- Phylogenetic analysis by AP-PCR, PFGE and ribotyping showed considerable similarity between human and fish isolates of pathogenic and pandemic *Vibrio parahaemolyticus,* which reflected the public health risk potentials of fish.
- Molecular characterization by PCR-SSCP analysis of *Oesophagostomum* spp. revealed two distinct conformers with six nucleotide polymorphisms (SNPs) among male and female parasites.
- Probability of evolutionary changes among prevailing genotypes of *Oesophagostomum* spp. was predicted by analysing secondary structure of RNA.

QUALITY CONTROL AND PRODUCTION OF VETERINARY BIOLOGICALS

- 223,500 doses of RD 'F' strain vaccine; 166,000 RD 'Mukteswar' strain vaccine; 132,500 doses of fowl pox vaccine; 184,195 doses of lapinized swine fever vaccine; 423,800 doses of tissue culture sheep pox vaccine; 160,600 doses of PPR vaccine; 57,925 doses of *Brucella abortus* strain-19 (live) vaccine; 3,640 ml HS adjuvant vaccine; 73,500 doses of tuberculin PPD; 50,000 doses of Johnin PPD; 25,250 doses of mallein PPD; 55,250 ml of *Brucella* agglutination test antigen; 6,900 ml of *Brucella abortus* Bang Ring antigen; 19,860 ml of Rose Bengal Plate Test antigen; 219 ml of *Brucella* positive serum; 9,000 ml of *S.* Pullorum coloured antigen; 45 ml *S.* Pullorum positive serum; 87 ml *Salmonella* 'O' serum and 3,750 ml of *S.* Abortus equi 'H' antigen were produced and quality tested.
- 7.32 million monovalent doses of FMD vaccine were produced at Bengaluru campus.
- 435,100 doses of PPR vaccine were produced; similarly, 33 PPR c-ELISA, 22 PPR s-ELISA and 100 doses of goat pox vaccine were supplied by Mukteswar campus.

Sero-monitoring of Important Equine Diseases

During 2009–10, equine disease sero-survey was conducted in Rajasthan, Haryana, Punjab, Uttar Pradesh, Uttarakhand, Madhya Pradesh, Chhattisgarh and Sikkim. A total of 176 out of 1,045 equines (16.84 per cent) had antibodies to equine influenza virus while 28 (2.67 per cent) were positive for EHV-1 and 107 (10.43 per cent) were positive for antibodies to Japanese encephalitis virus. Antibodies to *Babesia equi* were detected in 320 (38.09 per cent) of 840 equines, while *T. evansi* antibodies were detected in 49 (5.4 per cent) out of 906 equines tested. One case of equine infectious anemia was detected in Uttarakhand. EIA-positive equine was reported after a gap of 11 years from India. None of the 1,045 equines tested showed antibodies to *Brucella* and *Salmonella* Abortus equi 'H' antigen.

Japanese Encephalitis (JE) Virus Isolation

JE, a mosquito-transmitted disease caused by JE virus, is causing disease in equines, pigs and humans. It is prevalent in eastern and southern Asia. Serum samples from clinically affected horses and pooled mosquito vector samples from the surrounding field were found positive for JEV RNA by RT-PCR. Virus was isolated from one of the horses. This is the first report of isolation of JEV from a horse in India. The phylogenetic analysis of the envelope gene (1500 bp) sequence of the isolated virus, JE/eq/India/H225/2009 indicated that the isolate belonged to genogroup III of JEV, and that the equine isolate clustered together with Vellore group of JE isolates from India.

Camelpox Virus Zoonosis

Camelpox virus (CMLV) causes skin eruptions in addition to severe generalized exanthema mostly in camels and outbreaks occur frequently in camel-rearing north-western region of the country. However, their zoonotic potential has not been proven conclusively. The disease was confirmed on the basis of isolation of the virus from diseased camels, detection of the CMLV antibodies in the serum of diseased camels and human patients by serum neutralization test and amplification of CMLV-specific three full-length envelope protein genes (A27L, H3L and D8L) and one partial gene (C18L) of the isolated viruses from camels and human specimens. The phylogenetic analysis indicated that CMLV isolates cluster closely with variola virus (VARV) and vaccinia virus (VACV). The CMLV zoonosis indicates the changing pattern of host specificity of poxviruses and certainly points towards a declining cohort immunity against Orthopoxviruses in humans, which could be a public health concern. This is the first confirmed report of camelpox virus zoonosis.

Cysteine Proteinases from Trypanosoma Evansi

Parasite derived cysteine proteinases were proposed as potential targets for immunotherapeutic, chemotherapeutic and diagnostic reagents for parasitic diseases. The proteinases of camel isolate of *T. evansi* were analysed by gelatin substrate SDS-PAGE containing collagen antigen confirming the presence of cysteine proteinases in *T. evansi*. Some of these cysteine proteinases were reactive with experimental/field serum sample from equines with *T. evansi* infection. This indicated the potential of these enzymes in serodiagnosis of this economically important disease of livestock.

8

Livestock Production and Management

THE ENVIRONMENTAL IMPACT OF INTENSIVE LIVESTOCK PRODUCTION

The intensification of livestock production has had not only a significant effect on animal welfare, but also a considerable impact on the environment. In this regard, the disposal of the enormous quantities of animal waste produced by animals confined in intensive livestock production units has created the greatest environmental problems. The reduction of manpower levels and the need for efficient animal management have meant that 'factory' farmers have often abandoned the use of straw as bedding in livestock houses; the animals instead stand on slatted or solid concrete floors. The faeces and urine which these animals produce must therefore be disposed of differently.

When straw bedding is used, manure is collected in dung heaps, exposed to air it gradually ferments, bacteria are thus to some extent destroyed and the excreta eventually becomes fertiliser to be spread on the land. Slurry, on the other hand, must be stored in tanks, where it may decompose and sometimes produce a variety of noxious smelling gases. This method does not kill pathogens quite as effectively as the traditional method.

While straw bedding most closely resembles the natural surface on which most animals lie—and is thus promoted as the ideal floor surface on which animals should be kept—it can provide the conditions under which disease organisms can flourish. If poorly maintained, such bedding when excessively dirty and wet not only causes distress to the animals, but can also lead to environmental mastitis in dairy cattle, 'farrowing fever' in pregnant sows and an increased exposure of poultry to the dangerous *E.coli* bacteria.

Slurry can be returned to the land as fertiliser, but its storage and application can be particularly problematic. It cannot be spread when the ground is too wet and must, therefore, be stored in huge tanks, sometimes for several months. The failure of these storage tanks would have a significant impact on the local environment, most particularly if pathogens, such as salmonella, *E.coli* and leptospira, which are dangerous to both human and animal health, were still present in the slurry. To reduce the risk of infection, farmers have been warned against spreading slurry on grazing land.

If they must do so, they are advised to allow the slurry to adequately ferment and to leave the land on which it has been spread unpastured for at least six weeks afterwards. The main problem with modern farms is that they do not always have enough land to absorb the waste that the animals produce. The waste must, therefore, often be transported elsewhere at considerable expense to the livestock holder. As Johnson suggests, waste disposal can be the main factor which determines how many animals can viably be kept in a livestock production unit, rather than animal welfare. To illustrate this problem, he describes the crisis faced in the Netherlands during the late 1980s.

The great concentration of factory farms had put so much pressure on land that by January 1987 the Dutch authorities placed a ban on the further expansion of livestock numbers. With farms producing around 100 million tonnes of effluent per annum, twice the amount of liquid waste accumulated than could safely be spread on the country's 2 million hectares of farmland. As a consequence, the transport of slurry to areas away from factory farms was encouraged and subsidised by the Dutch government and large-scale dumping of slurry was banned due to the potential pollution and health risks it presented. Dutch farmers are now required to pay the full costs of disposal through sewage plants if they are unable to find appropriate land to spread their farm waste on.

Like slurry, the production of silage—a grass crop harvested when green and then partially fermented to make animal fodder—can present serious environmental problems. The anaerobic fermentation which accompanies silage production can potentially produce a waste liquid which has been estimated to be 200 times more polluting than domestic sewage, particularly when the grass is wet. Such waste presents great pollution problems if it escapes the silos in which it is held and enters natural waterways. The aquatic life of rivers and streams can be either poisoned or starved of oxygen as a result of the introduction of this effluent.

The contamination of natural waterways is one of the most serious environmental consequences of modern agricultural practice. Aside from the problems posed by slurry and silage effluent, intensive livestock production contributes towards nitrate pollution. In addition to this, the additives used in animal feed and the antibiotics which the animals are routinely given are potential sources of environmental contamination. Although the blame for environmental pollution is often apportioned to industry, rather than agriculture, the intensification of farming methods—most particularly with respect to the waste it produces—presents a significant threat to the environment.

In addition to the phenomenal impact on animal welfare and the environment, the intensification of livestock production has created new opportunities for disease organisms to flourish and threaten the health of both animal and human populations. As this book has thus far clearly illustrated, our interactions with other animals have always brought with them the potential risk of infectious disease. However, the increasing exploitation of and dependency on animal resources, which has occurred through the intensification of livestock production, has meant that disease *risks* which did not exist, or were unthinkable, forty to fifty years ago are today commonplace. In short, the increase in stocking densities has led to the increase of opportunities for disease transmission.

The housing and living conditions of intensively farmed animals—most notably pigs and poultry—potentially provide significant exposure to bacteria and viruses, thereby enhancing the risk of disease. Disease risks can be greatly reduced with good stockmanship. On 'factory farms', however, animal management is primarily linked to maintaining or increasing productivity. In the past, stockmen could attend to animals more individually and were familiar with the small herds they tended.

Today, stockmen are not only fewer in number, but are also required to manage huge numbers of livestock as efficiently and cost effectively as possible. The reduction of manpower levels in livestock farming combined with the increasing numbers of animals kept on site can lead to inefficiency in disease control and hygiene which, in turn, may result in the outbreak of infectious disease amongst the animal stock and—depending on the nature of the disease in question—possibly even transmission to the humans who come into direct contact with them. This is, of course, not to mention the risk of contaminated animal protein entering into the food chain and putting human consumers at risk of infection.

One of the disease management techniques adopted in intensive livestock production is to accept the pervasiveness of infection. Antibiotic drugs and vaccinations are thus routinely administered in order to keep stock healthy and maintain productivity. When disease outbreaks occur, it may not be viable to cull stock given the size of the animal population housed. Similarly, farm buildings may have been too poorly constructed and cannot be properly disinfected or cleaned, the animal's excreta, as suggested above, creating potential hazards for both animal and public health.

Management of disease purely through drug intervention is far from ideal. Moreover, it arouses considerable consumer concern about the quality of the animal produce which ends up on our dinner plates. The use of growth hormones and the possibly harmful effects of veterinary drug residues in food have become an increasing source of consumer anxiety in recent years; as has the suggestion that the persistent administration of antibiotic drugs may ultimately lead to the evolution of pathogens resistant to such substances.

For fifty years, antibiotics have been added to animal feeds in order to improve feeding efficiency and growth rates. Since the 1960s, scientists have looked into the possibilities of resistance risks and by 1970 the European community had made moves to prevent the antibiotics that are used in human medicine being used as animal growth promoters. Today, the issue of resistant risks has again become the topic of political and scientific debate, stimulated in part by the development of a veterinary antibiotic called *avoparcin* which is closely related to an important human medical antibiotic known as *vancomycin.* There is a concern that such veterinary antibiotics may promote the emergence of bacterial strains that have a genetic resistance to antibiotics.

Although there is currently little microbiological consensus on this topic, some scientists have gone so far as to suggest that the widespread use of antibiotics in livestock production, particularly those used to treat infections, may well be responsible for resistant strains of salmonella and campylobacter that are being found in increasing numbers of food poisoning cases in humans. In recent years, the problem of animal disease control has also been tackled from another angle, that rather than seeking to alter the living conditions of the animals, instead seeks to alter the animals themselves. In this context, attempts have been made to rear strains of 'minimal-disease' animals which are kept in highly controlled environments. Poultry flocks today can be bred to be 'specific-pathogen free', which essentially means that particular disease-producing agents have been eliminated by breeding and rearing birds in sterile conditions. Such isolation, however, can render such birds vulnerable to other microorganisms that might otherwise have been innocuous.

Modern intensive livestock production is a risky business and sometimes it is human health that is at stake. In spite of the industry's endeavours to produce animal products that are safe for human consumption, as indicated above, food-borne infections still pose a threat to public health. The intensification of poultry rearing is the main culprit for bacterial infections in humans. Salmonella infection is one of the ever-present hazards of broiler rearing. It is in fact widely accepted that salmonella will persist in poultry stocks, although it can be kept at bay by good feeding and housing practices.

The pervasiveness of this bacteria in poultry is largely due to contaminated feed, although people and wild animals or birds can also be the vectors of transmission. Intensively reared poultry, like other farm animals which can carry the salmonella bacteria, are generally fed on high-protein feeds which can contain slaughterhouse waste and other animal by-products such as dried poultry manure. If such feed is inadequately sterilised, disease organisms can persist and can be passed on to other animals. Sterilisation is a costly process which occurs through heat treatment. However, irradiation of animal feed is not always efficient because strains of salmonella have emerged which are heat-resistant and continue to contaminate feed.

The transmission of salmonella requires 'the transfer of faeces or intestinal contents from an infected individual to the digestive tract of a susceptible one'.

Like animals, humans can become infected by salmonellosis through the ingestion of contaminated animal produce. The disease can cause gastroenteritis, septic shock syndrome and focal infection in humans, and can even be fatal to vulnerable individuals, such as infants or elderly people. It should be emphasised that good hygiene and effective cooking methods from the abattoir to the kitchen can reduce the risk of infection for the human consumer.

The modern systems of livestock rearing, mass animal transportation and conveyor belt style slaughterhouses, however, increase the possibilities of cross-infection and the contamination of human food. Although the most common bacterial infection, salmonella is not the only disease which may be contracted from the ingestion of animal produce. Bovine tuberculosis, brucellosis, trichinosis, campylobacteriosis, listeriosis, staphylococcal enterotoxicosis, clostridial toxicosis and toxoplasmosis may be transmitted through tainted animal food and cause human illness.

Aside from food-borne infections and intoxications, consumer health is generally not affected by the diseases which farm animals carry. The people at greatest risk from agricultural animals and their products are in fact those who work with them; in other words, farm workers, veterinarians and individuals who handle animal carcasses occupationally. In Western Europe today, the major zoonoses, such as brucellosis (contagious abortion), tuberculosis and anthrax, are uncommon. Other zoonotic diseases, however, present serious occupational hazards to those in the meat, dairy and animal product industry.

For example, leptospirosis—commonly known as Weil's disease—is a disease which can seriously affect dairy farmers, both economically and physically. It is a contagious bacterial disease which can produce a high abortion rate in cattle late in pregnancy; it can be responsible for high mortality in calves, reduced milk production and contaminated milk. The organisms that cause leptospirosis are transmitted through urine and live in water or moist soil. Cattle—and other livestock—become infected through being splashed with infected urine, grazing on contaminated pasture or by drinking infected water.

Other wild species, particularly rodents, are responsible for transmitting the disease to cattle, and humans can be infected by bathing in contaminated water or direct contact with infected urine. The effect of leptospirosis on humans varies from inapparent to acute febrile illness which can be accompanied by fever, hepatitis, jaundice and meningitis—amongst other maladies. Good animal management, vaccination and reduction of contacts between livestock and rodents reduce the risks of infection by leptospirosis. There are many more health hazards for those who come into routine contact with food-producing animals. Other occupational hazards for humans are presented by rickettsial diseases, fungal infections (*e.g.* ringworm), chlamydia infections and respiratory viral diseases, in addition to bacterial infections such as those described above.

The fact that intensive livestock production today happens within closed systems, away from public view, has much to do with the reduction of infection risks. Modern livestock housing is essentially designed to keep out disease-causing organisms and their vectors of transmission and to keep disease-free animals in. Such systems are far from foolproof; infection may, for example, occur through the drinking water or animal feed. Furthermore, farmers, stockmen and veterinarians may introduce disease to livestock. It is commonplace when one enters a livestock production unit for one to have to don special overalls and boots in order to minimise the possibility of introducing new pathogens to the animals from outside. However, keeping all visitors out, apart from stock and veterinary personnel, not only aids disease prevention and control, but to some extent prevents consumers from gaining a first-hand glimpse of the conditions under which their food is produced.

Modern consumer society demands a continual, cheap and dependable supply of animal products which are pre-packaged and ready to enjoy: yet never before have the majority been so divorced from the production of the animal proteins which sustain them. The urban consumer is estranged from the origins of the food he eats and the leather he wears and has limited involvement in or knowledge of the circumstances of their production. The consumer is very much reliant on the producers and retailers to offer fresh and uncontaminated animal products—at a low price—on demand. The 'farm fresh' animal products which are nowadays found in urban supermarkets are however unlikely to have been produced locally, as was formerly the case. Animal produce may originate from farms and factories situated hundreds of kilometres from the consumer's domicile. The transport networks and storage capacities which today so efficiently ensure the swift despatch and delivery of farm-fresh animal produce to the consumer were not possible until recently.

Previously the movement of livestock and carcasses was fairly restricted, both within and across provincial or national borders, not only through lack of adequate transportation and refrigeration, but also due to the fear and possibility of the spread of contagious disease. Once a sufficient level of understanding and knowledge of transmission and control through governmental intervention had been reached, disease could be better contained, cross-infection prevented and animals could be more safely moved from one location to another.

Yet although governmental regulations pertaining to animal disease control have existed both nationally and internationally since last century, it is still possible for contaminated animals to slip the net and infect others (be they animal or human). Given that infected livestock or tissue can easily and quickly be moved from one (international) location to another, new opportunities for disease-producing organisms can develop in areas and amongst animal-human groups previously unaffected. The risk of disease can be further exacerbated by the failure to properly disinfect vehicles used for animal transportation, and delays in the transfer of live animals or meat from one place to another facilitate contagion further. Since, on an international level, animal movements are more likely to increase than decrease as a result of political and economic accord, the potential problem of animal disease transmission could, in the future, be one of increasing significance.

VETERINARY REGIME DURING THE TWENTIETH CENTURY

As the above discussion suggests, the possibilities for the transmission of infectious diseases are very real, most particularly given the risks posed by intensive systems of agricultural animal management and animal transportation. It is largely for this reason that veterinary medicine has become so important during the twentieth century in maintaining and monitoring animal health, to keep the spread of infectious disease in check and to ensure that farmers adhere to and understand the guidelines and regulations concerning disease prevention and control. Yet while disease *risks* may today be far greater than ever before, the danger of humans *actually* contracting diseases from contact with animals or the consumption of animal produce is in fact relatively small.

Throughout the course of the twentieth century, the veterinary regime has continued to move from strength to strength. The state veterinary services that were instituted in the industrialised nations of Europe during the late nineteenth century have played an increasingly important role in the control and prevention of animal disease. Legislative measures to combat the effects, spread and introduction of contagious disease have been enforced with increasing vigour.

The successes of earlier campaigns and stringent legal controls to stamp out the animal plagues that devastated European livestock populations provided the basis for future legislation. For example, veterinary inspection and the rigourous enforcement of all regulations pertaining to the importation, transport, sale, slaughter and destruction of livestock successfully led to the eradication of contagious bovine pleuropneumonia in the Netherlands by 1887 and Britain in 1898.

The introduction of measures to financially compensate farmers for their losses through compulsory slaughter policies undoubtedly provided an incentive to abide by governmental regulations and allowed veterinary officials to perform their duties without hindrance. Eventual success in eradicating disease helped to persuade livestock producers to accept the ruthless and restrictive regulations that the state sought to enforce. As a consequence, state veterinary intervention has become institutionalised within modern livestock production. Moreover, during the twentieth century, the appreciation of the need for animal disease control and prevention has more or less been incorporated into the habitus of livestock producers. Scientific developments have, of course, played a crucial role in the prevention and control of animal disease. The development of the science of bacteriology, advances in pathology, parasitology, immunology and microbiology and so forth have provided important contributions to advances in animal and public health. By the turn of the century, European governments had become increasingly aware of the need for better scientific understanding of animal disease if they were to reduce the risks that it posed. In the Netherlands, for example, the State Serum Institute was founded in 1904-by the pioneering veterinarian and bacteriologist Jan Poels—with a view to steering the work of the established Dutch Veterinary Service towards more scientific means of preventing and curing livestock disease.

The Serum Institute experimented with the production of a serum against foot-and-mouth disease and an anthrax vaccine. Moreover, it conducted bacteriological and chemical investigations into milk and dairy produce. Interestingly, this new institute operated specifically outside the dominion of the veterinary school, being instead an organ of the state. In 1933, nearly three decades later, the State Veterinary Research Institute was established, primarily to study and develop a vaccine against foot-and-mouth disease—a highly infectious disease that has perhaps wreaked the greatest havoc on European livestock and the livestock industry during the twentieth century. By 1946, the Institute's director, H.S. Frenkel, had succeeded in developing a method by which a vaccine against foot-and-mouth disease could be produced in large quantities. This method, however, only proved effective until 1951, when a more virulent strain of the disease struck Dutch livestock.

From 1945 onwards, the structure of the state veterinary machine became increasingly more complex, as government-subsidised Animal Health Services were established in each province within the Netherlands. The earliest regional animal health service had in fact already been established in Friesland in 1919. Its early work led to significant progress in the battle against bovine tuberculosis. The new regional services were modelled on their Friesian predecessor and strove for the coordinated control of animal disease, involving not only scientific and governmental veterinary advisors, but also representatives of local farmers and dairy organisations and practising veterinarians.

The eleven regional Animal Health Services were in turn coordinated by the Industrial Board for Agriculture, who established an Animal Health Committee to deal with the allocation of funds and organisation of schemes for disease control. After the Second World War, the regeneration of the meat and dairy export industry became an important part of Dutch economic recovery. A key aspect of this was a national plan to rid Dutch cattle herds of tuberculosis, thereby ensuring the safety and quality of the national product. A considerable portion of the Marshall Plan funds that were made available to the Netherlands to assist post-war economic reconstruction was devoted to achieve this end. By 1956, the disease had been conquered due to the efforts and cooperation of the Animal Health Services, practising veterinarians and the meat inspection service. On the basis of this national plan's success, similarly successful attempts were made to control and eradicate brucellosis from the Dutch livestock population. By 1957, the Serum Institute and the State Veterinary Institute amalgamated to form the Central Veterinary Institute, which functioned under the auspices of the Dutch Ministry of Agriculture and Fisheries. The Veterinary Service was effectively given the governmental mandate, through the establishment of legislation such as the Livestock Act and the Veterinary Medicines Act, to prevent animal disease and preserve public health.

The work of the Central Veterinary Institute continued to grow and diversify throughout the 1960s and 1970s. In 1986, it consolidated and moved to new premises in Lelystad. By 1994, the CVI amalgamated with a number of other agricultural animal research institutes to create the Institute for Animal Science and Health.

Today, this institute employs over 600 people and conducts research into animal health, breeding and reproduction, animal welfare, stress and management, animal nutrition, and so forth. Moreover, it is responsible for diagnostic work, the production of vaccines, specific-pathogen free animals, diagnostic preparations, etc., and the control and standardisation of veterinary medicines.

The developments in state-directed animal disease control in the Netherlands can be mirrored in the rise of the State Veterinary Service in Great Britain. There too was an understanding for the necessity of improved scientific knowledge and diagnosis of animal disease. The first governmental diagnostic laboratory was set up in 1894 in response to an outbreak of swine fever. This was followed by the establishment of the Central Veterinary Laboratory at Weybridge, Surrey, in 1913. Initial research was conducted into spontaneous abortion, swine fever, scrapie, foot-and-mouth disease and diseases of poultry.

As farmers began to increase the size of their flocks, as the process of intensification of production began to take off, the laboratory found itself increasingly dealing with the effects and greater incidence of poultry disease. During the 1930s, the laboratory began to research and produce tuberculin, in addition to an anti-abortion vaccine. The Agricultural Act of 1937 facilitated further expansion of the laboratory. New research departments of bacteriology, biochemistry, parasitology, pathology and poultry disease were established and a further laboratory was set up in Scotland.

By 1960, the Animal Health Division successfully completed its programme to eradicate tuberculosis from the British cattle population. During the 1960s, the new challenges to animal health presented by the changing methods in livestock production gave extra impetus to the fight against animal disease. The intensive rearing of calves, for example, led to an increasing incidence of respiratory disease. This unforeseen consequence of new methods of animal husbandry required urgent bacteriological and virological investigation. Similarly, the development of artificial insemination and its widespread use in cattle presented new problems for animal disease control. The movement of inseminators between farms and the importation of and trade in semen presented new hazards for the transmission of infectious animal disease. In 1955, legislative measures were taken to prohibit importation of foreign animal semen without a licence and the Animal Health Division took steps to screen the health of bulls.

In addition to the Central Veterinary Laboratory, the Veterinary Investigation Service was established in 1922, with regional centres throughout Britain. This service essentially provided local laboratory facilities on which practising veterinarians could rely to help diagnose and investigate animal health problems in the herds that they attended.

Initially, the Veterinary Investigation Centres were attached to agricultural colleges, but in 1946 they were incorporated into the Ministry of Agriculture, Fisheries and Food. The interaction between local veterinarians, veterinary investigation officers, the researchers of the Central Veterinary Laboratory and legislators in fact provides a good illustration of the increasing complexity of animal disease control during the twentieth century. At the grassroots level, practising veterinary surgeons observe and report the local incidence of animal disease.

This in turn is investigated by veterinary investigation officers who conduct field investigations and epidemiological studies to gauge the extent of animal health problems. In the meantime, the Central Veterinary Laboratory attempts to understand the aetiology of the disease and explores how it may effectively be controlled.

These combined investigations may in turn lead to the inception of statutory control measures, vaccination, test and slaughter programmes to contain and eradicate animal disease. In Britain, this kind of cooperation eventually led to the eradication of brucellosis and is today responsible for the investigation and extensive monitoring of new pathogens and diseases, *e.g.* bovine spongiform encephalopathy. In October 1995, the Central Veterinary Laboratory and the Veterinary Investigation Service merged to form the Veterinary Laboratories Agency in a further bid to improve the collaboration necessary to ensure animal disease control and the provision of advice to the Ministry of Agriculture, Fisheries and Food and other veterinary health workers.

Throughout the twentieth century, the necessity for collaboration and cooperation has extended far beyond the activities of local and nationally organised veterinary organisations. International cooperation on matters of veterinary public health first began during the nineteenth century as a result of the devastation caused by the cattle plague. In 1924, the *Office International des Epizooties* was established in Paris. This intergovernmental organisation initially sought to promote and coordinate research into the control and surveillance of animal disease, and has remained active on this front to this day; its original membership of 28 having gradually been expanded to some 144 member states.

During the post-war period, additional international fora for dealing with the economic and health problems associated with animal disease were institutionalised. In Europe, the European Union—and its former incarnations, *i.e.* the EEC and EC—have provided the main political arena within which European governments can jointly discuss and legislate on matters pertaining to veterinary public health and animal welfare. The European Economic Community (EEC) was originally established in 1958- following the signing of the Treaties of Rome -with a view to achieving economic and political union amongst its member states, namely Belgium, Germany, France, Italy, Luxembourg and the Netherlands. In 1973, Denmark, Great Britain and Ireland became members of the EEC. In later years, Austria, Greece, Finland, Spain, Sweden and Portugal joined the ranks of states seeking European economic union; the organisation is set to expand yet further in future as former Eastern Bloc countries vie for membership.

Since its inception, the European Union has had a critical impact on agricultural practice and the trade in live animals and animal produce within Europe. The Common Agricultural Policy was instituted during the early 1960s with a view to increasing agricultural productivity, ensuring a fair standard of living for the agricultural community, stabilising markets, ensuring food supplies and providing consumers with food at reasonable prices.

It has been perhaps the most important Common Policy that operates within the Union and has served as an important precursor to the 'single market' that is today intended to guarantee the free movement of goods, services, capital and people between the fifteen member states. The Common Agricultural Policy greatly influenced the economics of livestock farming within Europe and initially encouraged farmers to intensify production.

Ultimately this led to massive overproduction and the creation of huge surpluses, which in turn depressed market prices. 'Butter and beef mountains' came to symbolise the crisis in European agriculture; livestock numbers needed to be reduced and milk quotas were introduced. The Common Agricultural Policy has consequently undergone significant reforms and today it even encourages the *extensification* of production methods, particularly in the beef sector.

Free trade between member states remains at the top of the European Union's agenda, and trade naturally includes the trade in livestock and animal produce. As the epizootics of the past clearly evidenced, the movement of animals across national borders presents potentially great hazards to both public and animal health. Each member state has its own internal legislation and procedures regarding the control and prevention of livestock disease; the efficacy of which is entirely dependent on the efficiency of and cooperation between livestock producers, veterinarians and governmental agencies. However, membership of the European Union has required that each country's system of animal disease control be brought in line with those of other member states, in order to ensure a harmonisation and uniformity of disease prevention. Moreover, a system of veterinary inspection and veterinary and live-stock regulations have necessarily been introduced in an attempt to ensure the health of animals and the safety of meat that are crossing national borders.

With regard to livestock, each animal that is traded between member states, or enters the Union from a non-member state, must be accompanied by a veterinary health certificate that has been signed by an official veterinarian. Two types of certificates exist: one for animals intended for breeding and production; and another for animals intended for slaughter. The health requirements for breeding animals are much more stringent than for those destined for slaughter. These certificates detail the animal's origins, destination, means of transport, age, breed, type, means of identification and whether they passed through a market or assembly point. Such certificates are valid for ten days after the animal has been loaded for transport. In this way, diseased animals can be traced along with those that they came into contact with during transport.

The health certificate states that the animal has been examined and has been passed as healthy by an official veterinarian and that it originates from a herd or region free from specified diseases. Animals may have to undergo further tests depending on their origin, type and destination. Traded livestock must meet a number of welfare considerations determined on a Union-wide basis. Likewise, the rules for the trade in meat have been harmonised throughout the European Union, in order that animal flesh is produced, preserved and transported according to rules of hygiene identical to every member state. Certificates issued by official veterinarians at the place of meat production provide a guarantee that the meat is fit for consumption and serve as a valid transport document which testifies to the safety of the product. Produce from outside the European Union is subject to similar checks at border inspection posts and is required to meet the standards deemed acceptable within the Union.

The veterinary regime has thus been further consolidated on an international level and has led—at least in principle—to the standardisation of the control of animal disease, welfare and health care and the quality of meat and other foods of animal origin within Europe. The legislators in Brussels have ensured that member states adopt uniform policies with regard to livestock production. European legislation, therefore, has increasingly come to determine the nature of national legislation.

For example, since 1988, EU countries have maintained a ban on all growth-stimulating hormones, although such products are still widely employed in the USA. Attempts to eliminate the illegal trade in such hormone products even led to the murder of a Belgian veterinary inspector in 1995. An illustration of the impact of the EU's veterinary regulations on member states can also be provided by events surrounding the bovine spongiform encephalopathy (BSE) crisis. The severity of the legislation has greatly disabled the British livestock and meat trade and has also had significant repercussions for other countries.

It seems likely that the public health issues and trade problems surrounding BSE will continue to feature prominently on the European economic, agricultural and veterinary agenda in Brussels in the future. Similarly, the way the 1997 outbreak of classical swine fever in the Netherlands has been dealt with has to a great extent been affected by EU regulations. While vaccination may provide the key to ending the crisis, the EU has committed itself to a policy of eradication of classical swine fever and, at present, forbids the use of a vaccine which may be effective against *the* disease. The Dutch pig industry has been crippled by this outbreak, export embargoes have been instituted and over three million pigs have been destroyed, not necessarily because they had contracted the disease, but to prevent overcrowding of pig stalls.

While animal disease control and prevention measures have become increasingly stringent throughout Europe during the twentieth century, leading to the virtual eradication of the majority of animal plagues that decimated livestock and played havoc with the European economy in centuries past, we cannot forget that the modern global economy and the routine movement of individuals—both human and animal—across and between continents also have a bearing on veterinary public health within the bounds of Europe. Diseases such as tuberculosis and classical swine fever, for example, are known to have been transmitted to western European livestock from live animal imports or animal produce from Eastern Europe.

Similarly, imports from countries where animal disease control is not so advanced or strict may still pose a threat to public health since spores of diseases such as anthrax, which have more or less been eradicated in advanced Western societies, can still be introduced through the import of hides, animal foods etc. from regions where the disease still persists. It must also be noted that while a highly effective veterinary regime exists within Europe, veterinary policing is not always as efficient as it should be and regulations are not always well enforced.

Essentially, it only takes one weak link in the chain of interdependencies involved in the process of transforming animals into safe food to lead to inefficiencies and failures in animal disease control. Again, the recent BSE crisis provides a clear example of this. Inadequate compliance with and policing of the meat and bone meal bans, first introduced by the British government in

1988 to prevent the feeding of mammalian protein to ruminants, appear to have allowed the epidemic to continue, when it might otherwise have been curtailed. One may thus conclude that any animal disease control measures that the legislators introduce can only be as good as their policing.

The control of animal disease today is far from simply a local or regional problem, it has instead become a global one. The involvement of veterinary experts in the policy making and execution of the World Health Organisation (WHO) and the Food and Agriculture Organisation (FAO) certainly attests to this. However, although advanced industrialised societies have the animal disease problem more or less under control—producing significant animal protein surpluses in spite of continuing population growth—the (rural) populations of developing nations across the globe continue to suffer both the economic and health effects of animal disease. Furthermore, animal diseases can affect livestock populations and food production to such an extent that they do not even return a profit. Livestock disease therefore not only generates economic problems, but also wastes resources which could help to solve the problem of world food shortages.

While the focus of this chapter is far from global, it is still important to be aware of the fact that the relationship between humans, animals and the environment in which they all live is a continually changing one which is not confined within national borders. The complexity and scale of interdependencies between humans and other animals is such today that failure to keep disease-causing microorganisms adequately in check could be potentially devastating for both the physical and economic health of world populations. The necessity of an effective and globally-oriented veterinary regime in present-day society is, therefore, manifest.

NEW FRONTIERS OF ANIMAL USE

The scientific developments of recent years often seem like the stuff of science fiction, rather than science fact. Almost daily one can leaf through the newspapers, or turn on the television, to be confronted with news of the latest advances that will purportedly provide new hopes for the treatment of disease or the manufacture of food. Whether it be for the production of a cheaper and more effective vaccine to treat hepatitis B, or for the manufacture of a tomato that contains fish genes so that it can be successfully frozen, 'genetic engineering'—as it has been popularly dubbed—makes for good headline news because it is a topic that captures the public and scientific imagination alike. Genetics has been heralded as the science that will transform agriculture, cure diseases and conserve other species, although it could potentially also be employed to more sinister and destructive ends.

Essentially, molecular genetics is a science that explores the mechanisms of heredity. Each individual, human or animal, possesses a unique genetic 'blueprint' that makes him or her different from any other living creature. This genetic blueprint can, however, be modified. This can be achieved through a recombinant DNA technique that involves the injection of a foreign gene into an organism in order to produce specific new characteristics that in turn will also go on to be inherited by future offspring. Organisms that have been modified in such a fashion are known as 'transgenic organisms'.

The science of molecular genetics has found an increasing number of applications with regard to food-producing animals and veterinary science. In recent years, the gene mapping of farm animals has become an important tool for breeding and has been employed to localise, isolate and characterise the genes that are responsible for specific traits important to both health and production. With the development of new molecular genetic techniques that will help to isolate and identify the DNA markers that are linked to the genes responsible for economically important production traits and disease resistance, animal breeders will in the future be better equipped to single out the animals carrying the most desirable genes.

Theoretically, once such genes are identified, animals may be genetically selected or engineered for disease resistance. Another related application of genetic engineering is the production of more effective medical and veterinary vaccines. This may offer livestock producers a viable solution to the serious problem of enhanced infection risks that has been caused by increasingly high stocking densities. Advances in DNA recombinant technology have created the possibility for live carrier virus vaccines' to be developed that can be tailored to veterinary needs. It is thought that such modified live viruses will be far cheaper to manufacture and more efficient in the prevention of animal disease than existing ones.

Likewise, biotechnology—a term that may be generally applied to the manipulation of organisms for commercial purposes—has found important uses within animal reproduction. For many years, artificial insemination has provided the basis for the systematic genetic improvement of animals. A more recently developed reproductive technique, *in vitro* fertilisation (IVF), may possibly be employed within livestock production in the years to come. IVF involves the maturation of the egg outside the body and its fertilisation in a test tube. The embryo is then implanted into the womb of a surrogate mother. Although IVF in cattle is still in its infancy, embryo transfer is a technique that is now widely employed in livestock species and has various advantages with respect to genetic improvement, health protection and productivity.

Biotechnology has also been applied to improve animal nutrition. It has been utilised at three levels: in the feed, in the digestive tract (i.e. in relation to the ruminal microflora that aids digestion) and in the physiological regulation of the animal. Genetic technology can, for example, be employed to alter rumen bacteria in order to manipulate the animal's digestive tract ecosystem. Thus far attempts to interfere with animal metabolism have met with considerable resistance. In this respect, it is the application of an active agent, the growth hormone bovine somatotrophin (BST), to increase milk yield in cattle that has caused perhaps the greatest consternation, raising serious questions about the lengths to which livestock producers may be willing to go in order to increase productivity. As mentioned in Chapter 5, the use of BST, which is currently produced from transgenic microbes, has been prohibited within the European Union.

One of the most controversial applications of recombinant DNA technology has been the creation of transgenic animals. The first experiments with transgenesis in mammalian species took place with animals used for laboratory experimentation. For instance, during the mid-1980s, a transgenic mouse, christened the 'oncomouse', was developed by biomedical researchers at Harvard University in their bid to understand and find a cure for breast cancer. The 'inventors' of such transgenic creatures have even gone so far as to patent their creations, leading to heated debates upon the nature of intellectual property rights and their application to living organisms.

Not surprisingly, such genetic technology has also found its way into the realms of livestock production. Since the mid-1980s, fledgling experiments with transgenesis have taken place involving food-producing animals, such as sheep, goats, pigs and cattle. In the first instance, the creation of transgenic farm animals has been seen as a possible means to increase and improve agricultural productivity, to meet increasing human demands for animal produce without incurring further detriment to the environment by requisitioning extra land for agricultural production.

However, some of these transgenic experiments with food-producing species have had quite different aims. One of the prime incentives to create transgenic farm animals has been to produce substances known as biopharmaceuticals in milk. Bio-pharmaceuticals are substances produced by human genes that are essential to fight off disease and keep healthy. When such genes are incorporated into the DNA of other organisms, e.g. bacteria or yeast, these organisms become able to produce biopharmaceuticals beneficial to human health upon a scale that is commercially viable.

A prime example of such experimentation with transgenesis and the production of biopharmaceuticals can be provided by Herman, the world's very first transgenic bull. Born in 1990-and the result of a unique collaboration between the biotechnological company Pharming BV and a Dutch governmental research institute—Herman was the product of a fertilised egg in which the DNA had been modified to include a synthetic gene identical to the human one that controls the production of lactoferrin, a milk protein that has strong anti-microbial properties and stimulates intestinal flora to provide protection against bacterial infections. Apart from this single gene, Herman is essentially no different to any other bull with respect to his genetic make-up. Herman was created specifically in order to sire female offspring who would inherit this foreign gene and consequently produce human lactoferrin in their milk.

The first mature milk from Herman's transgenic progeny became available early in 1996 and was found to contain the appropriate lactoferrin. The commercial implications of such successful biotechnological developments are potentially enormous, for they offer new possibilities for the production of pharmaceutical drugs to treat human conditions at a far lower cost. For example, Pharming, the biotech company responsible for Herman and his daughters, is currently attempting to develop a range of human health care proteins in transgenic cattle.

The human lactoferrin that they have developed is intended as a component in clinical nutrition that can be used in the treatment and prevention of bacterial gastrointestinal infections in patients who have been immuno-compromised. A further planned application is in the production of speciality medical formulas for newborn and premature babies. However, perhaps more importantly, the introduction of this new transgenic animal technology signifies a new role for domesticated animals within human society. It is likely that in the not-too-distant future, animals will not only continue to function as the providers of protein, but may also become living, breathing and walking 'pharmaceutical factories'.

Transgenic technology may also find other important human health applications in future years, particularly with respect to the use of animal organs for transplantation into human beings. Since organ transplant technology was first developed for the kidneys and liver and then first employed in human heart surgery by Christiaan Barnard in 1967, organ transplantation has become increasingly important in the treatment of life-threatening conditions. The continued success of surgical procedures such as kidney and heart transplantation has led to an ever-increasing demand for healthy human organs, yet—unfortunately for those awaiting such radical, though potentially life-saving, operations—there is a great shortage of suitable organs. As the population ages, it is inevitable that this demand will increase further still. In spite of numerous high profile campaigns to encourage people to carry donor codicils, the shortage of donor organs has remained critical, inspiring scientists to search for new alternatives to human organs for transplantation. Some have sought solutions in the form of artificial organs; others have begun to seriously investigate the possibilities of using animal organs for transplantation into humans.

Xenotransplantation, as such cross-species organ transplantation is known, has been heralded as a potential solution to the organ crisis, for it could provide a ready-made supply of healthy donor organs to the thousands of people who today, and in the future, will require transplant surgery. Because of the animals' genetic affinity to humans, scientists initially turned to primates—particularly baboons—in their attempts to transplant organs to humans. The very first attempt at clinical xenotransplantation in fact occurred as early as 1910 when a German medical scientist, Dr Unger, transplanted a monkey kidney into a uraemic girl.

More recently, in 1992, a baboon liver was transplanted into an HIV patient who was dying of hepatitis B in the USA. Primates, however, pose particularly high risks to humans with regard to the transmission of zoonotic disease, raising serious questions as to their suitability for use in xenotransplantation. The recent emergence of new highly infectious diseases such as Ebola and Marburg have further emphasised the potential role of primates as the reservoirs of viral diseases deadly to humanity. Further to this, the use of primates in medical experimentation has become increasingly controversial due to a growing sensitivity to the semblance between humans and primate species. In recent years, calls have even been made for the great apes to be accorded equal rights to humans.

In view of such developments, scientific attention has instead turned to the domestic pig in the bid to produce organs suitable for human transplantation. Although more distantly related to people, the pig has already proved to be the ideal provider of insulin to treat human (and pet animal) diabetes and valves to correct heart abnormalities. Porcine organs are also approximately the same size as human ones. Moreover, given that these animals breed most prolifically, they could potentially provide a regular and reliable supply of organs for transplantation. All this being said, there are major problems that would have to be overcome before pig organs could actually be transplanted into the human body. Organ rejection is probably the greatest obstacle. This commonly occurs in the transplantation of human donor organs and is generally counteracted by the life-long usage of anti-rejection drugs by the organ recipient; in xenotransplantation rejection is likely to be hyperacute.

One of the solutions to this problem that is currently being sought by scientists is to employ transgenesis in order to genetically modify the porcine donor's organ in a way that will trick the human recipient's body into accepting it. The hearts of transgenic pigs that have been bred by researchers have already been transplanted into monkeys with some degree of success, arguably giving some indication of future possibilities for human application. The question of disease transmission through such transplant technology is also here at issue.

Some porcine pathogens may indeed pose risks to human health, though it is more than likely that pigs intended for xeno-transplantation will be bred to be specific-pathogen free. In fact, some may even be inclined to argue that the option of transplanting animal organs carries *less* disease risk than the transplantation of human donor ones. The animals used would necessarily have been genetically modified, bred and kept under strict clinical conditions; unlike the human donor who may carry all manner of diseases that may go on to infect the recipient once the organ has been transplanted. Another recent development in the science and technology of genetics that may find its most ideal application with regard to producing transgenic animals for both transplantation and biopharmaceutical purposes is cloning. Once relegated to the realms of science fiction, cloning has today become very much a reality. In February 1996, the public and scientific world were stunned by the announcement that two genetically identical lambs, named Megan and Morag, had been born at the Roslin Institute in Scotland. These sheep had been cloned by the process of nuclear transfer from a cultured cell line, originating from different cells of the same embryo. Morag and Megan provided concrete proof of the possibility of creating animal life without the need for male sperm

In February 1997, the even more astonishing news broke that a lamb had been cloned—by the very same scientific team—from the cells of an adult sheep. To create Dolly, as she was christened, DNA from a single cell had been taken from another sheep's udder, making her the very first mammal to have been created from the non-reproductive tissues of an adult animal. Dolly the sheep became a celebrity overnight and her very existence precipitated fervent debates about the ethics of cloning and its potential application to humans. A month later, hot on the hooves of this story, came the news that Australian researchers had managed to create almost five hundred genetically identical cattle clones. These were created from egg cells that had been taken from cow ovaries. At the time of writing, it remains to be seen whether these embryos can be advanced through successful pregnancies to birth.

The potential consequences of cloning are enormous. In the future it may well be possible to produce hundreds of copies of an adult animal. Such cloning would certainly assist the production of biopharmaceuticals in milk. Moreover, it could be feasibly used to create animals that grew faster for meat production and that were more resistant to disease; though one may imagine that a herd of identical animals may indeed be at a far greater risk of disease. Presumably, cloning could also play a role in protecting the bio-diversity of species by ensuring that breeds would not die out through keeping the appropriate cells and embryos in cold storage; though at the same time it poses an enormous threat to the genetic diversity of animal species.

From the point of view of both the consumer and livestock producer, cloning could theoretically ensure that meat and milk of a reliable and standard quality could be produced and sold at a standard price and would turn a standard profit. In this sense, cloning is an extremely attractive proposition for the livestock industry; as is transgenesis also—presumably these technologies will also be combined. They are, however, scientific phenomena with which many already feel extremely uncomfortable. Such technology necessarily involves direct human interference in—what is often perceived to be—the natural order and raises serious questions about the ethics and morality of such scientific enterprise. At present, national governments and scientific regulatory bodies have chosen to err very much on the side of caution when it comes to biotechnological experimentation. Often it is only when it can be proved that genetic experiments with animals can provide important benefits to humankind—as has been successfully argued for the transgenic production of lactoferrin—that they are permitted under stringent controls. Legal prohibitions on the application of such genetic technology in human beings are already firmly in place, particularly with regard to the cloning of human embryos. It remains to be seen whether either animal cloning or the production of transgenic animals will gain widespread acceptance or become commonplace in the new millennium.

Clearly, the scientific developments of recent years and the impending future present a colossal moral quagmire through which scientists, law makers and ethicists alike will be forced to wade; undoubtedly, there are also likely to be significant implications for animal welfare. However, there are also a whole host of practical problems that would have to be overcome before such technology could be successfully and safely applied. The practice of genetic engineering, for instance, carries with it an inherent risk because of pleiotropy; that is a phenomenon which occurs when a gene that does one specific thing may have other, possibly unrelated, effects upon or could interfere with the behaviour of other genes.

One of the major and commonly espoused objections to the genetic technology that has been developed in recent years is that it necessarily involves the deliberate and wholesale interference with 'nature's design'. Further to this, the idea of genetically modifying other life forms has also met with considerable hostility for it reeks of eugenics; the implications of which are still fresh in our collective memory from the darker events of the twentieth century. History has already taught us that the calculated 'improvement' of humans can be a highly dangerous business with the most sinister of consequences. Yet leaving the issue of the application of genetic technology to humans aside and focusing upon animals alone, it is pertinent to consider just how well grounded our objections to and fears of the genetic modification of animals actually are.

Although the genetic technology of recent times has often been portrayed as an extremely new and outlandish development, the fact remains that there is—at least in principle—very little new about it; most certainly as far as animals are concerned. The genetic modification of animals can logically be viewed as an extension; or rather a continuation of the process of domestication and selective breeding that has already been going on for the past 10,000 or so years. As this book has sought to illustrate, for thousands of years, humankind has actively attempted—in ever-increasingly successful measures—to alter and improve the behaviour, physical appearance and productivity of animals in order to service its needs. Whether it be to produce a more tractable animal or one that will produce more milk or larger calves, animals have persistently been deliberately and selectively bred to achieve highly specific ends. In this sense, people have been practising the art of 'genetic engineering' since animals were first domesticated. Traditionally, new varieties of animal species have been produced by cross-breeding. Until very recently, breeders had no detailed knowledge of the genetic mechanisms involved in producing such new varieties. Animal husbandry was a purely empirically based occupation. Farmers gradually learned that if, for example, cattle that produced a high milk yield were bred with other high-milk-yielding cattle, they would produce offspring that would go on to produce even more milk.

This kind of genetic manipulation is, however, far from an exact science. Moreover, such selective breeding can take an extremely long time before it produces the desirable result. Recombinant DNA technology has expedited this process, for it has provided the means by which the genetic make-up of animals can be modified in order to produce specific traits that—by traditional means—would have taken many generations to introduce, or would never have occurred at all since they involve the insertion of DNA foreign to the recipient animal.

Likewise, although many tend to balk at the mention of it, there is little new about biotechnology. Biotechnology can essentially be defined as 'the application of scientific and engineering principles to the processing of materials by biological agents to produce goods and services'. Whilst this may sound like a new-fangled idea, the fact remains that humankind has been using biotechnological techniques for many thousands of years in order to manufacture products such as beer, wine and bread.

To make such products, microorganisms have been routinely and deliberately manipulated for highly specific ends. The role of yeast in bread baking is perhaps the best example of this. Again, it has been the development of recombinant DNA technology that has increased the capabilities of biotechnological production, rendering it an important and efficient technique for the future. As has already been suggested, it is likely that biotechnology will play an increasingly important role in veterinary medical science, particularly with respect to the improvement of animal health and nutrition, the production of veterinary vaccines and prevention of animal disease. As a corollary to this, biotechnology may also be crucial to the maintenance of public health as it may contribute greatly to the control of zoonotic disease.

There are, however, good reasons why we should be cautious about the application of recombinant DNA technology to domesticated animals. Selective breeding, in the traditional fashion, has already proved to have had serious consequences for animal health and welfare. Take, for example, the current state of broiler chickens and turkeys that have been bred to grow so fast and produce such grossly enlarged breast muscle tissue that their legs become crippled; or the double-muscled Belgian Blue calves that necessarily have to be born by caesarean section. Such breeding practices have been developed in the constant bid to increase productivity to meet consumer demands and to produce higher profits, generally at the expense of animal well-being. But, as Chapter 6 illustrated, it is not only the food-producing species that have fallen foul to irresponsible breeding practices.

The health and well-being of pet animals have also been compromised severely through breeders' continued attempts to produce new varieties and to emphasise specific animal characteristics. One could argue that the application of genetic technology could plausibly solve many of the problems created by both intensive farming practices and pet-keeping. For example, it may in the future be possible to create a race of hornless cattle, thus obviating the need for dehorning altogether. Likewise, transgenic chickens might be produced that lack nerves to their beaks and could thus be debeaked without causing them any pain. Whilst such developments may indeed be possible, they will nonetheless do nothing to improve the general conditions in which intensively farmed livestock live and are bred. Similarly, transgenic cats could be produced without claws and dogs with modified vocal cords for the convenience of their human owners.

It remains to be seen whether the genetic modification of animals will create serious new problems with regard to animal welfare. The application of recombinant DNA technology to animals is so recent that it cannot yet be said with any great certainty whether, for example, the inclusion of a single human gene to cattle in order to produce biopharmaceuticals will endanger animal health and well-being in any fashion. Nor for that matter whether it may affect the health of humans who consume such genetically modified milk or its pharmaceutical derivatives. It is likely that such cattle will essentially continue to function just as and be treated like any other dairy animal. In that sense, transgenesis may change little with regard to how animals will live within human society.

Creating transgenic pigs (and possibly other species also) for xenotransplantation, on the other hand, will most likely entail that the animals be kept in extremely sterile, clinical environments that are a far cry from the conditions, for better or worse, that they now experience. What is, however, clear is that with the introduction of genetic engineering, our manipulation and exploitation of animals has entered a brand new phase. The production of new animal varieties can now take place over the course of one single generation, rather than many. Moreover, and perhaps more importantly, recombinant DNA technology means that animals can be modified using genetic material that is completely alien to the recipient animal species. Needless to say, this signifies a crucial new development in the age-old practice of animal husbandry. It remains to be seen to what extent the practice of genetic engineering will be continued and commercially employed, or whether this will be deemed ethically and biologically unacceptable.

As the new millennium approaches, it is clear that the role of domesticated animals in human society is continuing to diversify, rather than diminish. Whilst it appears that many people are becoming more sensitive to the lot and rights of other animals, humanity as a whole seems to be perpetuating the tacit belief that animals may be exploited as a natural resource. In this respect, the message of this book is rather bleak for those who seek to reduce or rid the world of animal exploitation. For the foreseeable future, this is unlikely to happen; though it does not necessarily exclude the possibility that animals could well be treated in a more humane fashion whilst being used by humankind.

A world where animal flesh is no longer eaten, nor animals exploited for human ends in any way, remains steadfastly within the realms of fantasy and science fiction. The Utopian vision of *Star Trek* where humanity has attained the pinnacle of civilisation by no longer breeding and killing animals for food, instead obtaining its nutrients and raw materials through synthetic replication, is far removed from the stark realities of the twentieth—and probably twenty-first—century. Contrary to what many may hope, the human exploitation of animals has greatly increased and looks set to increase still further in the years to come as new uses for animals are dreamt up along with the new technologies that can achieve them.

The final question remains of just what these developments will entail for the veterinary regime. It seems likely that as the frontiers of animal exploitation continue to expand, so will the importance of veterinary medical science to society also. Veterinary expertise has thus far been crucial to the development of transgenic animals for the production of biopharmaceuticals and xenotransplantation. The close interaction between veterinarians and scientists from other academic spheres will presumably continue as such technologies are further refined or expanded. Scientific developments seem likely to require that in the future veterinarians will be obliged to be complicit in the activities of the biotechnologists, as much as they have become in the activities of the livestock industry during the post-war period.

If biotechnology turns out to offer all that it today seems to promise, then it seems certain that the efficacy of the veterinary regime will improve further still. The prospect of new and more effective vaccines to combat animal disease and the improvement of animal nutrition have great implications for both the health and well-being of animals and public health. If the damage caused by animal disease is reduced and the methods of controlling it become cheaper, the consumer will undoubtedly also reap the benefits of veterinary advances through being able to buy safer, cheaper and better quality animal produce. However, at present, many consumers are skeptical about the advantages that genetic technology purportedly offers and the impact that it will have upon their lives. Likewise, policy-makers and legislators have responded cautiously to recent scientific advances and have been keen to exercise their regulatory powers upon the burgeoning biotechnological industry. In this respect, as we enter the twentyfirst century, the state is likely to tighten its grip still further and continue its imposition of the veterinary regime with increased vigour, particularly within an international framework. Throughout the course of the late twentieth century, the importance of an effective veterinary regime has become increasingly evident. Calamities such as the recent BSE crisis have explicitly highlighted the potential consequences of our increased exploitation of animals and the necessity of intensifying measures for animal disease control. As the new millennium dawns, it seems more than likely the burden of beasts for human society will continue to be a heavy one.

EVALUATE CURRENT AND PROPOSED LIVESTOCK MANAGEMENT PRACTICES

To maximize forage productivity, it is best to combine or alternate various livestock breeds on a range. Their differing food preferences can help to keep plants productive by minimizing overgrazing of a particular favoured area and allowing less preferred plant species time to mature. It is prudent to make superior forage available to those animals with the highest needs. When forage is limited, livestock managers may decide that young and milk-producing animals must have first access to new pastures and ranges with a wide variety of abundant forage. Managers should investigate the value of various systems of rotating livestock. Rotation allows land to be grazed continuously throughout the year. Livestock can be rotated between fields or ranges to prevent the buildup of disease and to vary grazing pressures. Through either fencing or herding, they can be relocated into croplands to consume crop residues. Assessment of seasonal grazing patterns should include potential impact on soils.

Dry-season grazing can benefit the land by breaking up crusted soil and working seeds into the ground. By contrast, grazing on moist soil can cause considerable soil compaction, which reduces the soils ability to absorb moisture and can result in increased erosion from run-off during the rainy season. Many of the environmental impacts from livestock production are associated with particular practices of livestock managers.

Thus it is critical to understand current practices and how the proposed project might alter these practices or promote new ones:

- Who are the local community's livestock managers?
- What practices do a family or community use to control the size and composition of livestock herds?
- How do livestock managers currently control livestock movement? Will the proposed project change these movements in a way that might harm the environment?

- Does the proposed project require the construction of fences? If so, will they interfere with wildlife migration or transit of livestock belonging to other communities? Could the fences lead to overgrazing and land degradation? Will the fences be built with local materials? Would living fencing be practical? Would solar-powered electric fencing be technically and economically feasible?
- Are streams and riverbanks currently protected from livestock damage? If the proposed project will open new areas to grazing, will water supplies need to be protected?
- Must steps be taken to prevent new livestock and associated animals (*e.g.*, dogs) from transmitting disease to wildlife? Is there a vaccination/ animal disease control programme available for this purpose?
- Will the project involve construction of improvements (*e.g.*, boreholes or other infrastructure)? Could these lead to unplanned changes in herding patterns and overgrazing?

ASSESS DEMAND AND USE OF LIVESTOCK PRODUCTS

- Who is marketing livestock and livestock products?
- Is the demand for livestock products coming from local or outside populations? How rapidly is it increasing or decreasing? How stable is the demand?
- In preparing livestock products, are people using technologies which reduce impacts on the environment, open additional markets, or improve health and nutrition (of people and animals.)

ASSESS LIVESTOCK ECTOPARASITE MANAGEMENT

Epidemic and endemic diseases continue to be a major constraint to livestock productivity in large parts of the developing world. Although vaccines have controlled many of the epidemic diseases, they continue to cause severe economic losses through morbidity and mortality.

These diseases include the infections caused by vector-borne haemoparasites and helminths. Existing technologies, such as chemotherapeutic agents and live vaccines that were previously successful in controlling these diseases, are no longer effective—because of acquired resistance or weakened delivery services. Appropriately designed alternatives are often lacking.

CONSIDER POPULATION PRESSURE AND DISEASE BURDEN

Two factors that may affect the outcome and impact of projects population growth and fatal or debilitating epidemic diseases. Population growth may lead to conflicting uses of grazing lands or reduce individual farms or rangeland to sizes that cannot sustain livestock.

Fatal or debilitating epidemic diseases may weaken effective dissemination or replication of proper livestock management techniques.

- What is the current and projected population growth rate in the project area? How might this affect project sustainability in the future?
- What is the current extent of the HIV/AIDS epidemic in the region? How might this affect the composition of the population (size, ethnic makeup, age/gender distribution) and family structures necessary for project sustainability? How will development and livestock technical support services be affected?
- Are there other epidemic diseases in the region, such as sleeping sickness, that might adversely affect project implementation?

ROLE IN INDIAN POULTRY PRODUCTION

Remarkable growth of poultry sector in the Indian sub-continent has been essentially chicken dominated; dependent on intensive system of production involving high technology with high external inputs. This is not necessarily appropriate and sustainable to the socio-economy of a densely populated country facing routine shortage of conventional poultry feed resources. The prevailing socio-agro-economic scenario of this sub-continent calls for some kind of cafeteria approach for poultry production; scientists should develop and offer a broad spectrum of poultry alternatives to meet the different local requirements. The availability of such diversification possibilities would maximize the returns from a given level of inputs and also minimize the risks and hazards to the environment. In this context, importance of alternate poultry species is being rediscovered. Production of alternate poultry may never rise enough to compete with commercial chicken, but these birds could become a significant source of food for the masses and also a source for substantial supplement income.

Guinea fowl and turkeys are interesting gallinaceous birds being farmed for centuries to meet poultry production requirements of diversified eco-agriculture mellieu. The traditionally raised guinea fowl and turkey ensure reasonable returns at insignificant cash inputs, while quail production is a distinct diversification offer. Intimate understanding of the basics of domestication; disease control, nutrition and genetics has helped farm production of these avian species. Present paper covers the present and future importance of these alternate poultry germplasm resources available in the country.

GERMPLASM RESOURCES

Guinea fowl were introduced into the Indian sub-continent during the slavery era of mediaeval centuries. Preliminary survey of guinea fowl showed its distinct popularity with marginal farmers and other vulnerable groups as small-scale poultry enterprise. The indigenous germplasm seems well adapted to the diversified agro-climatic conditions prevailing in semi-arid regions. Guinea fowl population rank third after chickens and ducks. It is referred to by different regional names in this sub-continent 'Titari in northern plains; and 'Chittra' in the Majority of guinea fowl are raised in\ semi-arid pockets of Punjab,Uttar Pradesh, Bihar, Rajasthan, Madhya Pradesh, Maharashtra, Saurashtra, Karnataka, Andhra Pradesh, Orissa and Tamil Nadu. Present production status in this country is best summed up as widely spread and highly fragmented.

The descriptions of guinea fowl breeds/ varieties are essentially based on the plumage colour variations. Pearl, Lavender and White are the three main varieties recognised world over. In India, Pearl guinea fowl are most common among village stocks and usually referred as the 'local' breed. Pearl individuals are characterised by dark-gray feathers with uniformly distributed white spots that present a pearl like appearance. Pearl plumage, colour is also regarded as the 'wild' mutant. Levender birds are identified by their light gray distinctly spotted feathers. White variety possesses completely white plumage. Guncari guinea fowl stocks developed at the Central Avian Research Institute, Izatnagar are selected for high early body weight and breed true to their plumage characteristics. Some minor plumage mutant varieties, *viz.*, Violet, Lilac, White breasted pearl and White breasted lavender also available.

Ecogenetic Potentials

The ecogenetic characterisation is identification of species specific potentials to survive under non-conductive harsh environments. Evolved in tropics, guineafowl is considered to possess relatively higher level of tolerance for rigourous tropical climatic conditions. Although no detailed studies are reported but preliminary observations indicate existence of better potentials.

Body conformation and plumage cover are the two important factors in determination of heat tolerance potentials. Light in body weight, long-legged guinea fowl present greater surface area for heat dissipation per unit body weight. Low feather weight estimates (7.8 per cent) at 12-week to 16-week age are comparatively lower than the average values reported for different domestic fowl classes. Long neck devoid of feathers, poorly feathered thighs, exposed and highly vascular nares, cere, wattles and helmet play important role in thermoregulation. Its more fluffy monotype plumage covering and potential to elevate feathers permit better circulation of air through plumage, more efficient dissipation of internal heat.

Normal respiratory actions and stress panting with outstretched wing help dissipation of internal heat through vaporisation movement. The normal respiratory frequency for guinea fowl is not different from that of chicken (18-30 breaths/min). Experiments revealed comparatively little polypnea during moderately high ambient temperature (85°C - 110°F) among guinea fowl vis-à-vis chicken. Mycotoxin contamination of feed grains is more common in tropical and sub-tropical countries, particularly when coupled with poor handling operations. The higher aflatoxin tolerance of guinea fowl indicates it potential to utilise lower quality feed ingredients.

Production Systems

Guinea fowl may be raised in the modern intensive system characterised by high input, good husbandry and hygiene and supply of balanced feed, however this system is confined largely to Government/ public sector farms. The traditional extensive system of rearing is most popular in rural areas. A limited survey showed that birds are maintained in complete harmony to the village environment; housing and management is completely traditional and birds are raised on free range. Average flock size varies widely (15-200) in guinea fowl producing pockets. Droving is the most popular form of husbandry. Birds offer no management problems, primarily due to their flocking instincts. In semi-extensive system, birds are maintained in pens communicating with spacious well-fenced enclosures. Management is entirely through incorporation of family labour; two persons can easily manage even large flocks. In the traditional rearing system, the nutritional requirements of voraciously omnivorous guinea fowl are met through its catholic feeding habits; birds accept fallen grains, leaves, weeds, root-bulbs, fruits, diversified insect fauna and flora and even carrion. Supplementary feeding, if any, usually consist of waste grains, household waste and crop residues.

Strategies of combining guinea fowl raising with arable crops, fruit plantations and pastures for ruminant production contribute towards the production economy as well as stability of the system.

Its advantages include:

- More complete utilisation of land,
- Control of waste herbage, weed growth and the insect pests,
- Increased land fertility and
- The plantation environment reduces heat stress to birds during foraging activities.

Several reports indicate that guinea fowl may be beneficially integrated with crops of rice, ragi, grapes, apple and ruminant farming, etc., because of its potentials to minimize the harmful insect pests and weeds. Our observations strongly suggest complete economic viability of guinea fowl production. A study of inputs and returns showed reasonable net gains for guinea fowl production when 2-3 crops of at least 500 heats are raised every year on foraging: particularly when mortality was kept under control. In view of the increasing demands for more and more diversity in poultry meat, and elite class consumer preferences for the native free range fed poultry products, there seems and ample scope for the guinea fowl produce.

Germplasm Resources

Although, turkeys were introduced in this subcontinent continent several decades back by the Christian missionaries, but till date turkey farming has achieved very little progress as well popularity. Possibly high cost of production, long generation interval and less market demand are the main reasons for its low popularity. A study conducted at HAU, Hisar showed that production price of 1 kg live weight of turkey was more than 1.5 times higher that that of broilers. Generally speaking, turkey production has failed to generate popular interest in most of the tropical countries outside Latin America. Recently, it has atso started gaining popularity in several African countries. Bronze, White Holland Bourbon Red, Narangansett, Black, Slate and Beltsville Small White (BSW) are the standard breeds. Among a dozen or more non-standard varieties, Broad Breasted Large White (BBLW) and Broad Breasted Bronze (BBB) are more popular. The white-feathered turkeys are mutants of the original Bronze birds. White birds are preferred over coloured ones because coloured birds contain a black pigment in their quill feathers. In India, Beltsville Small White type turkeys are being maintained at most of the Government/university experimental farms. At few centers, Broad Breasted Bonze and Broad Breasted Large White units are also being kept.

Organised turkey farms under private sector are practically non-existent, limited to few small units at Government/public sector farms. Domestic turkey maintained at private small-scale units scattered throughout the country are mainly the decendents of semi-improved Broad Breasted Bronze turkey imported from Europe and North America during the early ears and it is referred to as local breed. Few flocks of Norfolks and Cambridge turkeys, may be also there. The total turkey population is small, may be one lakh birds are raised annually. These unimproved turkeys are produced as a cash crop for meeting the market needs for Christmas dinner, birth days, wedding or holidays treats.

Ecogenetic Potentials

Evolved from a truly tropical bird, whose natural habitat was open forests and wooded areas, the present domesticated farms have largely retained its self-reliance, ruggedness and adaptability. They can be reared virtually anywhere. The hardy 'local' turkey stocks are able to thrive under hot climate to the arid conditions, from sea level to mountains with near temperate climates and from rain forests to deserts. Turkeys walk rather than fly and find almost all their food on ground when kept on range. They possess excellent foraging and scavenging abilities.

Production System

The turkey may be reared under intensive as well as extensive system of rearing. The intensive rearing system is largely restricted to the Government/ public sector. The intensive management is almost similar to that of chicken and is basically of deep litter type. One sq ft of floor space per poult is required during the first 3-4 weeks and thereafter the floor space requirement per bird is 1.5 sq ft upto 8 th week, 2 sq ft upto 12 th weeks of age, 2.5 sq ft upto 16 th weeks of age and 305 sq ft after 16 th weeks of age. Free range system of rearing is most popular for rearing the local stocks of turkey. They are natural foragers and scavengers and always range farther. Indeed, they thrive best where they can rove about freely feeding on seeds, fresh grass, other herbage and insects which include locusts, cicadas, crickets; grasshoppers, worms, slugs and snails, etc. The self reliant 'local' stocks cat do well with little management and can still fly short distances to avoid predators.

ORGANIC PRODUCTION OF POULTRY

Any number of pasture-based systems can be classified as organic if the producer chooses to follow the certification process which requires that feed, pasture and processor be certified organic. In terms of organic production, the USDA'S National Organic Programme requires outdoor access and organic feeds produced without synthetic chemicals and prohibits the use of antibiotics (USDA NOP, 2006; Fanatico, 2007). In Florida, there are a number of producers that follow organic standards for raising their flock but have chosen not to become certified organic for various reasons. For more information on organic standards in Florida contact Florida Organic Growers Association (www.foginfo.org). The main drawback of organic production is the elevated cost of feed and the lack of processing facilities that offer services for certified organic producers.

Opportunities of Pastured Poultry Production

According to the National Center for Appropriate Technology (NCAT), there is a strong demand for pasture-raised poultry products when marketed directly to consumers. These types of operations can be managed by one person and producers can start small and grow as they become proficient in producing and marketing poultry and/or eggs. There is minimal initial capital investment compared to most agricultural enterprises. However, poultry products can also attract customers for other products. Moreover, pastured poultry production is arguably a more sustainable system because the nutrients in poultry excreta are being deposited into the soil. These deposits serve as a source of organic matter and nutrients that help build soil fertility, assist water retention and increase forage production. This can be especially beneficial in areas with deep sandy soils and low fertility.

Challenges of Pastured Poultry Production

The challenges associated with pasture-raised poultry are labour, seasonality and processing. This is a labour intensive operation: the birds will need to be fed, watered and moved at least once–and sometimes twice–daily. Meat production in North Florida is mostly seasonal with production and harvest ranging from February to June and August to December in order to avoid extremes in temperature. The climactic conditions will dictate how early or late in the season birds can stay in the pasture because weather related stress can affect feed conversion and meat quality. On the other hand, laying flocks can cope with varying climactic conditions if adequate housing is provided. Predation pressure also presents challenges, especially for new producers because it takes time to develop predator control strategies to minimize losses.

Meat Birds: Breed Selection

Unlike with laying hens there are limited breed options for meat birds. The options include the Cornish cross hybrids which are selected for commercial production systems. Cornish crosses are considered a fast-growing broiler. These broilers reach a market weight of five to six pounds in approximately seven to eight weeks. Due to the fact that Cornish cross broilers are harvested at a younger age their meat tends to be more tender and juicy and does not exhibit a strong flavour like heritage birds.

An emerging trend in pastured poultry production is the use of dual purpose heritage breeds to help preserve the genetic pool of poultry in the United States. Heritage breeds are breeds recognised by the American Poultry Association prior to the mid 20th century. Heritage breeds include barred silvers and dual purpose breeds like New Hampshire, Buff Orpington and Barred Rocks. With dual purpose poultry breeds males are selected for meat production and females for egg production. The dual purpose chicken will not provide a large carcass as a meat bird, nor lay as many eggs as an egg layer. These birds tend to grow slowly, eat more feed and may take up to 12 weeks to reach market weight. Heritage breeds are also considered to have a stronger flavour because the age at which the birds are killed and processed is an important factor that contributes to flavour. Breed selection will be based on your preferences and in some instances the preferences of your market and consumer demand.

Fig. Buff Orpington Hen.

Selecting Laying Hens

Commercial hatcheries offer a wide variety of options when it comes to laying hens. A common misconception is that hybrids are man-made breeds. In reality hybrids are a cross between known breeds whose offspring will usually grow faster, provide flock uniformity and increased egg production. This is often referred to as hybrid vigour. Some examples of hybrids include red sex links, production reds and white leghorns. Once again, personal preference and market will drive the breed selection process.

Heritage breeds are also an option in a layer operation. A benefit of using heritage breeds is the conservation of genetic diversity. Some of the breeds in this category include New Hampshire, Rhode Island Red, Buff Orpington, Minorca and Ancona. Additionally, heritage breeds have adapted to changing environmental conditions and are natural and efficient foragers. Using this type of bird in a pasture-based production system can also be used as a marketing tool.

Some producers prefer to select a breed based on the colour of eggs desired. A rule of thumb for brown and white egg layers is the colour of the bird's ear lobes will, in many cases, correspond to the colour of the eggs which they will lay. White Leghorns are among the most popular and productive layers of white eggs whereas Rhode Island Reds are the second most popular and produce large brown eggs. If more unique shell colour or "designer eggs" are desired then the Araucana and Ameracana breeds will provide these results.

Starting the Birds

In a pasture-based poultry system newly hatched chicks will not go directly into the pasture. During the first couple of weeks an enclosed structure is needed in order to provide temperature regulation, feed and water. This enclosed structure is more commonly known as a brooder. In the case of broilers, the chicks will remain in the brooder for approximately three weeks or until they are sufficiently feathered to withstand climactic conditions. In contrast, chicks that will later become layers will need to remain in the brooder for approximately six weeks. These first couple of weeks will determine the long term survival, development and health of the flock in the pasture.

The brooder should meet the following requirements:

- One-half of a square foot per bird.
- A heat source (infra red lamps are recommended, have two in case one burns out).
- Three inches of dry bedding such as wood shavings, hay or straw. Try to avoid slippery surfaces like newspaper.
- One quart of water for every 25 birds.
- Clean feeders. The feeder should be placed near the heat source, but not directly underneath it.
- Some form of ventilation. Chicks need draft protection, but proper ventilation during the brooding period is very important.

During the first week, the temperature in the brooder should remain a constant 95 degrees Fahrenheit. After the first week the temperature should be lowered five degrees per week for the next two to three weeks, after the third week the birds are typically ready to be moved into the pasture. The temperature in the brooder can be decreased by raising the heat source. It is very important that fresh water and feed be offered ad libitum (free choice) daily because it is vital for their growth and health once they are in the pastures. Care should be taken to insure that the feed and water are not allowed to become too warm because this will result in the chicks not eating and drinking as much as they need.

On delivery day, you need to have the brooder up and running 24 hours before the chicks arrive in order to detect any problems beforehand. Also make sure that the hatchery sends the chicks via overnight mail to avoid losses and to get the chicks to starting drinking and eating feed. When placing the birds in the brooder each bird should be introduced to the water source by dipping their beak directly into the water.

U.S., consumers are increasingly interested in products they perceive as naturally-produced or environmentally-friendly, that provide a high level of nutrition, good flavour, improved welfare for the birds, and provide information about how the food was produced. These interests have led to specialty markets for poultry produced in alternative systems such as free-range or organic. In Florida, the producers of pastured poultry can meet the demands of this growing segment of consumers. However, the lack of processing facilities is the main challenge that producers will need to overcome either by building their own facilities, retrofitting existing ones or working collaboratively to gain access to mobile processing units.

9

Animal Breeding

INTRODUCTION

Animal breeding, controlled propagation of domestic animals in order to improve desirable qualities. Humanity has been modifying domesticated animals to better suit human needs for centuries. Selective breeding involves using knowledge from several branches of science. These include genetics, statistics, reproductive physiology, computer science, and molecular genetics. This article discusses the basic principles of how populations of animals can be changed by application of these principles, and a brief discussion of molecular genetics, immunogenetics, and newer reproductive technologies is included. The fundamental biological principles underlying animal breeding are discussed in the articles heredity and animal reproductive system.

The science of Animal breeding is defined as the application of the principles of Genetics and biometry to improve the efficiency of production in farm animals. These principles were applied to change animal populations thousands of years before the sciences of genetics and biometry were formally established. The practice of animal breeding dates back to the Neolithic period, when people attempted to domesticate wild species such as reindeer, goats, hogs and dogs. Domestication was performed through controlled mating and reproduction of captive animals which were selected and mated based on their behaviour and temperament. Judging from cave paintings that have survived, selection was also applied to some qualitative traits such as coat colour and the absence or presence of horns.

Without written records, there is no certain knowledge of the evolution of animal breeding practices, but written documents dating back more than 4000 years indicate that humans appreciated the significance of family resemblance in mating systems, recognized the dangers of intense inbreeding, and used castration to prevent the reproduction of undesirable males. Progress in the performance of domesticated animals through these selection practices was very slow; improvements were mainly due to animals adapting better to their environments. Robert Bakewell, an English animal breeder of the 18th century, is considered the founder of systematized animal breeding. He was the first to emphasize the importance of accurate breeding records, introduced the concept of progeny testing to evaluate the genetic potentials of young sires, and applied inbreeding to stabilize desired qualitative traits. He also promoted concepts such as "like begets like," "prepotency is associated with inbreeding" and "breed the best to the best." Bakewell and his contemporaries in Europe pioneered the development of diverse breeds of Beef cattle, Dairy cattle, Sheep. Colour, conformation, geographical origin and some production characteristics were the main factors that differentiated these breeds. Wide geographical redistribution of animal populations was also an important factor in the formation of new breeds, as invading armies, migrating people and traders transported livestock to new lands.

Animal breeding as a modern science belongs to the 20th century. Although numerous geneticists and biometricians have made significant contributions to the development of this science, J.L. Lush of Iowa State University is considered the father of the modern science of animal breeding. Lush and his students developed major scientific procedures applicable to the genetic improvement of farm animals.

BREEDING STOCK

"Breeding stock" is a term used to describe a group of animals used for purpose of planned breeding. When an individual is looking to breed animals, he or she is looking for certain valuable traits in purebred stock for a certain purpose, or may intend to use some type of crossbreeding to produce a new type of stock with different, and presumably superior abilities in a given area of endeavor.

To breed chickens, a typical breeder intends to receive eggs, meat, and new, young birds for further reproduction. Thus the breeder has to study different breeds and types of chickens and analyse what can be expected from a certain set of characteristics before he or she starts breeding them. Accordingly, when purchasing initial breeding stock, the breeder seeks a group of birds that will most closely fit the purpose intended.

Backyard Breeding

The term backyard breeder is a general term, sometimes considered derisive, used to describe people who are "home breeders", and breed only a few animals for personal use or occasional sale. However, while some backyard breeders pay strict attention to quality, the term usually describes those who allow animals, particularly dogs or horses, to procreate regardless of physical, genetic, and/or emotional health. It is also used to describe unreputable commercial breeders. In the process of careless breeding, many backyard breeders produce genetically weak animals that can be predisposed to genetic disorders or debilitating physical deformities, particularly those that result from inbreeding. In dog breeding, these individuals focus on only a limited aspect of the dog, while ignoring the original function and temperament of the breed. When such breeding is carried out on a large scale, the venue is called a *puppy mill* or *puppy farm*. Backyard breeding is popularly blamed for the proliferation of aggressive dogs for the sports of dog baiting and dog fighting. Dog fanciers generally believe that such ill-bred dogs are the reason for the bad public reputation of certain dog breeds, and the increasingly frequent enactment of breed-specific legislation. Because of the time and expense of feeding and caring for horses, which produce one foal per year, there are fewer horse breeders who produce animals *en masse,* though some individuals do engage in animal hoarding and breed far more animals than they can support.

In the horse world, over breeding of grade animals that cannot be sold except for very low prices raises concerns that such animals will be slaughtered for horsemeat. Some, though not all, colour breeders make errors similar to puppy mill breeders by focusing on only one characteristic to the detriment of health, soundness, and quality conformation. In addition, industrial breeders, such as those who produce *pregnant mare urine* for the manufacture of premarin, allow many mares to be bred indiscriminately, simply so that they are able to produce the urine laced with reproductive hormones that forms the base for the drug. Their foals are generally considered an unwanted byproduct, particularly males (colts), and sold off, frequently to the slaughter market.

Purebred Breeding

The idea of *breed purity* may strike an unpleasant chord with some individuals because it is reminiscent of nineteenth-century eugenics notions of a "superior strain" in humans, supposedly exemplified by members of an aristocratic social class or of certain races. The application of theories of eugenics had far-reaching consequences for human beings, particularly in cases such as The Holocaust. However, the idea of a superior strain in animals, such as that incorporated in the breeding of Thoroughbred race horses, was less controversial. By "breeding the best to the best," employing a certain degree of careful inbreeding, considerable culling, and selection for "superior" qualities, one could develop a bloodline or "breed" superior in certain respects to the original base stock or landrace which had been produced primarily by natural selection.

This process forms the basis for a "true-breeding" line of animals, which today is generally called a "purebred" line. Once created, a purified line is "closed" and preserved from dilution and possible "debasement" by unrelated stock. However, this theory is not without some negative consequences. When the ideal of purified lineage or purely aesthetic type is seen as an end in itself, without consideration of usefulness or vigour, a breed can suffer over time.

Also, if a breed registry is closed with too small a gene pool, over time, degradation due to inbreeding depression may occur. For many purebred breeders, the need for a periodic outcross is recognized and incorporated into an overall plan. The observable phenomenon of hybrid vigour stands in contrast to the notion of breed purity. However, on the other hand, indiscriminate breeding of crossbred or hybrid animals may also result in degradation of quality, particularly when animals are mass-produced without concern for overall health or longevity.

COW BREEDING

DIFFERENT BREEDS FOR DIFFERENT NEEDS

Dairy cattle are breeds that have been developed to produce milk whereas beef cattle are bred and selected primarily for the production of meat. Dual-purpose breeds do exist and they have been selected for both meat and milk production. In the last 20 years the genetic make up of the UK dairy herd has changed and is still changing. Cows are larger, earlier maturing, are producing far more milk and are more efficient at converting feed into milk.

As it is the birth of the calf that causes milk production, it is critical for the farmer that the cow becomes pregnant again during lactation and produces the next calf one year after the previous one. The average length of productive life of a dairy cow is between four and five lactations, although this varies from one or two to over ten. Since the fifties artificial insemination (AI) has increasingly replaced natural service. Embryo transplants (ET) are also being developed. A surrogate mother may have difficulty delivering, since the foreign calf is often larger than nature intended. When this happens, a rope or chain is attached to the calf's forelegs, and, with a winch-like device the newborn calf is pulled from the womb. Unnecessary suffering for both cow and calf.

Regular Calving

A high annual milk production per cow is dependent on regular calving at about 12- month intervals. Poor reproductive efficiency leads to longer calving indices and a reduction in milk sales per cow per annum. It also leads to an increase in the number of cows culled for failing to conceive. On average, in the UK, only about 85 cows calve per annum for every 100 cows available for breeding. Over thirty-five per cent of all cows culled are disposed of because of reproductive disorders.

Obtaining Semen

Photographer Henry Heap visited the Milk Marketing Board stud bull operation in Mid Wales. He provided Turning Point magazine with a complete set of photographs and commentary of his visit. "A large, handsome, tan bull stood stock still, tethered by the ring tethered through his nose to what looked like a giant circular clothes line or some grotesque merry-go-round. He was kept so motionless by a rubber hood that covered his eyes, which had a similar effect to that of hooding a falcon. A worker commented that the bull would stay still all day if they left the blindfold on. However, this bull was not the source of the semen, but instead what is known as a 'teaser'. The stud bulls are led out by the nose one at a time to mount the motionless hooded bull. After they have mounted the teaser their penis is diverted into a leather vagina which is used to collect the semen. As soon as the process is complete, another bull is led out. The hood keeps the teaser bull stationery for each successive beast to mount. These stud bulls never even see females.

During embryo transfer an embryo can be flushed from the uterus of a genetically superior cow and implanted in a receiver cow of so-called 'poorer quality' for foetal development. Embryos are also taken from dead cows at the slaughterhouse. Hormone treatment is used to induce multiple ovulations and the cow is then inseminated. A few days later several embryos can be flushed from her. For both embryo collection and implanting the cow receives a local anaesthetic injection in the tail. Embryos of the desired breed can then be used. ET maximises the number of offspring from selected breeding females and maximises profit. Do-it-yourself embryo transfer is now taking place in the UK. Courses are run to train in direct embryo transfer techniques using the quick-thaw method for frozen embryos. Although vets are not required to be present at the time of transfer, they must check each animal is suitable for implantation. Farmers genetically improve the herd by using techniques such as MOET (multiple ovulation embryo transfer) which involves inducing a genetically 'superior' cow to produce up to 20 embryos rather than one.

For ease of management, farmers prefer cows to calf at the same time. They, therefore, need to detect when cows come into oestrous. The oestrous detection rate (ODR) is the number of cows observed in oestrus during a 3 week period, in relation to the number of cows expected to be observed. The average ODR on farms is about 60 per cent. One alternative is to synchronise the oestrous period of several animals by injections of prostaglandin or administration of progesterone as an impregnated intraveginal coil implanted for 9-12 days.

ARTIFICIAL INSEMINATION

Artificial insemination (AI) now accounts for around 75 per cent of British cattle births. The cow is tied up and one hand of the inseminator manipulates the cervix through the rectum wall whilst the other discharges semen into the vagina and cervix using an inseminating gun. A regional manager with AI firm Genus the problems associated with the procedure. "Placing a quarter of a cc of diluted semen into an area as small as a penny coin, when the target area is inside 500kg of unco-operative cow, requires a high degree of expertise. The more the cow is able to move about, the harder it becomes to hit the target.

Poor handling facilities cause stress, the cow, the inseminator and both our bank balances." There are many invasive procedures carried out on farm animals that the general public rarely witness. One of many includes making sure the cow is in calf. This is carried out using an ultrasound device much like a small TV set with a long thick wire attachment. A vet will insert their hand and arms length into the vagina of the cow.

The animal is held in a holding device to keep her still. The screen shows the small calf in the middle of a dark shaped uterus. Herds usually calve at the same time (due to human intervention) so all the cows on the farm will usually be checked at the same time.

Jersey Cattle

A small, honey-brown breed of dairy cattle, the Jersey is renowned for the high butterfat content of its milk, as well as a genial disposition. The Jersey cow is quite small, ranging from only 800 to 1200 pounds (360 to 540 kg).

The main factor contributing to the popularity of the breed has been their greater economy of production, due to:

- The ability to carry a larger number of effective milking cows per unit area due to lower body weight, hence lower maintenance requirements
- High butterfat conditions, and to thrive on locally produced food.

Bulls are also small, ranging from 1200 to 1800 pounds (540 to 820 kg), and are notoriously very aggressive. Castrated males can be trained into fine oxen which, due to their small size and gentle nature make them popular with young teamsters. Jersey oxen are not as strong as larger breeds however and are generally out of favour among competitive teamsters. Due to the small size, curious docile character and attractive features of the Jersey cow, small herds were imported into England by aristocratic landowners as props for aesthetic landscapes. As its name implies, the Jersey was bred on the British Channel Island of Jersey.

It apparently descended from cattle stock brought over from the nearby Norman mainland, and was first recorded as a separate breed around 1700. Frank Falle has speculated (on the basis of DNA evidence) that the oldest settlers in Jersey were Danish vikings who had been to Nantes with Hatain, where it is recorded that tribute was given by the King of France to Hatain of 500 cattle to leave that area, whereupon they settled in the Normandy area. The marked resemblance between Jersey cattle and the Nantaise cow would seem to bear this out. While the breed is isolated from outside influence today, this was not always the case. Before 1789 cows would be given as dowry for inter-island marriages between Jersey and Guernsey. This was, however, not widespread.

Since 1789, imports of foreign cattle into Jersey have been forbidden by law to maintain the purity of the breed, although exports of cattle and semen have been an important economic resource for the island.

The restriction on the import of cattle was initially introduced in 1789 to prevent a collapse in the export price. The United Kingdom levied no import duty on cattle imported from Jersey. Cattle were being shipped from France to Jersey and then being shipped onward to England to circumvent the tariff on French cattle. The increase in the supply of cattle, sometimes of inferior quality, was bringing the price down and damaging the reputation of Jersey cattle. The import ban stabilised the price and enabled a more scientifically controlled programme of breeding to be undertaken.

SHEEP BREEDS

The domestic sheep is a multi-purpose animal, and the more than 200 breeds now in existence were created to serve these diverse purposes. Some sources give a count of a thousand or more breeds, but concrete documentation has not been produced for such a number. Almost all sheep are classified as being best suited to furnishing a certain product: wool, meat, milk, hides or a combination thereof (i.e. dual-purpose breeds). Other features that are important when classifying sheep include face colour (generally white or black), the length of the tail, the presence or lack of horns, and the topography for which the breed has been developed.

This last point is especially stressed in the United Kingdom, where breeds are described as either upland (hill or mountain) or lowland breeds. A sheep may also be of a fat-tailed breed, which is a type of dual-purpose sheep common in Africa and Asia with larger deposits of fat within its tail, or the meat from such a breed. Ovine breeds are also grouped based on how well they are suited to producing a certain type of breeding stock. Generally, sheep are thought to be either "ewe breeds" or "ram breeds". Ewe breeds are those that are hardy, and have good reproductive and mothering capabilities—they are for replacing breeding ewes in standing flocks. Ram breeds are selected for rate of weight gain and carcass quality, and are mated with ewe breeds to produce meat lambs. Many breeds, especially rare or primitive ones, fall into neither category strictly. Lowland and upland breeds are also crossed in this fashion, with the hardy hill ewes crossed with larger, fast-growing lowland rams to produce a type of ewe called a mule, which can then be crossed with meat-type rams to produce prime market lambs.

Sheep breeds are categorized by the type of wool they exhibit, irrespective of whether they are favoured for meat or wool production. Fine wool breeds are those that have wool of great crimp and density, and are consequently preferred by the textile industry. Most of these breeds were derived from Merino sheep, and Merinos continue to dominate the world sheep industry.

The record for the most valuable sheep ever belongs to an Australian Merino ram which was sold for $16,000 AUD.

Medium wool breeds have wool between the extremes, and are typically fast-growing meat and ram breeds with dark faces. Some major medium wool breeds, such as the Corriedale, are dual-purpose crosses of long and fine-wooled breeds created for high-production commercial flocks. Long wool breeds are the largest of sheep, with lengthy wool and a slow rate of maturation. Long wool sheep are most valued by modern agriculture for crossbreeding to improve the attributes of other sheep types. The American Columbia breed was developed through crossing Lincoln rams—a long wool breed—with fine-wooled Rambouillet ewes.

Coarse or carpet wool sheep are those with a medium to long length wool of characteristic coarseness. Breeds traditionally used for carpet wool show more variability in general traits than other types, the chief requirement being a wool that will not break down under heavy use (as would that of the finer breeds). As the demand for carpet-quality wool declines, some breeders of this type of sheep are attempting to reposition a few of the traditional breeds to alternative purposes. Others have always been primarily meat-class sheep.

A last group of sheep breeds is that of fur or hair sheep, which do not grow wool at all. Hair sheep are similar to the very early type of domesticated sheep kept before wooly breeds were developed, and are primarily raised for meat and pelts. Some modern breeds of hair sheep, such as the Dorper, are the result of crosses between wooled and hair breeds. For producers uninterested in selling wool, hair sheep are highly advantageous, as those who keep hair sheep need not go to the expense of shearing. Hair sheep are also resistant to parasites and hot weather, and are consequently increasing in popularity among a subset of producers.

With the modern rise of corporate agrobusiness and the decline of localized family farms, many breeds of sheep are in danger of extinction. Preference for only those breeds with uniform characteristics and fast growth rates has pushed heritage (or heirloom) breeds to the margins of the sheep industry. Those that remain are maintained through the efforts of conservation organizations, breed registries, and individual farmers dedicated to their preservation.

SELECTING SIRES FOR A SINGLE HERD

Most of the factors involved in sire selection have been covered in earlier sections, but one major question remains. Will the sire which has been proven to be superior in other herds also be superior in my herd? Older breeders will remember the term "nicking," which is not too often heard anymore. If a sire did well in one herd and poorly in a second, he was said to nick well in the first, poorly in the second. Today, researchers look at the importance of nicking by studying the importance of sire by farm interactions. Such interactions arise if there are real changes in the relative performance of sires (usually measured by the milk yield of their daughters) as they are used on various farms. Should there be an appreciable number of sires which rank high in some herds or farms, and low in others, and vice versa for other sires, this would show up as a sire-by-herd interaction.

The importance of this question is apparent; if there were such interactions, a dairymen would be unable to rely on USDA sire proofs, since he could not depend on sires with superior proofs to be superior in his herd. This was an area that obviously needed research, and during the past several years considerable research has been published on the subject. A classic study by North Carolina scientists involved sire-by-region interactions. The two regions involved were North Central and Southeastern United States and the question essentially was, "Do AI sires used in these two areas rank about the same, or are there some who are outstanding in one region but poor in the other, and vice versa?"

Much of the semen used by Florida dairymen is from sires bred, house and proven in the Eastern and North Central United States, so this question was of immediate concern.

Results showed that this particular type of interaction was not an important factor, and indeed was almost nonexistent. The authors stated, "provided adequate unbiased progeny information is utilized, a bull may be safely selected and used for artificial insemination in one region, even though he was evaluated in the other region." Studies on the importance of sire by herd interactions have been in very close agreement, with one being conducted by Florida researchers on records of Florida DHIA Guernsey, Holstein and Jersey first-calf heifers. Everyone recognizes that even the best bull may have a certain number of poor daughters and that a dairyman who has only a few daughters of a particular sire may have been unlucky, by chance, and gotten one or more of them. The opposite can happen also with poor sires.

The problem is to determine how important the interactions are overall, realizing that a certain amount of shifting in rankings of sires can occur by chance, particularly if a sire is represented in one or more herds by only a small number of daughters. The research found that under Florida DHIA herd conditions; these interactions were of little or no practical or theoretical consequence for milk and fat yields. There was some evidence among Guernsey's and Jerseys of an interaction for fat percentage; this has been suggested in other research studies also, but many of the scientists who have studied interactions have not looked at fat percentage. In summary, it appears that sires proven superior for milk and fat yields in other herds can be used with confidence in Florida dairy herds, without concern for possible sire by herd interactions. The possibility exists for the presence of interactions with fat percentage.

FINDING SUPERIOR FEMALES

Measurements for most of the economically important traits can be made directly on females. It frequently takes a long time to get the measurements of interest, however. Mature body size, would have to wait until the cow is 6 to 8 years old. An estimate of milk yield must wait until the first lactation is completed (at about 3 years of age), unless one accepts an estimate of the lactation based on the first several months.

These points are mentioned because genetic progress is influenced by the time at which measurements are made, and their reliability. Take milk yield as an example. An estimate of the milk producing ability of a cow comes with her first milk record. Certainly, however, one knows more about her when there are two records rather than one, and even more with three records. Generally, the more records available, the more reliable the estimate of her real ability; but it takes longer to get more records. An estimate of the producing ability of a cow was called "Most Probable Producing Ability" years ago by Dr. J.L. Lush, well known animal geneticist. This measure takes into consideration the idea of repeatability, and shows how the number of records a cow has should influence the estimate of her producing ability. If a heifer produces more than her herdmates during her first lactation, we will certainly guess that she is a better than average animal.

If the average of her first two records is above herdmates, one would be even more confident that she is truly above average, and her good performance did not just happen from some favourable accident which probably will not happen again. The calculations are simple and can be applied to any trait of interest. Although the concept is rarely mentioned today, since availablity of computers makes more sophisticated procedures feasible, MPPA serves as the basis for modern methods. State the milk record, or any other measurement, as a deviation from herdmates. If the cow has two measurements, state each as a deviation, and take the average. MPPA then can be calculated as:

$$\text{MPPA} = (\text{Average deviation}) \times [(nr)/(1 + nr - r)]$$

where n = the number of measurements and r = the repeatability. How much better two measurements or records are than one, as far as reliability, is determined by a similar equation.

$$\text{Gain in accuracy} = 1 - [(1 + nr - r) / n]$$

with n and r as in MPPA. The more records one has on a cow, the more one can accept her records at face value. If she has six records averaging 2000 pounds better than her herdmates, the best guess of her MPPA is 1714 pounds better.

Which cow should be selected as a dam for a future herd sire, a cow with two records averaging 3000 pounds above herdmates, or one with four records averaging 2400 pounds above? The two estimates of MPPA are close, being 2000 pounds for the cow with two records and 1920 pounds for the older cow. One probably would look at other characteristics of the two animals, particularly fat per cent but possibly others, before making a decision. The more sophisticated and useful application of this concept, of course, is that it is incorporated into calculations of ERPA.

GENETIC CHANGE IN MILK YIELD

The theoretical upper limits of genetic change in milk yield have been known for quite some time. Knowing the various repeatabilities, heritabilities, generation intervals and selection pressures (intensities), one can calculate what would be expected to happen. This tells nothing about what is actually going on, of course, but it does tell what one can expect under the best conditions.

Potential

If milk yield is given maximum emphasis in the selection programme, with little or no pressure being exerted on other traits, and efficient use is made of an AI sire proving programme, the maximum attainable genetic change in milk yield is slightly over 2.0 per cent per year.

Most of this, over 90 per cent, theoretically would come directly or indirectly from sire selection. In a Holstein population averaging 15,000 pounds per cow, this would amount to 300 pounds per year. Though this estimate has appeared disappointingly small to some breeders, it is more than the actual total change which has been attained in the United States.

Observed

It is only in recent years that it has been possible to obtain reliable estimates of the actual genetic change which has occurred in milk production. A major problem in the actual measurement of such progress is that environmental and genetic changes are confounded in such a manner over time as to make their separation extremely difficult. With the availability of high-speed large-capacity electronic computers, many of the problems can now be handled. Use of the animal model has improved the accuracy of the estimates. As far as progress arising from selection of males and females, theoretical estimates has been right on target.

About 90 to 92 per cent of all genetic change in milk yield has occurred from sire selection; this represents the selection of males and females to produce herd sires and the selection of males to produce herd dams. The remaining 8 to 10 per cent of genetic progress arises from selection of females to produce herd dams. The fact that we must raise nearly all female calves born alive in a population to maintain population size, and the difficulties of evaluating the true genetic merit of dams, precludes much selection pressure from selection of dams. Average annual genetic change in milk yield of tested dairy cattle has been variable but seems to be about 0.7 per cent of their average levels. This would amount to about 105 pounds per cow in a 15,000 pound herd, or 140 pounds in a 20,000 pound herd. In individual herds, one would expect to find genetic changes which were considerably higher or lower. If 2.0 per cent annual genetic change is possible, as theory suggests, dairymen are progressing genetically only about one-third as fast as possible.

SELECTION AND EXPECTED RESPONSE

Traits which are heritable presumably can be changed by selection, but the degree of heritability is only one of three factors which determine the progress which can be made. The other two are the culling level (intensity of selection) and the variability of the trait. These are logical since the more severely we can cull, the more progress we can make. Also, the more variable a trait is, the higher the population average will be for the selected individuals for any given culling percentage.

It turns out that the amount of progress which can be made in one generation is the product of intensity of selection, heritability for that trait, and the observed variability. Note that if any of the three components is zero, selection progress will be zero also. Good estimates are available for how much progress can be made from selection for most of the economically important characteristics of dairy cattle. Of course, there are several traits which are important, so there is a need to look further than anticipated results from selection for a single trait.

Correlated Response

Knowing what the genetic relationships are among the important traits, plus a few other things, permits us to predict what will happen to a trait such as fat percentage if we ignore it completely and select for milk yield, or vice versa. Though these results pertain to Holsteins, they are similar to results for other breeds. The changes expected in the trait being selected for; values off the diagonal are changes expected to happen in other traits. Heritabilities and variabilities used in these calculations agreed with those generally accepted by dairy cattle geneticists. Intensity of selection in the estimations was fairly conservative. If more severe culling were practiced, greater progress would be made. If milk yield alone is selected for at a reasonable level of intensity, an increase is expected in genetic merit in the next generation of 607 pounds. Since milk yield, fat yield and protein yield are genetically fairly closely related (that is, their genetic correlations are high and positive), an increase is expected in fat and protein yields, even though these were ignored in the selection programme.

Because of the negative genetic correlations between milk yield and fat percentage, and milk yield and protein percentage, these two percentages are expected to go down. Selecting for milk yield and ignoring the other four traits would be expected to result in an increase per generation of 607 pounds of milk, 23.2 pounds of fat and 13.7 pounds of protein, but decreases of 0.036 per cent fat and 0.018 per cent protein. If one attempted to increase fat percentage genetically in his herd by selecting for it and ignoring other traits, he could increase fat percentage. A reasonable increase per generation to be expected is 0.190 per cent. Along with this, however, there would be an actual decrease in milk yield of 287 pounds, although fat yield would increase 24.1 pounds.

Present pricing systems for milk in the United States are based mostly on its weight and fat percentage, although additional factors such as protein percentage may be included also. So the dairyman needs to decide what combination of milk yield and fat percentage and other changes is most profitable for him. He also needs to produce milk that passes minimum legal standards; this is an important factor in herds which already are near the minimum. Beyond this, there are a number of other important traits which must be considered. Dairymen, of course, mentally put all of these factors together in their selection programmes and, depending on their present herd status and present milk pricing, give each trait a certain amount of emphasis. In other words, they weight each characteristic they think is important. Geneticists do the same thing and attempt to establish numerical values for the weights; a numerical index, called a selection index, then can be formulated for each economic situation.

Milk Yield and Composition

Just about every index formulated for the United States gives yield more emphasis than any other trait, with enough emphasis placed on fat percentage to meet legal standards. Protein has become more important in pricing systems in recent years.

Reproductive Performance

Efficient reproduction is very important economically because a cow must first reproduce to produce. However, most measures of heritability are zero or close to zero, suggesting that little or no improvement can be made by selection. There already is considerable selection pressure being exerted on improving reproductive performance without any conscious effort being made by the breeder. This occurs because sterile males and females leave no offspring, and those with low efficiency leave a lot fewer than those of higher reproductive performance. So within a given herd culling for reproductive failure would have little effect on the genetic merit for reproduction of the herd. Now that large numbers of records are being obtained on reproduction of daughters of AI sires, however, it may turn out that this information can be taken advantage of in selecting which proven AI sires to use in the herd. Even though heritability of most single measures of reproduction are close to zero, estimates of average genetic merit of a large group of daughters for reproductive efficiency are much more reliable.

Desirable Body Type

Desirable type has appealed to dairymen for centuries, and major changes in appearance in dairy cattle have occurred over the years, accompanied by increases in milk yield. This fact does not itself mean that the two traits have (or ever have had) an appreciable, positive genetic correlation. Regardless of what has gone before, the breeder must concern himself with the present facts of life in his selection programme. The genetic relationship between type and production seems close to zero and may even be negative. There have been a number of changes in the breed association type classification programmes and policies in recent years, and evaluation of all such changes cannot yet be made. A number of scoring systems also have been developed by others such as universities and AI studs. Yet the final value of desirable type must be in its relationship with production and longevity of the cow.

Except for the breeder who derives an important fraction of his income from sale of breeding stock, dairymen should realise (according to research) that only 2 to 6 per cent of their income is related to variability in body confirmation. USDA researchers estimated several years ago that giving equal emphasis to type and production would reduce the possible increase in milk yield by 29 per cent. Though perhaps the final judgment on emphasis to be placed on desirable type cannot yet be made, it appears that if major attempts are made to improve type, they will be made at the sacrifice of possible genetic increases in milk yield.

Longevity

Longevity may be defined as the ability to avoid death or culling, the former being mostly for disease or accident and the latter for reproductive failure, low milk yield, mastitis, etc. Few cows die simply from old age. Of the various possible reasons for loss, some have a genetic basis but few are highly heritable. In any event, high producing cows remain in the herd longer than low producers. Natural selection pressures also work in favour of the ability to resist death or culling. Selection for high milk yield, plus natural selection, seem to be exerting sufficient selection pressure for increased longevity such that the breeder need not give it major emphasis in his programme.

Body Size and Weight

Within each dairy breed, large cows give more milk, on the average, than small cows, but they also require more feed. Research suggests that the additional costs involved with larger cows about equal the additional income derived from them. In the case of two cows of equal production but different size, one actually would expect the smaller cow to be more profitable. Most measures of size (weight, height, etc.) show appreciable heritability and variability, so larger cows could doubtless be developed if desired.

However, most research now suggests that milk yield in dairy cows can be increased considerably without there being any increase in size. Selection for increased body size could be successful, but the correlated response in milk yield though probably positive would be close to zero. Though North Carolina researchers indicated that it might still be wise to check cows considerably above or below the breed average in body weight for productivity and longevity, they concluded that "undue emphasis on larger cows in the short run could lead to the development of less profitable animals."

Other Traits

Many others are economically important and are heritable to some degree. Many qualitative (simply inherited) traits, ranging from lethals on one extreme to aesthetic traits on the other, can be important under some circumstances. Artificial insemination, particularly frozen semen, has caused concern because of possible widespread dissemination of undesirable genes. The genetics of the situation have been studied, however, and with present knowledge potential dangers apparently can be avoided. Traits other than those mentioned also are important. Mastitis resistance, is of extreme economic value and has measurable heritability. A certain amount of direct pressure for resistance will be exerted from culling of mastitic animals, although clinical cases may remain in a dairy herd for years under proper treatment.

Present knowledge of genetics of dairy cattle suggests that major selection emphasis should be placed on milk yield, with enough selection pressure on fat percentage to maintain composition of milk at legal standards. Other components of milk may become important if pricing systems change to emphasize them. Care should be taken to include as few traits as practical in a selection programme, since progress in milk yield may be sacrificed to a degree such that total profit per cow may decrease. This is especially true with traits of low heritability and only slight genetic relationship to milk yield. Reproductive efficiency and resistance to mastitis may appear as economically important traits in future sire proving systems.

The Animal Model

The Animal Model does many things notably different from previous evaluation systems. These include simultaneous evaluation of all sires and cows. This means all information on all relatives is used to increase the accuracy of any one animal's evaluation. This also means that cow families are directly included in the evaluation of all sires and cows, effectively eliminating the basis for previous objections by some breeders. All female ancestors and descendants are used in the evaluation of both sires and cows, weighted by how closely they are related. Since all known relatives of both registered and grade cows are used, computing the evaluations for the Holstein breed requires simultaneous solution of about 50 million equations. Other changes include correction for genetic merit of mates, use of a new genetic base, new ways of using records on cows missing first records and cows changing herds, and definition of different management groups for registered and grade cows.

There is some new terminology with the Animal Model emphasizing that new methods are truly different. Predicted Difference (PD) changes to Predicted Transmitting Ability (PTA), or one-half the breeding value. This is the prediction of what the sire transmits to his offspring. Cow Index changes to Predicted Transmitting Ability (PTA).

Repeatability (RPT) changes to reliability. Repeatability always has been a misnomer as used by the USDA in this context anyway. Since reliability includes contributions from additional relatives, reliability figures should be higher than the old repeatabilities. The biggest difference is for sires of cows with few daughters compared to those with many. A sire with no daughters would have a reliability of 37 per cent. Percentile is sire's ranking based on the active AI bull population; for cows it will be based on cows with recent lactations. The rankings are based on an index including milk, fat, and protein evaluations. Previously they were based on an index including only milk and fat. As good as the new Animal Model system is, it cannot correct for preferential treatment of daughters of certain sires. No genetic evaluation system can. Random AI sampling of young sires and wide daughter distribution will continue to be important in establishing accurate first evaluations. This also is true for sires with relatively few daughters.

Economic Evaluation of Sires and Their Semen

Predicted transmitting ability for dollar value (PTA PRO$) is based on milk, fat, and protein evaluations. Current national prices are used. For the July 1989 run these were $11.20 per hundredweight of milk containing 3.5 per cent milk fat and 3.2 per cent protein. Hence the equation for estimating PTA PRO$ is PTA PRO$ = $.0242 (PTA milk) + $ 1.54 (PTA fat) + $ 1.09 (PTA protein). Note that the dollar value is computed from pounds fat and protein produced, not percentages. Percentile rankings are based on this index.

In Florida, the milk price is higher than almost everywhere else in the United States, and the dollar values of superior sires should be proportionally higher. The ranking of bulls by evaluations would not change much, but the net value over their semen price would. Taking advantage of considerable previous research, Marsha Wilcox and coworkers developed techniques to aid dairymen in the selection of semen to purchase for herd use. The concept is that of net present value (discounted net income) of semen. Although the method is computationally elaborate, availability of microcomputers to dairymen will permit each to develop a system of evaluating semen based on a set of selection criteria tailored to individual needs. Information necessary for semen selection include sire proof for milk, fat, protein, type (PDT), interest rate, and such management parameters for the individual herd as conception rate, calving interval and female mortality rate. Outcome of this linear net merit index for semen is Present Value Dollars (PV$). These researchers then evaluated a number of possible economic and management situations to determine the effects on sire selection; considerable shifting of rank occurred with changes in criteria. For one sample selection programme (all selection pressure on milk, none on desirable type), rank correlations of PV$ and PDM, and PV$ and PD$ + semen price, were.56 and.52. Such procedures should prove useful and profitable to dairymen who are able to characterize their own management parameters and to define their selection goals. Thus, the concepts of proving sires are not difficult. Efforts to obtain better estimates, thus increasing rates of genetic improvement, continue to make the exact calculations and formulas very complex.

10

Animal Feeding

Animals with rumen cannulae have proved to be one of the most useful tools for evaluation of feed resources and especially for determining the effect of a feed on the rumen environment, which has a major influence on the processes of digestion. The technique can be criticized on the grounds that it is an interference with the normal functioning of the animal and that if not carefully maintained the fistula and the cannula can be a source of stress. *In vitro* methods of gas production promise to be an appropriate alternative for measurement of the fermentation potential of a feed and of the relative effect of anti-nutritional compounds present in many tropical feed resources.

PRINCIPLE OF THE METHOD OF THE ANIMAL FEEDING

It is not necessary to starve animals prior to surgery; in fact it is desirable to have the rumen relatively full. A tranquillizer given prior to the actual surgery is an advantage in the case of cattle. The animals should be accustomed to handling by attendants and should be docile and easily led. The surgery is carried out with the fed animal standing in a crush or even restrained in the corner of a yard by a moveable gate. The animal is tranquilized by intramuscular injection (*e.g.*, with Rompun) but this is not absolutely necessary. A 350 kg cow requires about 0.5 ml of Rompun to be sufficiently sedated.

Approximately 15 minutes after the injection of the tranquillizer, surgery may commence. The area of incision should be closely clipped or shaved and cleansed with a mixture of alcohol or alcohol and iodine. The incision should be made high on the left side in the anterior dorsal abdomen. The site of the incision is identified by marking a triangle from the point of connection of the last rib with the spine and moving the same distance along the spine from the last rib. The area between the last rib and where it connects with the spine should be sufficiently large to take the external flange of the cannula. In general the closer to the spine on the flank the incision is made the better; but only experience will allow accurate placing and estimation of size of the incision.

Before starting such operations cannulae of different sizes (5-12 cm internal diameter) should be on hand. Analgesia of the incision area can be produced by paravertebral anaesthesia; however, this requires experience and skill. A more practical approach is to inject a local anaesthetic in a series of subcutaneous and intramuscular injections immediately above and along the site of the incision. Approximately 25 ml of Zylocain is injected into a steer of 250 kg and 15 ml into a sheep.

PRINCIPLE OF THE METHOD

A metal clamp is applied to a fold of the rumen wall exposed by an incision into the body cavity. The clamp holds the fold of rumen wall outside the body, occludes the blood supply and causes the damaged area below the fold to adhere to the body wall. In ten to fourteen days, the clamped piece of rumen 'sloughs off' leaving a fistula through which a cannula can be readily introduced and secured.

FACILITIES AND EQUIPMENT

Only minimum facilities are needed; a simple crush or some method of restraining the animal in a standing position (cattle only), a minimum of surgical equipment (scalpel, forceps, etc.), a tranquillizer and local anesthetic. The clamp consists of two brass rods 11 cm long and 0.6 cm in diameter (for sheep) and about twice this size for cattle. Each brass rod has two holes 2.5 cm from either end of the rods. The holes in one rod are threaded to take a brass screw which is fitted through a hole in the other rod so that, when the screws are turned, the two rods draw together forming a clamp.

COLLECTION OF RUMINAL FLUID BY OESOPHAGEAL TUBE

For sheep a plastic tube of 10 mm internal diameter and some 90 cm long is suitable. The rumen sampling tube should be moistened and the sheep's mouth opened by placing a thumb in the region without teeth. The tube is then passed over the back of the tongue and into the oesophagus. Test for its presence in the rumen by checking for the smell of rumen fluid, and the lack of respiratory air movements along the tube. A vacuum pump is used to apply suction to draw the rumen liquid into the sampling bottle. With cattle, a larger tube is required (15 mm internal diameter and 150 cm long). The rumen fluid can be obtained by lowering the animal's head until fluid runs from the tube. Move the tube in and out a few centimeters in taking the samples. Filter rumen fluid through gauze. Note that samples obtained in this way may be contaminated with variable amounts of saliva.

RUMEN INCUBATIONS WITH NYLON BAGS

This method is given first priority for researchers in developing countries, as the most appropriate tool for providing information on:

- The nutritive value of a feed for ruminants;
- The efficiency of the rumen ecosystem.

It generates useful information from the point of view of both the carbohydrate and protein status of a feed; and the degree to which it will be digested in the rumen or escape to the intestines. The method is described in detail for this reason.

CHARACTERISTICS OF THE BAG

The bags should be prepared from a nylon or other synthetic fibre material with a pore size of between 20 and 40 microns. The pore size is a compromise between minimal loss of small particles and making sure that microbes, including protozoa, can enter the bags uninhibited; and also that gas can escape from the bags. When gas does not escape the bags may float on top of the solid phase of the rumen and give very variable results.

The bags should be sewn with polyester or nylon thread with double seam and close stitching. Overall dimensions for cutting out should be 17 × 10 cm to give an effective length of about 12-14 cm. Smaller bags can be used if samples are smaller. It is not necessary to introduce a draw-string in the neck of the bag, as they can bc closed with a separate length of nylon thread (*e.g.*, fishing linc), and/or attached to a long nylon string (*e.g.*, baler twine) or a plastic rod. The bags can be reused as long as there are no holes in them; each time they should be checked for breakages.

MANUFACTURE OF RUMEN CANNULAS

The use of the one-step fistulation technique means that animals (cattle, buffalo, sheep or goats) can be prepared for use in almost any research laboratory. The lack of availability of manufactured cannulas has often been the reason for not preparing fistulated animals. It is relatively simple to devise rumen cannulae for both cattle and sheep.

AVAILABLE MATERIALS

In most countries PVC conduit is available with diameters from 13 to 300 mm and with a wall thickness of 3-5 mm. In the Dominican Republic, PVC tubing has been used for cannulae which were placed in the fistula of cattle which were subsequently under experimentation for more than two years without apparent problems. Rubber tubing is also available in most countries (*e.g.*, car radiator manifolds) and has been used to prepare cannulae. Car tyres or the protective band from inner tubes usually provide a suitable rigidity for retaining flanges for the cannula.

CONSTRUCTION OF CANNULAE FROM RADIATOR TUBING

Flexible rubber cannulae are preferred since these can be easily compressed to introduce them into the oval fistula that results from the method of cannulation. For sheep, the components of the cannula can be constructed from a section of radiator hose and a round flat piece of rubber cut from a truck tyre. These two parts may be sewn together with nylon thread. Insertion of this cannula into sheep is facilitated by twisting a section of the retaining flange into the tube. When the cannula is in position the retaining flange may be pushed out of the tube of the cannula to allow it to assume its normal shape but inside the rumen.

THE USE OF RUMEN AMMONIA CONCENTRATION

The level of rumen ammonia is critical for efficient microbial fermentation of feed. Rumen ammonia concentration can therefore be used to diagnose a deficiency of fermentable N in a diet. This will indicate when urea supplements are required. The critical ammonia level in the rumen for efficient microbial growth on different substrates is likely to vary according to the fermentability of the substrate.

As a rule of thumb, rumen ammonia nitrogen should be at least 15-20 mg/ 100ml rumen liquor. Where rumen ammonia is to be used as a diagnostic tool then the times of sampling of rumen fluid are critical. It is necessary to synchronize the availability of ammonia with the fermentation of the carbohydrate. The ammonia level at 4 to 6 hours post feeding or following the commencement of grazing is the recommended time for sampling. Ammonia concentrations in rumen fluid must be above the critical level for prolonged periods on fibrous diets which are only slowly digested in the rumen. For this reason the concentration immediately before feeding may also be an index of the need to supplement.

ASSAY FOR BY-PASS PROTEIN IN A SUPPLEMENT

Wool growth in sheep is highly dependent on the quality of amino acids absorbed from the intestines, in particular the sulphur amino acids. However, these amino acids are not absorbed other than from the protein (dietary and microbial origin) digested in the intestines. Increased wool growth rate in response to ingestion of a protein supplement is directly related with the content of by-pass protein in the supplement.

PROCEDURE

Mixed sex, cross-bred wool sheep (the original work was done with 1-year-old Merino x Border Leicester sheep) are housed in individual pens and given *ad libitum* a basal ration of low-nitrogen roughage (*e.g.*, rice straw) plus a mineral mixture and 1% urea (to ensure adequate fermentable-N in the rumen).

The sheep are randomized to the treatments (protein-containing supplements) with 10 animals per treatment. In the validation of the method, these were 60g/d untreated casein and 0, 20, 40, and 60 g/d of formaldehyde-treated casein (HCHO-casein) prepared as described later (formaldehyde treatment leads to almost complete protection of the casein from degradation in the rumen). Wool growth was estimated by clipping and weighing the wool from a 10 cm square patch on the flank of the sheep every three weeks.

Preliminary studies indicated that the carryover effects of diet on wool growth were reduced to insignificant levels after 3 weeks. Trials are therefore carried out for a six week period and only the wool growth in the final three weeks is measured and related to the amount of protein in the supplement. In subsequent studies, the sheep are re-randomized before being allocated to treatments. The data in Figure show the results obtained with different levels of formaldehyde-treated casein.The amounts of wool clipped from the patch were linearly related with the level of HCHO-casein added to the basal diet. When soluble casein was added to the diet, wool growth rate was only slightly increased over the control animals indicating that this protein had no by-pass characteristics.

Selected results from the use of this assay to evaluate a number of plant protein. The wool growth represents the level of by-pass protein relative to formaldehyde-casein. Meals that had received most heat treatment gave the highest wool growth and were therefore the best sources of by-pass protein. Sunflower seed meal was a poor source of by-pass protein, especially when the oil had been extracted by the expeller system. The better by-pass characteristics of the protein in meals produced by the solvent extraction process are because these meals are usually "toasted" at 120°C after the oil is extracted.

Similar temperatures may be reached in the expeller process but the results are more variable, as the temperature is produced by friction in the press and this varies with the kind of oilseed being processed. Feeding trials ranked the protein meals in the same order as indicated by the wool growth assay (Leng R A, unpublished data). In this case the criteria were feed intake and liveweight gain, both of which are good indicators of the by-pass protein status of a supplement when added to a low-protein diet.

BIOLOGICAL TEST OF SOIL FERTILITY

Feed resources are generated from the land and specifically from soil and water. To maintain the fertility of soil and, even better, to enrich this basic natural resource, is one of the most important indicators of sustainability. This parameter is obviously a complex one and will depend on many factors both chemical and physical.

To analyze the individual components is costly and time-consuming and is not technically nor economically feasible in the majority of tropical developing countries. There are therefore good reasons for developing a simple method which can be performed with minimum inputs and which gives a broad assessment of the capacity of the soil from a given cropping system to support the subsequent growth of another crop.

About 10 kg of soil (cores taken to a depth of 30cm) taken from the test plot are air-dried and placed in 0.5 litre plastic bags arranged in blocks in the open. Samples are replicated three times in each of four blocks in a random design. One maize seed was placed in each bag and the biomass harvested 42 days later for fresh and dry weight determination. Adequate moisture was ensured by frequent irrigation. No plant nutrients were used. It would be useful to include a "standard" soil type in future tests so as to have a reference point that would facilitate comparisons among sites within and across countries. This could be based on sand with three levels of, say, excreta from laying hens or from cattle.

CHEMICAL ANALYSIS OF FEED AND FAECES

PREPARATION OF SAMPLES

Samples of material to be analyzed must be oven-dried at 65°C and then ground to pass through a 1 mm screen. Further drying to constant weight may be necessary to remove residual moisture. Dried material may be stored at room temperature in sealed vessels or plastic bags, preferably under nitrogen gas.

MOISTURE

A sample containing the equivalent of about 2 g dry matter is dried to constant weight at 95-100°C for 24 hr. Use an aluminium dish or porcelain crucible. Calculate percentage moisture from the loss in weight (to first decimal place).

ASH

Weigh a 2 g sample into a weighed porcelain crucible and place in a temperature-controlled furnace preheated to 600°C. Take care to avoid loss of material by convection currents. Hold at this temperature for 4 hr. Transfer crucible directly to desiccator, cool and weigh immediately. Calculate percentage ash (to first decimal place).

DISTILLATION

This is the same as for ammonia estimation. Turn on the heater under the steam generator and increase the heat to boil the water steadily (not violently) and turn on water to condenser. Put the empty digestion flasks on the collector tubes and, with the alkali stopcock closed and steam directed into the apparatus, run steam through the assembly and collect the condensates in 100 ml beakers for several minutes. This serves to warm up the apparatus, and flush out any residual alkali. When the apparatus is preheated, open the alkali stopcock and direct the steam into a water drain.

Place samples in the distillation apparatus and place 100 ml flasks containing 20 ml 2% (w/v) boric acid (containing indicator) under the condenser stem. Be sure the tip of the condenser stems are below the surface of the boric acid solution. Admit alkali solution through the alkali stopcock (about 5 ml alkali for 1 ml of H_2SO_4 used in the original digestion) and close the alkali stopcock.

Then turn steam on through the apparatus and allow steam distillation to proceed for 6 min. Near the end of this period, lower the receiving beaker so that the distillate washes any remaining ammonia solution from the tip of the condensing units. When the distillation is completed, turn steam stopcock into the position which diverts the steam to sink waste and another opens the distillation flasks to atmospheric pressure. Remove distillation flasks and turn steam stopcock to the off position.

MENKE' GAS PRODUCTION METHOD

Apparatus

The apparatus used in the gas production technique may vary slightly from one laboratory to another. At the International Feed Resources Unit (IFRU), the apparatus used is simple. It consists of glass syringes of 100ml capacity which are incubated in a water bath where the temperature is accurately controlled with a water stirring heater. The following procedure is based on the apparatus used in our laboratory.

SYRINGES

Good quality syringes are essential (syringes of Hliberle Labortechnik, 7901 Lonsee Ettlenschieu, Oberer Seesteig 7, Germany are recommended). The syringes and their pistons should be numbered with a permanent (water-proof) dye starting for example with number 1. A few extra syringes are left as replacement for broken ones.

BUFFER SOLUTION

Stocks of the main elements solution (pH 6.8), the buffer solution (pH 8.1), the resazurin solution and the trace element solution can be prepared and stored in dark bottles. The reduction solution must be freshly prepared. The preparation of all solutions is as described on page 9 of the paper by Menke and Steingass. The pH of the buffer mixture solution (*i.e.* main elements + buffer solution + resazurin solution + trace element + reduction solution) should be about 7.1 ± 0.15.

PREPARATION OF SAMPLE

The samples are milled through a 1.0 mm screen and their DM content determined. Before weighing the samples, the position of the syringes in each run should be planned. Ideally there should be 3 replicates of a blank and a standard roughagein each run. The triplicates of the blank or the standard roughage should be dispersed among the syringes. Thus one of each is incubated as the first (No 1) and second (No 2) syringes, the second blank and standard will be in the middle, while the third replicate of the blank and standard roughage should be the last 2 syringes. Samples are normally run in duplicate and a run is usually repeated 3-4 times.

CHEMICAL ANALYSIS OF RUMEN AMMONIA CONCENTRATION

The concentration of ammonia in the rumen is one of the most important factors that determine the rate and efficiency of digestion of fibrous feeds. There are two methods of measuring rumen ammonia which are relatively simple, one of which can be used under field conditions (*i.e.*, the use of an indophenol-dye to produce a colour reaction with ammonia). This method is used by extension officers in Queensland, Australia to predict when urea supplementation through drinking water is likely to be effective in increasing productivity of grazing ruminants.

CHEMICAL PRINCIPLES

Ammonia reacts with free chlorine to form chloramine which then condenses with two phenol molecules to form an indophenol dye (strongly reducing compounds in rumen fluid are oxidized by the hypochlorite). Since excessive amounts of both salicylate and hypochlorate are present, the amount of dye produced depends on the amount of ammonia present. The relatively high concentration of salicylate was chosen to "swamp" the effect of any phenolic compounds which may be present in the rumen fluid.

ACETATE CLEARANCE

Hypothesis

It is proposed that in, cattle and sheep, acetate clearance rates reflect the balance of nutrients available for metabolism for a given productive state and that acetate clearance from blood will be directly related with feed intake. Accumulation of acetate in blood indicates that there is a shortage of co-factors (NADPH) that are required to incorporate the acetate into lipids at sites of deposition. These co-factors are derived from glucose or its precursors, or from amino acids. If the clearance rate of acetate from the basal diet is slow, this indicates that supplementation is necessary with "by-pass" protein or with glucose precursors.

METHOD

Cattle (about 150 kg) are injected with about 2.5 mM sodium acetate per kg liveweight. The injection can be done *via* an in-dwelling cannula in the jugular vein or injected directly into the vein. The injection should be done slowly over about 4 minutes. Blood samples (10 ml) are taken at intervals post injection for analysis of acetate or total VFA.

INJECTION SOLUTION

Dissolve 30 g of sodium acetate in 300 ml sterile double-distilled water. Inject directly into the jugular vein.

BLOOD SAMPLES

When an in-dwelling cannula is used, this is normally filled with dilute heparin solution (100 units/ml of 0.9% saline) to prevent it being blocked with coagulated blood, and then sealed with a nail. Prior to taking samples, remove the heparin solution and 5 ml of blood from the cannula and discard. Take a further 20 ml of blood into a bottle containing two drops of heparin (3,000 units/ml). Flush the cannula with the dilute heparin solution and seal it. Take samples at 10, 20, 30 and 40 minutes post injection.

FEED COMPOSITION AND ITS COLLISION ON ANIMAL MANUFACTURE

In order to meet the nutritional requirements of livestock a precise knowledge of feedstuff composition is necessary. Such information is particularly vital when trying to achieve the high levels of production required in today's competitive markets (where there is growing interest in quality, efficiency and the environment) through the preparation of balanced diets for the animals.

Knowledge is also essential in order to mix the right proportions of various ingredients for the manufacture of valuable compound feeds. Ultimately, knowing the chemical characteristics and the nutritional value of raw materials is fundamental in the planning of forage production on the farm so that crop yields can be balanced with the animals' requirements. The general aspects of this have been reviewed by Topps.

From a political point of view, data on feedstuff composition and nutritional value provide essential information to policy-makers that allows them to develop competitive and sustainable agriculture. Feedstuffs are usually described by their chemical composition as well as by their nutritional value. Information on the composition of a feedstuff is acquired through chemical analysis, and most commonly includes such parameters as dry matter, protein, fibre fractions, organic matter and fat contents. The nutritional value of a feedstuff is assessed from experiments with animals (*in vivo* digestibility/*in situ* degradability, etc.) and provides information on how feedstuffs are digested and metabolized by the animal, mainly through interpreting differences between the input and the output of a series of components. Among these, the digestible/degradable organic matter and the digestible or metabolizable energy values are the ones most frequently represented in databases and used to predict energy supply from feeds and diets. To provide both chemical and nutritional information, feedstuffs have been analysed in laboratories and animal-based studies and the resulting data collected in tables of feed composition for more than 200 years.

The first recorded publication was that of Thaer. More recently, collections of data have been organized into computer programmes, creating databases of feedstuffs composition. The first database available on a computer was produced in 1963 at Utah State University in the United States. Today, modern databases are powerful tools with which information can be rapidly retrieved, sorted, updated and printed.

FEED COMPOSITION FOR CATTLE AND SHEEP

Nutrition research spanning more than 100 years has defined the nutrients required by animals. Using this information, rations can be formulated from feeds and ingredients to meet these requirements. Animals fed these rations should not only remain healthy but be productive and efficient. The ultimate goal of feed analysis is to predict the productive response of animals when they are fed rations of a given composition. this is the real reason for information on feedstuff composition.

TABLE VALUES FOR FEEDSTUFF COMPOSITION

Feedstuffs vary in composition. Unlike chemicals that are "chemically pure" and therefore have a constant composition, feeds vary in their composition for many reasons. Actual analysis of a feed to be used in a ration is more accurate than tabular data. Obtain and use actual analysis whenever possible. Often, however, it is either impossible to determine actual compositional data, or there is insufficient time to obtain an analysis. Tabulated data are the next best source of information. When using tabulated data, remember that feeds vary in their composition. the organic constituents (*e.g.*, crude protein, ether extract, crude fibre, acid detergent fibre and neutral detergent fibre) can vary as much as 15 per cent, the mineral constituents as much as 30 per cent, and the energy values at least 10 per cent, from table values.

Therefore, the values shown can only be guides. For this reason they are called "typical values." they are not averages of published information. Some judgment was used in arriving at some of the values in the hope that the values will be realistic for use in cattle and sheep rations. Feeds can be chemically analyzed for many things that may or may not be related to the response of the animals to which they are fed. In the accompanying table, certain chemical constituents are shown. the response of cattle and sheep when fed a feed, however, can be termed the biological response to the feed in question. this is a function of its chemical composition and the ability of the animal to derive useful nutrients from the feed.

The latter relates to the digestibility or availability of a nutrient in the feed for absorption into the body and its ultimate efficiency of use in the animal. this also depends on the nutrient status of the animal and the productive or physiological function being performed by the animal. Ground fence posts and shelled corn may have the same gross energy value in a bomb calorimeter, but have markedly different useful energy value (TDN, digestible energy, net energy) when consumed by the animal.

That means that the biological attributes of a feed have much greater meaning in predicting the productive response of animals. However, they are more difficult to accurately determine because there is an interaction between the chemical composition of the feed and the digestive and metabolic capabilities of the animal being fed.

USING INFORMATION

Feed Names

The most obvious or commonly used feed names are given in the table. Feeds designated as "fresh" are feeds that are grazed or fed as fresh cut materials.

DRY MATTER

Typical dry matter (DM) values are shown. However, the moisture content of feeds can vary greatly. thus DM content can be the biggest reason for variation in the composition of feedstuffs on an "as fed basis." For this reason, the composition of chemical constituents and biological attributes of feeds are shown on a DM basis. Because DM can vary greatly, and because one of the factors regulating total feed intake is the DM content of feeds, ration formulation on a DM basis is more sound than using "as fed basis." To convert the values shown to an "as fed basis," multiply the decimal equivalent of the DM content times the compositional value shown in the table.

PROTEIN

Crude protein (CP) values are shown for each feed. Crude protein is determined by taking the Kjeldahl nitrogen times (100/16 or 6.25). (Proteins contain 16 per cent nitrogen on average.) Crude protein does not give any information on the actual protein and non-protein content of a feed. Digestible protein has been included in many feed composition tables, but because of the large contribution of body protein to the apparent protein in the feces, digestible protein is more misleading than CP. Calculate digestible protein from the CP content of the ration fed to cattle or sheep by the following equation: per cent DP = 0.9 (per cent CP) - 3, where per cent DP and per cent CP are the ration values on a dry matter basis.

Rumen "by-pass" protein, or undegraded intake protein (UIP), represents the per cent of protein that passes through the rumen without being degraded by rumen microorganisms. Like other biological attributes, these values are not constant. By-pass values for many feeds have not been determined. Reasonable estimates are difficult to make. Degradable intake protein (DIP) is used to meet the nitrogen requirements of rumen micro-organisms. Nitrogen sources such as urea are the most economical sources of DIP. Balancing DIP and UIP sources provides a more accurate way of meeting the metabolizable protein needs of ruminants.

CRUDE, ACID DETERGENT AND NEUTRAL DETERGENT FIBRES

After more than 100 years, crude fibre (CF) is declining in popularity as a measure of low digestible material in feeds. the major problem with CF is that variable amounts of lignin, which is not digestible, are removed from various feeds in the CF procedure.

In the old scheme, the material removed was called nitrogen-free extract (NFE) and was thought to be more digestible than CF, even though many feeds have been shown to have a higher digestibility for CF than NFE. One reason CF remained in the analytical scheme for feedstuff analysis was its requirement for the determination of TDN.

Newer procedures have developed an alternate analytical scheme, namely, acid detergent fibre (ADF) and neutral detergent fibre (NDF). ADF is highly related to digestibility in the animal. NDF is related to voluntary intake of the feed and the availability of net energy from digestible energy. Both measures relate more directly to predicted animal performance, so they are more valuable than CF. Also, if TDN is replaced by other measures of energy value, there will be little use of the CF content of feeds. As more complete data on the ADF and NDF content of feeds are developed, CF will be dropped.

MINERALS

Values are shown for only certain minerals. Calcium (Ca) and phosphorus (P) are important minerals in most feeding situations. Potassium (K) becomes important as the level of concentrate increases in the ration, or when non-protein nitrogen is substituted for intact protein. Sulfur (S) also becomes more important as the level of non-protein nitrogen increases in the ration.

VITAMINS

Vitamins have been omitted from the table. Only vitamin A is of general practical importance in cattle and sheep feeding. the vitamin A and carotene in feeds depend largely on maturity and conditions at harvest and the length and conditions of storage. therefore, it is probably unwise to rely entirely on harvested feeds as a source of vitamin A. Where roughages are being fed that contain good green color or are being fed as immature fresh forages (*e.g.*, pasture), there will probably be sufficient vitamin A.

ENERGY

Four measures of the energy value of feeds are shown in the table. TDN is shown simply because there are more TDN values for feeds, and because this has become a standard system for expressing the energy value of feeds for cattle and sheep. there are several technical problems with TDN, however. there is a poor relationship between crude fibre and NFE digestibility in certain feeds.

TDN also overestimates the value of roughages compared to concentrates in producing animals. Some have argued that energy is not measured in pounds or per cent, so TDN is not a valid measure of energy. However, this is more a scientific argument than a criticism of the predictive value of TDN.

Digestible energy (DE) values also are shown. Many studies have shown there is a constant relationship between TDN and DE: there are 2 Mcals of DE per pound of TDN. Obviously, DE can be calculated by multiplying.02 times the per cent TDN content. Because DE is measured in calories, it is technically preferred over TDN. With greater emphasis on ADF and NDF as replacements for CF and the use of the bomb calorimeter to measure DE directly, use of TDN should gradually decrease.

It should be apparent, however, that the ability of TDN and DE to predict animal performance is exactly equal. Interest in the use of net energy (NE) in evaluating feeds for cattle and sheep was renewed with the development of the California net energy system. the main reason is the improved predictability of results depending on whether feed energy is being used for maintenance (NEm) or growth (NEg). the major problem in using these NE values is predicting feed intake and, therefore, the proportion of feed that will be used for maintenance and growth.

Some use only the NEg values in formulating rations. this suffers the equal but opposite criticism mentioned for TDN — NEg overestimates the feeding value of concentrates relative to roughages. Others use the average of the two NE values, but this would be true only for cattle or sheep eating twice their maintenance requirement.

THE PROCESS OF QUANTIFYING THE AMOUNTS OF FEED INGREDIENTS

Feed formulation is the process of quantifying the amounts of feed ingredients that need to be put together, to form a single uniform mixture (diet) for poultry that supplies all of their nutrient requirements. It is one of the central operations of the poultry industry, in view of its role in ensuring good nutrition. Feed costs account for more than 70% of the total production costs for most types of poultry, so it is important that returns are maximised through use of adequate diets. Feed formulation is a central operation in poultry production, ensuring that feed ingredients are economically used for optimum growth of chickens.

It requires a good knowledge of poultry and feed ingredients. Most large-scale poultry farmers depend on commercial feed mills for their feeds, to obviate the need to do their own formulations or feed preparation. It is therefore essential that formulations are accurate, to ensure a large number of flocks are not adversely affected.

TYPICAL FORMULATION

Feed formulation is both a science and an art, requiring knowledge of feed and poultry, and some patience and innovation when using formulae. Typical formulations indicate the amounts of each ingredient that should be included in the diet, and then provide the concentration of nutrients (composition) in the diet. The nutrient composition of the diet will indicate the adequacy of the diet for the particular class of poultry for which it is prepared (*e.g.* egg layers, meat chickens or breeders). It is common to show the energy and protein contents of the diet but comprehensive information on concentrations of mineral elements and amino acids may also be provided.

Since lysine is a key amino acid for poultry, the concentrations of the other amino acids may be related to it. Because some nutrients interfere with the utilisation of other nutrients, relationships between nutrients, particularly amino acids, also show if each of the nutrients will be used efficiently.

One of the reasons for formulating diets for poultry has already been identified, *i.e.* to produce a single uniform feed that can be delivered efficiently to poultry. The production rates expected on modern poultry farms also dictate that the dietary requirements of poultry be carefully identified and precisely met. Diet formulation enables the poultry industry to maintain some uniformity in levels of production. Industry nutritionists tend to use the same formulations over relatively long periods of time, so that the quality of the product (meat or eggs) remains stable over time. The product quality can also be easily predicted if the same diet formula is used and all other factors remain unchanged. It is worth stating that chickens are able to select from different ingredients, placed separately, in order to grow and lay normally. Such a practice is known as choice feeding, and is practised on a limited scale by some poultry producers. However, choice feeding is cumbersome and cannot be economically applied in large-scale modern commercial situations without modifications to the infrastructure.

SOURCES OF DATA ON FEEDSTUFFS COMPOSITION

Information on feedstuffs composition and nutritional value plays a fundamental role at the farm, feed manufacturer and government levels, so the data inserted into databases need to be reliable and of high quality. The quality of the data depends, in the first place, on the origin of the information, which may have been extracted from the literature, collected from laboratories, obtained from other databases or generated specifically from animal studies and analyses of feeds.

DATA ASSEMBLED FROM THE LITERATURE

Feedstuffs information in the literature usually concentrates more on the data value obtained and does not give a full description of the feedstuffs concerned. When feedstuffs are insufficiently or inappropriately described there is a risk of confusing their real origin. A typical example of this occurs when publications generally refer to the product "distillers' dark grains", which is a by-product of the whisky distilling industry that, depending on the type of distilling process, may be derived from barley (in the malt distilling process) or wheat and/or maize (in the grain distilling process).

In this case a true and existing genetic variability is ignored. It is evident that the compositions of the two products will differ, so the provision of additional information on the origin of the feedstuff is essential; it is not enough merely to refer to it as "distillers' dark grains". In the absence of additional information, when trying to obtain information on the protein content specifically of barley distillers' dark grains, a value obtained from the average protein content of samples of both barley dark grains and wheat/maize dark grains would be retrieved from the database, and this value would be meaningless.

Data obtained from the literature can therefore often be inappropriate, particularly when they use averaged data which may incorrectly merge different feedstuffs. Data from the literature are recommended only when no other sources are available.

DATA COLLECTED FROM LABORATORIES

When compiling databases of feedstuffs information, the data used have usually been produced in a number of different laboratories. When assembling collections of data, it is necessary to have specific information regarding the analytical procedures used to obtain the data, in order to guarantee that those used to create a database have been obtained through standardized methods, are expressed in standardized units and can, therefore, be compared.

The standardization issue is particularly important for data on the nutritional value of feedstuffs, obtained through *in vivo* digestibility or *in situ* degradability studies. In this case, together with the strictly analytical issue, the laboratories and research institutes involved should have standardized procedures for conducting animal studies. These relate to the numbers of animals and the species used (sheep, cattle, pigs, etc.) and to the feeding and collection protocols employed. However, when conducting *in vivo/in situ* studies, results will always be substantially influenced by variability among individual animals, no matter what level of standardization is achieved.

The standardized analytical procedures used to collect data from laboratories need to be monitored continuously, and the data they produce checked for accuracy and relevance. For this purpose, ring tests among participating laboratories should be carried out in order to evaluate the precision of standard or newly developed techniques.

MERGING OF NEW AND OLD DATA

It is common practice to merge new databases with old ones. In these cases it is important that the analytical procedures used to obtain all of the data are understood in order to generate information that is expressed on a common basis and is, therefore, comparable. The evolution of analytical methodology is a substantial issue.

Old data may have been obtained with obsolete methods or with surviving analytical procedures that, in the past, produced data that were less accurate than those produced today. Thus, the merging of old data with new may be inappropriate and may reduce the overall quality of the information.

DATA PRODUCED SPECIFICALLY FOR USE IN DATABASES

The data generated specifically for use in databases are likely to be those of the highest quality as all the necessary information will be included and the methods standardized and checked. The chemical composition of feeds can be analysed using the most up-to-date and standard methods and animal studies can be designed appropriately, by deciding which animal species to use and in which physiological state the animals studied should be. Both feeding and collection protocols can also be set with the aim of meeting the specific requirements of the database. Ideally, an entire database would be produced by analysing specifically the composition of feedstuffs and studying their nutritional value through animal studies. In this way the variability arising from extrinsic factors such as analysis methods and animal protocols would be reduced to the minimum. On the other hand, when using data from a single source it is critical that the techniques and procedures used have been certified for precision and are representative. However, if it is to contain representative information, the database has to be established with an appropriate number of samples.

The costs involved in analysing feedstuffs and, particularly, performing animal measurements are so prohibitive that it is rare for feedstuff information to be produced specifically for the purpose of assembling a database. Moreover, much time is required to obtain the data, particularly those derived from animal studies. Therefore, the data are usually extracted from the different sources listed previously. Ultimately, it is the responsibility of the database manager to judge which information to include in order to generate a meaningful database.

SOURCES OF DATA VARIABILITY ON FEEDSTUFFS

Information on feedstuffs may be associated with different sources and types of variability. Variability can be "intrinsic" and caused by real differences among feedstuffs, such as the genetic origin of the feed or the processes that have been applied to it. Variability can also be "extrinsic", that is caused by differences in sampling and analysis procedures.

INTRINSIC VARIABILITY OF DATA

Each feed has particular chemical and nutritional characteristics which distinguish it from other feeds. However, the same feed can also be derived from different cultivars or influenced by its geographical origin. Moreover, an identical raw material may have been processed in a variety of ways to produce very different products. The wide range of diverse wheat-related materials provides a clear illustration of the need to distinguish among different feedstuffs derived from the same raw material. Terms for wheat-based materials include wheat, durum wheat, winter wheat, spring wheat, processed wheat meal and wheat distillers' dark grains. All of these products and by-products are wheat-based, but they vary substantially in terms of both chemical composition and nutritional attributes and it is therefore expected that they will have different effects when fed to animals.

When assembling a database it is necessary to make sure that the real variability arising from the intrinsic characteristics of the feedstuffs is correctly recognized in order not to affect the quality of the data. Information from feedstuffs databases is most commonly retrieved in the form of averages (or "consolidated data"), so the feeds have to be merged judiciously in order to obtain meaningful information. To achieve reliability, a precise description and the correct naming of feedstuffs are essential, and for this purpose all available information on the type, origin, processing, etc. of the feedstuffs is needed.

One important attribute that is often omitted is information on the forage's stage of maturity. As shown in the example illustrated in the Figure, the content of both neutral detergent fibre (NDF) and starch relate strongly to the stage of maturity of the forage maize, and this is clearly reflected in the dry matter (DM) content of the plant, which increases with maturation.

In this case, therefore, it would be incorrect to describe material harvested at different stages simply as forage maize; the different maturity stages should be indicated and, on the basis of these, the different types of product classified. With respect to the processes applied to the feeds, when introducing data on, for example, rice bran meal, it is necessary to know how the meal was obtained and which processes were involved in its production.

In the example shown in Table, at least two different types of products (expelled and extracted rice bran meal) need to be differentiated because they are characterized by different chemical compositions and, consequently, by different nutritional values. The process that produces the extracted rice bran meal leads to a material with a much lower fat content compared with the expelled one (7.3 g/kg compared with 90.0 g/kg DM), and this produces a direct effect on the metabolizable energy value. It is apparent that if, instead of specifying the type of process used, the values of the two types of rice bran meal had been averaged, the resulting values would have been meaningless.

Another, more simple, example is related to the on-farm processes applied to forages. Hay is obtained by drying the forage, but the drying process may be achieved by "sun curing" or "barn curing" the material. The different processes can lead to different compositions; for example, related crude protein content varies from 122 g/kg DM for barn-cured hay to 99 g/kg DM for the sun-cured variety. In this case also, information on the type of process applied to the forage is essential in order to maintain the two feeds as distinct entities and, thus, avoid erroneous mean values.

EXTRINSIC VARIABILITY OF DATA

Differences in analytical procedures may exist among different laboratories and also within the same laboratory. If not identified, differences can compromise the information retrieved from the database as it will compare values provided on different bases. In this respect, an important limitation of databases is related to the type and quality of the data they make available to their users.

In general, the diversity of analysis methods adopted by different laboratories is responsible for most of the extrinsic variability found in databases. It is safe to compare or mean parameters only if they have been grouped according to the method of analysis used to obtain them. An example in which incorrect mean values may result is in the estimation of lignin using potassium permanganate, according to the method of Goering and Van Soest, and using acetyl bromide, according to the method of Morrison.

These two procedures for measuring lignin not only give different results but are also expressed in different units and it is therefore incorrect, if not impossible, to merge and compare the two values. Recently, Beever and Mould highlighted large differences in the crude protein and starch contents of maize silages submitted to nine different laboratories. The values observed for one maize silage ranged from 57 to 119 g/kg DM, for crude protein, and from 165 to 272 g/kg DM, for starch. Investigations uncovered the fact that different methods were being used and that some were predicted by near infrared reflectance spectrophotometry (NIRS). Variability also often arises in data on the nutritional value of feedstuffs that have been obtained through *in vivo/in situ* experiments using animals.

The animal species may be a source of this inherent data variability. In particular, it is common practice to classify the nutritional value of feedstuffs on the basis of data obtained through measurement among both sheep and cattle under the general definition of "ruminants".

Differences in the passage rate of digesta and, therefore, of feed utilization between these two species are often substantial. Other generalizations commonly made arise from the lack of differentiation among the different physiological states of animals of the same species (dry and lactating cows, for example, or growing and finishing pigs). Such physiological differences may have a great impact on the data obtained, and when not recorded will compromise the quality of those data. Moreover, the use of non-standardized feeding and faecal collection protocols can affect the results obtained substantially. The *in situ* method used extensively to measure the rumen degradability of feeds is subject to many sources of variability. Huntington and Givens reviewed many of these, and the *in situ* method has been the subject of a number of ring tests. For example, Madsen and Hvelplund reported on the protein degradability of five concentrate feeds evaluated *in situ* at 23 different European institutes. Although the mean values obtained were similar to published values, the reproducibility among centres was poor. In particular, the ring test underlined the need to standardize the material used to make the bags and the washing procedures followed.

A last source of variability, in terms of mean data, is related to the number of samples taken of a feedstuff which can vary widely, resulting in consolidated data that are more or less representative. In conclusion, in order to produce a high-quality database on feedstuff composition and nutritional value, the intrinsic variability generally needs to be recorded and differences among materials need to be clearly identified. This is vital if effectively different feeds are not to be merged and averaged but maintained distinct. On the other hand, extrinsic variability should be kept to a minimum because, when a few standard procedures are each applied to a large number of samples, the resulting data will be more representative and useful than when a large number of different methods have each been applied to only a few samples.

EFFECTS OF DATA VARIABILITY FROM FEEDSTUFF

The use of information obtained from feedstuff databases characterized by low-quality data (with high extrinsic variability and low intrinsic variability) may have various effects depending on the end user of the database (*e.g.* animal nutritionist, compound feed manufacturer, farm manager, policy-maker). In this respect, in general, the quality of the data is more often influenced by the extrinsic than the intrinsic variability.

FOR THE ANIMAL NUTRITIONIST

Analysis of feedstuffs composition is costly and time-consuming. Nutritional values acquired through studies with animals are particularly labour-intensive.

The nutritionist can therefore not normally afford to analyse specifically each of the feedstuffs involved and often needs to base assessments of nutritive value on more readily available and less expensive information. The animal nutritionist is in charge of designing the animals' diet. The first consequence of unreliable information on feedstuffs, resulting from high extrinsic variability, may be the formulation of a nutritionally unbalanced diet. This may have different effects, depending on nutritional, environmental and economical factors; these effects are highly interrelated since the optimization of feed utilization leads to the optimization of animal output and financial returns.

Nutritional effects. Today, nutritional information on feedstuffs is used to predict animal performance (often with the use of complex models) through estimation of the energy supply to the animal. The incorporation of low-quality data in these models will produce unreliable estimates of animal performance. Furthermore, it is important to maintain the data resulting from *in vivo* measurement of nutritive value separate from those values predicted from chemical composition.

Environmental effects. The most common environmental effect derived from unbalanced diets is caused by nitrogen loss. Particularly for ruminants, the capture of nitrogen depends on the rate of production of microbial protein and is related to the availability of carbohydrates and energy in the rumen. It is therefore important to know the exact amount of degradable nitrogen (together with the carbohydrate and energy availability) when characterizing a feedstuff, in order to capture nitrogen for anabolic purposes and avoid the release of ammonia into the environment.

An example of how the efficiency of nitrogen conversion in the rumen is affected by the level of nitrogen in the diet. In the Table example, the differences in nitrogen intake are due only to differences in the crude protein content of the grass part of the diet.

The balance among nitrogen absorption in the intestine, its conversion into milk protein and its loss in faeces and urine is highly dependent on the capacity of the rumen to capture this nutrient. In the example, increasing the nitrogen intake by approximately 25 per cent resulted in nitrogen losses in the urine increasing by 80 per cent. It is therefore evident that even small differences in the concentration of nitrogen in different feedstuffs may give rise to substantial losses of nitrogen to the environment.

Economic effects. If, for example, maize gluten is fed at the rate of 5 kg per cow per day to a herd of 100 cows at a metabolizable energy (ME) level of 1.8 MJ/kg DM, it would supply the herd with 5 310 MJ ME/day. At a level of 12.8 MJ/kg DM, the daily ME supply to the herd would be 5 760 MJ. Over a 200-day period, the difference in ME supply would be 90 000 MJ, which represents a saving of some 8 tonnes of compound feed, or some US$2 500. Unbalanced diets may also produce an economic loss in terms of animal health, feed conversion efficiency and, ultimately, the output of animal products. The wrong proportion of a nutrient in the diet is reflected in milk production (in terms of both quantity and quality) and has a direct effect on the economics of production. Such effects can be seen extremely rapidly in fast-growing poultry, where even small changes in the nutrient balance can lead to drastic economic consequences.

FOR THE COMPOUND FEED MANUFACTURER

Compound feeds are composed of a blend of various raw materials and/or by-products. In designing these products, the feed manufacturer needs to find the balanced combination among different materials that will satisfy the requirements of specific animals in a specific physiological state.

In order to achieve this, the manufacturer initially needs to base purchasing decisions on the available information about the chemical composition and nutritional value of feedstuffs. It is therefore very important at this stage that the information available is representative and reliable, not least so that the economic value of the material can be assessed and compared with its market price. Subsequently, the composition of the different ingredients will be measured and the nutritional value of the product estimated, but failure to get the initial values correct will ultimately be reflected in a higher cost of production and will reduce sales of compound feeds to farmers.

FOR THE FARM MANAGER

The farm manager needs to plan crop production in order to meet the animals' requirements. Managers therefore need to obtain information on the supposed chemical and nutritional characteristics of various ingredients in advance and, based on this information, plan the correct combination and quantity of different forages. If the information is inaccurate, the crops produced will provide unbalanced proportions of nutrients (*i.e.* too much protein, too few available carbohydrates, etc.). The immediate consequences will be an inappropriate diet for the animals, with direct effects on production, and the necessity to compensate the lack of some nutrients by purchasing from the market.

Bibliography

Amarjeet Singh: *Animal Husbandry*, Rajat Publication, Delhi, 2008.

Aruna T. Kumar: *Handbook of Animal Husbandry*, Indian Council of Agricultural Research, 2008.

Ashok Kumar: *Animal Husbandry*, Discovery Publication, Delhi, 2006.

Bibek Ghosh: *Crops Residues and By-Products in Animal Feeding*, Gene Tech Books, Delhi, 2007.

C.V. Singh: *Animal Breeding and Genetics*, New India Publishing Agency, Delhi, 2015.

Devendra Thakur and D.N. Thakur: *Tribal Life in India, Vol. 3. Tribal Agriculture and Animal Husbandry*, Deep and Deep Publication, Delhi, 2009.

Dinesh Arora: *Biotech's Dictionary of Animal Husbandry*, Biotech Books, Delhi, 2005.

Gyan Deep Singh: *Animal Husbandry*, Anmol Publication, Delhi, 2008.

L. Lal: *Animal Husbandry (Objective Fundamentals)*, Agrotech Press, Delhi, 2010.

M P S Bakshi and M Wadhwa: *Animal Feeding Concepts and Practices*, Satish Serial Publishing House, Delhi, 2016.

Malcolm B Willis: *Dalton's Introduction to Practical Animal Breeding*, Blackwell Science, 2004.

Manoj Kumar Rai: *Handbook of Animal Husbandry*, Oxford Book Company, Delhi, 2012.

Manoj Kumar Rai: *Textbook of Animal Husbandry*, Oxford Book Company, Delhi, 2012.

Nishamani Kar: *Animal Husbandry and Rural Development : Restructuring and Planned Development of the Bovine Economy*, Deep and Deep Publication, Delhi, 2002.

P K Saraswat: *International Encyclopaedia of Dairy Technology and Animal Husbandry (Vols 1 and 2 Set)*, Anmol Publication, Delhi, 2009.

P Mathialagan: *Textbook of Animal Husbandry and Livestock Extension*, International Book Distributors, Delhi, 2007.

P.P. Joglekar and Pankaj Goyal: *Animal Husbandry and Allied Technologies in Ancient India (From Prehistorical to Early Historical Times)*, Pentagon Press, New York, 2015.

R Renaville and A Burny: *Biotechnology in Animal Husbandry, Vol. V*, Springer Publication, 2008.

Rajendra Reddy and J.P. Abhay Shankar: *Animal Husbandry*, Commonwealth Publication, Delhi, 2008.

Shagufta Jamal and H P S Arya: *Participatory Rural Appraisal in Agriculture and Animal Husbandry : A Training Manual*, Concept Publication, Delhi, 2004.

Sohrab Ali: *Practical Animal Breeding*, Arise Publication, Delhi, 2008.

Vir Singh, Ashoka Kumar and R.S. Jaiswal: *Animal Feeding and Production in India*, Daya Publication, Delhi, 2011.

William J.A. Payne and R. Trevor Wilson: *An Introduction to Animal Husbandry in the Tropics*, Wiley, India, 2013.

Yasin Khan: *Animal Husbandry and Farming Management*, Agrotech Press, Delhi, 2013.

Index

I

L

M

N

O

P

R

S

T

V

W

Z